AF326843

Neutron Scattering Applications and Techniques

Series editors

Ian S. Anderson, Oak Ridge National Laboratory, Oak Ridge, TN, USA
Alan J. Hurd, Los Alamos National Laboratory, Los Alamos, NM, USA
Robert L. McGreevy, ISIS, Didcot, UK

More information about this series at http://www.springer.com/series/8141

Gordon J. Kearley · Vanessa K. Peterson
Editors

Neutron Applications in Materials for Energy

 Springer

Editors
Gordon J. Kearley
Vanessa K. Peterson
Australian Nuclear Science and Technology
  Organisation
Lucas Heights, NSW
Australia

ISSN 1868-0372   ISSN 1868-0380   (electronic)
ISBN 978-3-319-06655-4   ISBN 978-3-319-06656-1   (eBook)
DOI 10.1007/978-3-319-06656-1

Library of Congress Control Number: 2014949356

Springer Cham Heidelberg New York Dordrecht London

Printed on acid-free paper

Springer is part of Springer Science+Business Media (www.springer.com)

Preface

Neutron scattering has grown from being a physics and chemistry centred technique in which lattice dynamics and crystallography were the mainstays, to a more interdisciplinary field which includes areas as diverse as engineering, archaeology, food, plastics and all manner of nanostructured materials. The study of sustainable-energy materials is highly cross-disciplinary in nature, with the diverse activities frequently having some background in neutron-scattering techniques, which reflects the innate applicability of neutron scattering in this area. It is not surprising to find that many neutron-scattering centres are supporting an "energy project" of some description that brings together the in-house and user activities. It is refreshing to see scientists with widely-varying expertise making a joint approach to understanding and improving energy materials. Progress is made through a number of avenues:

1. Increasing performance of neutron sources and their associated instrumentation.
2. Improvement of specialised instrumentation, sample environments, and ancillary equipment (such as in situ cells), aided by collaboration, workshops, and conferences.
3. Computational resources and algorithms adapted to modelling structure and dynamics in increasingly complex materials are being both validated and used in the study of sustainable energy with neutrons.
4. Studentships and fellowships in the field are increasing, bringing fresh ideas and new approaches to the way neutron scattering is used and the data analysed.
5. Increasing awareness of the importance of sustainable energy in society, and the role that neutron techniques of analysis plays, helping to increase the resources that are allocated to this area.

This book brings together some of the core aspects of sustainable-energy materials that can be studied with neutrons, but there are obviously many other important neutron-based studies in the area that fall outside this core, for example in the fields of wind, hydro, and biomass. Similarly, a large number of non-neutron techniques are used to study the materials that we discuss in this book, and it is frequently the combination of information from a number of techniques that leads to the final understanding.

Although the book does not contain dedicated theory or instrumental sections, the volume is aimed at those who have little or no knowledge of neutron-based techniques of analysis, and instead we refer to an earlier volume in this series in which these topics are presented in detail. The contents of this book are aimed at professionals at all levels in the field of sustainable energy, to show the types of question that can be addressed using neutrons. Some chapters in the book take the form of a review, whilst others use case studies to provide a more targeted approach. The book is structured chronologically, beginning with energy generation, moving onto storage, and then to use, although each chapter can be read independently of the others. The loose theme of application of sustainability-materials in transport applications that runs through many of the chapters is rather artificial, because each chapter has at least some bearing on stationary applications.

We are sincerely grateful to all the authors for their willingness to find the time to provide their chapters. They are all active in research and can only make their contribution to this book by taking time away from their research projects. We do understand.

Lucas Heights, NSW, Australia, October 2014   Gordon J. Kearley
                                                               Vanessa K. Peterson

Contents

Part III   Energy Use

Contributors

David G. Carr Australian Nuclear Science and Technology Organisation, Lucas Heights, NSW, Australia

Deanna M. D'Alessandro School of Chemistry, The University of Sydney, Sydney, NSW, Australia

Anita Das School of Chemistry, The University of Sydney, Sydney, NSW, Australia

Joseph A. Dura Center for Neutron Research, National Institute of Standards and Technology, Gaithersburg, MD, USA

Juergen Eckert Department of Chemistry, University of South Florida, Tampa, FL, USA

Antonio Faraone Center for Neutron Research, National Institute of Standards and Technology, Gaithersburg, MD, USA

Lucas A. Haverkate Faculty of Applied Sciences, Reactor Institute Delft, Delft University of Technology, Delft, The Netherlands

Hervé Jobic Centre National de la Recherche Scientifique, Institut de Recherches sur la Catalyse et l'Environnement de Lyon, Lyon, France

Mark R. Johnson Institut Max von Laue-Paul Langevin, Grenoble, France

Maths Karlsson Chalmers University of Technology, Göteborg, Sweden

Christian A. Kaufmann Helmholtz-Zentrum Berlin Für Materialien Und Energie, Berlin, Germany

Gordon J. Kearley Australian Nuclear Science and Technology Organisation, Lucas Heights, NSW, Australia

Sangcheol Kim Materials Science and Engineering Division, National Institute of Standards and Technology, Gaithersburg, MD, USA

Michael Law Australian Nuclear Science and Technology Organisation, Lucas Heights, NSW, Australia

Wiebke Lohstroh Maier-Leibnitz Zentrum, Garching, Germany

Fokko M. Mulder Faculty of Applied Sciences, Reactor Institute Delft, Delft University of Technology, Delft, The Netherlands

Kirt A. Page Materials Science and Engineering Division, National Institute of Standards and Technology, Gaithersburg, MD, USA

Vanessa K. Peterson Australian Nuclear Science and Technology Organisation, Lucas Heights, NSW, Australia

Brandon W. Rowe Materials Science and Engineering Division, National Institute of Standards and Technology, Gaithersburg, MD, USA

Susan Schorr Freie Universität Berlin, Institut Für Geologische Wissenschaften, Berlin, Germany; Helmholtz-Zentrum Berlin Für Materialien Und Energie, Berlin, Germany

Neeraj Sharma School of Chemistry, The University of New South Wales, Sydney, NSW, Australia

Christiane Stephan Freie Universität Berlin, Institut Für Geologische Wissenschaften, Berlin, Germany

Sven C. Vogel Los Alamos National Laboratories, Los Alamos, MN, USA

Marnix Wagemaker Faculty of Applied Sciences, Radiation, Science and Technology Department, Delft University of Technology, Delft, The Netherlands

Mohamed Zbiri Institut Max von Laue-Paul Langevin, Grenoble, France

Chapter 1
Neutron Applications in Materials for Energy: An Overview

Vanessa K. Peterson and Gordon J. Kearley

Abstract Creating a global energy-system that is both environmentally and economically sustainable is one of the largest challenges currently facing scientific and engineering communities. Alternative energy-technologies and new materials have risen as a result of the combined needs for energy and environmental sustainability, with the focus moving increasingly away from fossil fuels. Neutron-based techniques of analysis play a role in almost all aspects of sustainable-energy materials research, and the chapters of this book will enlarge on these studies using examples and case studies to illustrate research approaches, methods, and outcomes.

1.1 Introduction

Research on renewable materials of relevance for environmentally benign energy-technologies is one of the most rapidly-growing research areas in materials science. The primary challenge in this research is the development of materials for such technologies that are viable in competition with existing energy-technologies, responding to application requirements such as efficiency, durability, and cost. Understanding the fundamental properties of materials and their functionality at the atomic and molecular level is crucial in addressing the global challenge of cleaner sources of renewable energy.

This book is divided into three main parts: materials for energy production, storage, and use. The central theme is identifying where the energy carrier is in the material and its interaction with its immediate environment so that these can be tailored to increase the concentration and/or transport of the carrier, which may be electrons, ions, atoms, or molecules. The theory of neutron scattering and analysis techniques, as well as the

V.K. Peterson (✉) · G.J. Kearley
Australian Nuclear Science and Technology Organisation,
Lucas Heights, NSW 2234, Australia
e-mail: vanessa.peterson@ansto.gov.au

G.J. Kearley
e-mail: gke@ansto.gov.au

© Springer International Publishing Switzerland 2015
G.J. Kearley and V.K. Peterson (eds.), *Neutron Applications in Materials for Energy*, Neutron Scattering Applications and Techniques,
DOI 10.1007/978-3-319-06656-1_1

associated instrumentation, are explained elsewhere in the "Neutron Scattering Applications and Techniques" book series [1] and we particularly refer the reader to Chaps. 2 and 3 , respectively, in the "Neutron Applications in Earth, Energy and Environmental Sciences" edition of this series [2] which are available on-line [3].

1.1.1 The Need for Neutrons

Despite the superficial similarity of the application of neutrons with those of photons, neutrons have some key differences to photons that enable neutron-based techniques to play a particularly important role in the sustainable-energy area. A full account of neutron theory and techniques is given elsewhere in this series [1] with key methods prevalent in this book outlined here. For our purposes we can regard all of the advantages and disadvantages of neutron scattering as having a single origin: neutrons interact with, and are sometimes scattered by, atomic nuclei. In all neutron-scattering methods this leads the following advantages:

i. The scattering characteristics of each type of isotopic nucleus are well known, but vary almost randomly from one isotope type to another (scattering-lengths are shown in Fig. 1.1). This provides considerable scope for measuring light nuclei in the presence of very heavy nuclei, and also changing the scattering length by using a different isotope of the same element. For materials such as lithium-ion battery electrodes, where lithium must be probed in the presence of transition metals (see Chap. 7), this is a considerable advantage over other techniques where the scattering arises from electron density.

ii. Incoherence arises when scattering from the nuclei do not interfere constructively. The random relation between the nuclear spins of hydrogen and neighbouring atoms contributes to the extreme incoherent neutron-scattering cross-section of the ^{1}H nucleus, which can be turned off by simple deuteration. Hydrogen is probably the most important element in sustainable energy-materials and it is very convenient that neutron scattering provides this selectable sensitivity for this element.

iii. Neutron-scattering cross sections are in general quite small, so neutrons are relatively penetrating, where measurements (reflection is an exception) occur for the bulk of the sample. This penetration alleviates the need for special window materials in difficult sample environments and in situ studies. There are good examples of this in Chap. 7 for lithium-ion batteries in which the composition of the electrodes can easily be measured in an operating battery.

iv. Neutron-absorption cross sections are also normally quite low, contributing to the high penetration of neutrons, but some isotopes have very large absorption. Lithium is perhaps the second most important element in sustainable-energy materials, so it is again convenient that ^{6}Li has high absorption, whilst ^{7}Li does not. Consequently, either radiography (and other bulk techniques) or normal neutron-scattering experiments can be made using the appropriate isotopic composition.

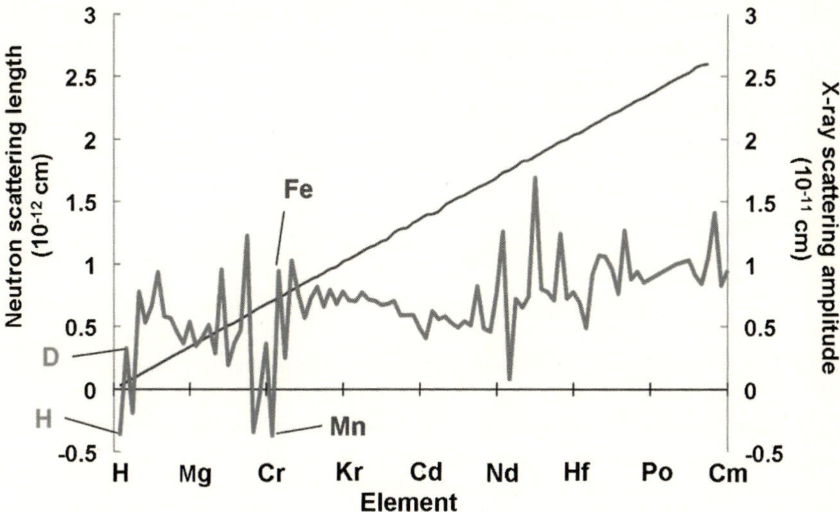

Fig. 1.1 Neutron-scattering lengths and X-ray scattering amplitudes for various isotopes (neutrons based on naturally-abundant isotopes unless specifically identified) and elements (X-rays)

The list above shows that neutron scattering is well suited to the study of sustainable-energy materials, but there are other important considerations to be made before undertaking a neutron experiment. Firstly, neutrons are difficult to produce in high quantities with the correct energies for the studies in this book, and consequently, virtually all experiments require a central facility and associated logistics. Secondly, even at the most powerful sources the neutron beams are weak when compared with photons or electrons, so the samples and the counting times for neutrons are correspondingly greater. Consequently, neutron scattering is rarely used when the desired information can be obtained by another means, and maximum use of complementary techniques (most frequently X-ray) and computer-modelling methods is common.

1.2 Neutron-Based Analysis of Energy Materials

In the next sections we will outline the basics of the neutron techniques of analysis that underpin the chapters that follow. Neutrons have the same principle attributes as photons for the study of a wide range of materials. Neutrons can be diffracted giving information about atomic position, scattered inelastically giving information about atomic (or molecular) motion, and neutrons can be absorbed giving spatial information concerning material composition through radiography and tomography. The instrumentation for photons is well known, but for neutrons there is an almost analogous group of techniques that together cover length scales from fractions of an

Å to microns (and up to many centimetres for radiography) and timescales that cover from femtoseconds to hundreds of nanoseconds. The generic properties of neutrons lead to the recurrent use of particular neutron scattering and neutron-based analysis throughout this book, and this section explains the rudiments of these.

1.2.1 Structure

Structure is determined using neutron diffraction (ND). The structure factor, $S(Q)$, describes scattered neutrons of the beam in terms of the wave-vector transfer, Q, where $Q = 4\pi\sin\theta/\lambda$, and θ is the angle of the scattered neutrons with λ being the incident-neutron wavelength. For a single crystal, the scattering will consist of Bragg peaks. In an ideal powder sample, small crystallites are randomly oriented and scattering from a particular set of lattice planes corresponds to the scattering obtained by turning a single crystal. In powder samples, Debye-Scherrer cones are obtained in place of Bragg peaks, where intensity from the cones can be determined simultaneously using large-area detector arrays.

1.2.1.1 Neutron Powder Diffraction

The workhorse of ND is powder diffraction, which has been developed to the point where complete structural information can be obtained from polycrystalline samples. Even modest dynamical information, such as diffusion pathways, can be deduced from atomic displacement parameters. The technique relies on the well-known coherent-interference pattern that is scattered from well-defined lattice planes, which, due to rotational averaging in a powder, collapse to a simple one-dimensional powder pattern. The level of detail that is available from this pattern is largely dependent on the resolution and range of the diffractometer, plus the availability of refinement algorithms, all of which continue to improve. The technique is not limited to a single compound, and measured diffraction patterns are often used to establish the phase composition in complex materials, for example where doping is used to modify electronic structure in solar-cell materials (such as in Chaps. 5 and 6).

Although very detailed structural information can be obtained from powders, the comparative simplicity of the technique also makes it the prime candidate for in situ studies, which also profit from the penetration and isotope selectivity of neutron scattering. In the context of this book, the technique is commonly used to follow the evolution of structure with temperature, or composition, for example in charging and discharging electrodes (such as in Chap. 7).

The main constraint on neutron powder diffraction (NPD) is the larger samples required compared with X-rays, and the large incoherent neutron-scattering from the ^{1}H nucleus that causes a high background. In general this background can be eliminated by deuteration, with the added advantage that the crystallographic positions of these atoms can then be more-easily established.

1.2.1.2 Pair-Distribution Function Analysis

Disordered materials often provide mechanisms for improved diffusion and transport, which is desirable for many sustainable-energy materials. Pair-distribution function (PDF) analysis is becoming increasingly important in studying these materials at it provides local structure, interatomic distances, bond-angles and coordination numbers in disordered materials such as glassy and amorphous materials. The essential difference between conventional ND and PDF (linked to the neutron total-scattering experiment) is that while in ND only the Bragg peaks are considered, neutron total-scattering means that also the diffuse, weaker, scattering present between the Bragg peaks is analysed, where deviations from the average can be measured. It is this "extra" scattering that provides information about the structure on a local scale and is therefore of high importance for structural studies when the material is not fully periodic.

The PDF, or, $G(r)$, is obtained from the structure factor, $S(Q)$, via a Fourier transformation,

$$G(r) = \frac{2}{\pi} \int_{0}^{\infty} S(Q - 1)(Qr)QdQ \tag{1.1}$$

It is the analysis of this quantity that gives information about the local structure of the material. The Fourier transform requires data over a large Q range to avoid truncation effects, so the technique typically uses rather short-wavelength neutrons, either at the hot-end of reactor sources or at spallation sources.

1.2.1.3 Neutron Reflection

The structure and kinetics at, and close to, interfaces is of importance in many sustainable-energy devices (e.g. electrolytes and electrodes, see Chap. 7), but these properties are difficult to establish. Within certain constraints, the neutron reflection (NR) experiment can establish the scattering characteristics beneath a surface by measuring the reflected intensity as a function of angle. Above a critical angle (representing Q), total reflection occurs, but below this each layer interface produces an oscillating reflected amplitude with period $\Delta Q = 2\pi/T$, where T is the thickness of the layer. However, the measured reflected intensity is the total from all interfaces present, and because phase information is lost, the usual way forward is to fit the measured signal with models. In practice it is the variation of sample composition within the depth of the sample that is of interest, and this is characterized as the scattering-length density (SLD) profile, which is the sum over the number density of each isotope at a given depth, z, times its bound coherent neutron-scattering length. The essential advantages of neutron measurement of reflection is the variation of scattering length with isotopic nuclei, which allows contrast variations, measurement of buried layers, and favours the light elements in the presence of heavier ones found in energy materials.

The main constraints in NR are the comparatively large and atomically-flat surface that is required, and establishing suitable models for analyzing the results.

1.2.1.4 Small-Angle Neutron Scattering

Small-angle neutron scattering (SANS) is a well-established characterization method for microstructure investigations, spanning length scales from Å to micron sizes. Within the SANS approximation, Q simplifies to $Q = 2\pi\theta/\lambda$. The SANS Q range is typically from 0.001 to 0.45 Å^{-1}. For example, scattering from a simple spherical system in a solvent yields a SANS coherent macroscopic neutron-scattering cross section (scattering intensity in an absolute scale) of:

$$\frac{d\sum_c(Q)}{d\Omega} = \frac{N}{V}V_P^2\Delta_{\rho^2}P(Q)S_I(Q) \tag{1.2}$$

where (N/V) is the number density of particles, V_P is the particle volume, $\Delta_{\rho 2}$ the contrast factor, $P(Q)$ is the single particle form factor, and $S_I(Q)$ is the inter-particle structure factor. $S_I(Q)$ has a peak corresponding to the average particle inter-distance.

1.2.2 Dynamics

The dynamic structure-factor, $S(Q, \omega)$, describes scattered neutrons in terms of the wave-vector transfer Q and the neutron energy-transfer $\hbar\omega$, where $\hbar = h/2\pi$ and h is Planck's constant. The timescales and corresponding energies of the processes that are accessed by neutron scattering from energy materials are illustrated in Fig. 1.2. A particular strength of neutron scattering is that the size and geometry of the volume explored by the dynamic process is also available, and is shown on the horizontal axis of Fig.1.2.

This is particularly useful when measuring local or long-range diffusive processes, for example in fuel-cell electrolytes (see Chaps. 9 and 10).

Fig. 1.2 The frequency (E), time (t), and space (*Q*) domains in which the dynamics of energy materials are typically studied using neutron scattering

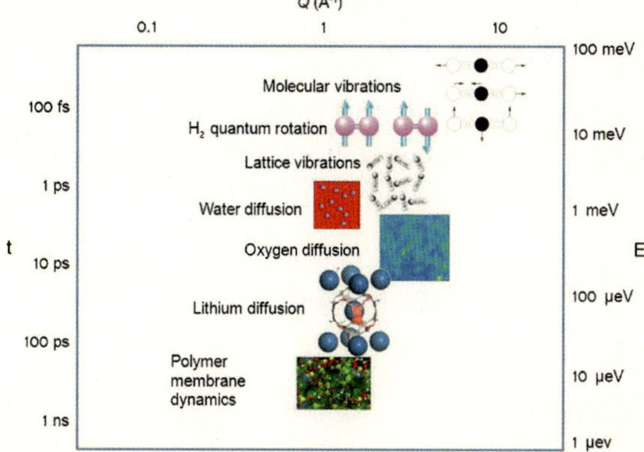

Fig. 1.3 Representative inelastic and quasielastic signals from some idealized motions that give rise to them. Reprinted from (H. Jobic and D.N. Theodorou, Microporous Mesoporous Mater **102**, 21 (2007)) [4]

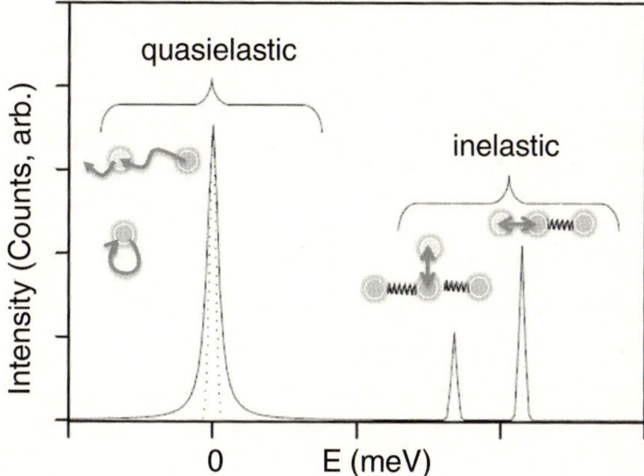

Inelastic peaks usually arise from a periodic motion, and the forces controlling this motion are stronger than those that would define a more diffusive motion. Correspondingly, inelastic peaks usually arise at higher energy than the quasielastic broadening, as illustrated schematically in Fig. 1.3.

1.2.2.1 Inelastic Neutron-Scattering

Inelastic neutron scattering (INS) is crucial in the study of hydrogen storage where the neutron can excite the quantum motion of the H_2 molecule, and measure the transition energies directly (see Chap. 8). However, in the more general case INS amounts to "vibrational spectroscopy with neutrons" and in the present book is used to study local structure, vibrational dynamics, and the nature of hydrogen-bonding interactions. The main strength of the technique arises from the large neutron-scattering cross section of hydrogen, which causes vibrations involving hydrogen to dominate the spectra. This domination, particularly when combined with selective deuteration, is very powerful for providing assignment of the observed peaks to specific vibrational-modes. In addition, it is now straightforward to calculate the INS directly from a molecular model which is not only an aid to assignment, but also a validation procedure for the model. The technique is comparable to infra-red and Raman spectroscopy, which have better resolution, but lack the hydrogen selectivity and simplicity of assignment. Incoherent INS has no selection rules and even modes that are silent in both infra-red (IR) and Raman can have significant INS intensity.

1.2.2.2 Quasielastic Neutron-Scattering

The term quasielastic neutron-scattering (QENS) can be used to describe any broadening of the elastic peak regardless of its origin, but use of the term in this book is always with reference to a stochastic or diffusive non-periodic motion. The technique is important for energy materials because it provides the relevant time and length scales on which the atomic-scale dynamics of protons and small molecules typically occur, for example in proton-conducting perovskites (Chap. 9). Various molecular processes can be distinguished from the data, which can be quite straightforward, although in systems of any complexity it is now common to use molecular-dynamics simulations from which it is now easy to produce a calculated QENS spectrum.

1.3 In Situ

Function is of great importance to the study of energy materials. At the heart of the application of neutron-based techniques of analysis to the study of energy materials is the understanding of structure- and dynamic-function relations. Comprehending the working mechanism, at the atomic and molecular scale, is the key to progressing alternative and sustainable-energy technologies, and fundamental to this is the study of the materials during operation. As such, in situ and even operando studies are commonplace and necessary in energy materials research. The in situ technique, often applied to materials under equilibrium, has been extended in recent years to operando studies, where the materials are studied under non-equilibrium conditions whilst performing their function. The advent of new-generation reactor and spall-ation neutron sources, as well as associated faster instrumentation, has greatly assisted in facilitating such research.

1.4 Perspectives

Whilst the difficulties of realising sustainable energy are decreasing, the difficulties of fossil-fuel based energy are increasing and the point will inevitably arrive when sustainable energy is not only socially and environmentally more favourable, but also makes economic sense in its own right. Sustainable-energy materials will develop over the long run, and the role of neutron-scattering techniques in the understanding of these is almost certain to develop in parallel. Generic improvement in neutron sources and instrumentation will enable smaller samples to be measured in shorter times, and this is part of the wider scientific agenda. However, there are also specific improvements that will benefit the study of energy materials.

In situ and operando experiments are crucial for "close to market" studies, and whilst these are not simple with neutrons, they are generally more straightforward than with other methods. Perhaps ironically, it is now possible to construct real lithium batteries that are optimised for operando neutron-diffraction measurements, where the optimisation may affect cost, but has little or no effect on the actual operation. Often the modification for operando neutron scattering amounts to deuteration of materials and neutron-scattering centres are increasingly housing specialised deuteration facilities, capable of deuterating complex molecules.

The complex systems that characterise the development of energy materials give complex neutron-scattering signals, from which it can be difficult to deconvolve unambiguous information. However, the rapid increase in computer hardware and software is enabling the experiment, data treatment, theory, and modelling to be brought together to provide consistent interpretation of the neutron-scattering data. Although at present this is the domain of specialists, considerable efforts are being made throughout the neutron-scattering community to bring this type of approach within the reach of non-specialist users. Although previously the multiprocessor computer hardware required for this type of work was only available at central-computing establishments, it is now becoming ubiquitous in universities and neutron-scattering centres where there is generally good local support.

The experimental programme at neutron-scattering centres has to strike a balance between scientific, societal, commercial, and national interests, the details of which depend on the strategy and "terms of reference" of the centre. Sustainable-energy materials are almost equally important in all aspects of this balance, which provides a unique opportunity for communication and collaboration across these aspects and between neutron-scattering centres. Although this type of initiative has yet to occur, there have been a large number of conferences and workshops at the purely scientific level that have been funded from a diverse range of sources. Larger gatherings, specifically highlighting neutron scattering, would provide an over-arching description of problems, bottle-necks, and resources, from industry, strat-egists, and through to experimentalists.

References

1. I.S. Anderson, A.J. Hurd, R. McGreevy (eds.), *Neutron Scattering Applications and Techniques*, (Springer, Berlin)
2. L. Liang, R. Rinaldi, H. Schober. (eds.), *Neutron Applications in Earth, Energy and Environmental Sciences* (Springer, Berlin, 2009)
3. http://www.springer.com/series/8141. Accessed 4 March 2014
4. H. Jobic, D.N. Theodorou, Microporous Mesoporous Mater **102**, 21 (2007)

Part I
Energy Generation

Energy generation is not only important for meeting the requirements of consumers and industry, but is crucial for national security and economic competitiveness. Environmental sustainability is a global issue that needs to respond to existing damage from direct emissions as well as unplanned future events such as spills and leakages. Neutron techniques of analysis play some role in progressing all technologies for renewable-energy generation, although this may be limited to structural materials for wind, marine and hydro energy generation, whilst neutron scattering in earth sciences plays a significant role in geothermal energy. Chapters in Part 1 concentrate on those aspects of energy generation that are mainstream for neutron-based methods, but are nevertheless relevant to the more general sustainable-energy technologies in energy generation.

Catalysis (Chap. 2) not only plays a central role in sustainable-energy generation where renewable feedstocks are used, but also plays a more general role in increasing efficiency, reducing energy consumption, and producing cleaner targeted products from fossil-based sources such as oil, natural gas, and coal. Active sites in catalysis are often present in only trace quantities, and characterizing these with neutrons is usually limited to model compounds. However, almost the whole range of neutron techniques of analysis have been used to help in the design, characterisation and optimisation of catalysts by measuring the structure of the catalysts themselves, and following the dynamics of the reactants and products. The hydrogen economy requires an efficient means of generating, storing, and using hydrogen, all of which involve some catalysis. However, because the most important role of catalysis is in energy generation, we gather all aspects of the topic in Chap. 2.

Although global CO_2 emissions threaten today's way of life, cost-effective methods to separate, capture, store, or use CO_2 from fossil fuels represent a major challenge. As existing technologies use solvents that impose a heavy energy-penalty (about 30 % of the energy generated by the plant), a key scientific challenge is the development of materials that can interact with flue-gas streams to capture and concentrate CO_2 with lower energy requirements. Porous materials such as coordination polymers offer a new way to address this in that they not only possess the highest surface area of any known materials, but they can be engineered to be selective for CO_2. Therefore, solid porous hosts represent one of the most promising technologies for separating and storing gases of importance in the generation and use of energy. Studying the uptake of CO_2 in such materials at the fundamental level is required to progress these towards commercialisation, with such studies allowing direct feedback into the synthesis of materials with enhanced CO_2 uptake, selectivity, and chemical stability. Neutron scattering is essential in this research with in situ studies involving pressure and/or temperature being of particular importance. Both structural and dynamical information is important in this area in order to establish the gas–host interaction, and forms the basis of Chap. 3.

Structural materials are important to all forms of sustainable-energy production and neutron scattering is increasingly being used to characterize and understand fatigue and failure, and in this context we concentrate on materials for nuclear-energy applications in Chap. 4. Nuclear materials are particularly demanding because they must meet the mechanical demands not only under pressure, temperature, and chemical environment, but also under the effects of irradiation. Neutron-based characterisation of such materials mainly takes the form of neutron diffraction to understand how the microstructure and crystal structure characterize bulk material-properties. Superficially, this is a straightforward measurement, but in practice we need to understand how crystal structure, microstructure, chemical composition, and orientation are all coupled, and how these can be controlled to obtain (or avoid) particular properties. Major neutron-scattering centres now have at least one instrument that is conceived specially to do these types of experiment.

One of the greatest contributions neutron scattering has made in the study of sustainable energy-materials is in solar cells, which are divided into inorganic and organic in Chaps. 5 and 6, respectively. Photovoltaics (PV) is required for each of them, and is the direct conversion of light into electrical energy, the first PV device having been built by Edmond Becquerel who discovered [R. Williams, J. Chem. Phys. **32**, 1505 (1960)] the PV effect in 1839. The operating principles of this effect are based on a sequence of light–matter interactions that can be summarized as follows:

(i) Absorption of photons with a given energy matching the semiconducting properties of the device (band gap and intrinsic coefficients);

(ii) Free charge-carrier generation in inorganic semiconductors and bound exciton (pair of electron-hole) creation in the case of the organic analogous and exciton [E.A. Silinsh, V. Capek *Organic Molecular Crystals -Interaction, Localization and Transport Phenomena* (American Institute of Physics, New York, 1994)] dissociation;

(iii) Charge transport via relevant pathways;

(iv) Charge collection at dedicated electrodes and photocurrent generation.

PV solar cells convert solar energy into electricity via the PV effect in a variety of strategies that can be classified as (Fig. 1): Multijunction, single-junction GaAs, crystalline Si, thin-film, as well as organic and emerging (including hybrids).

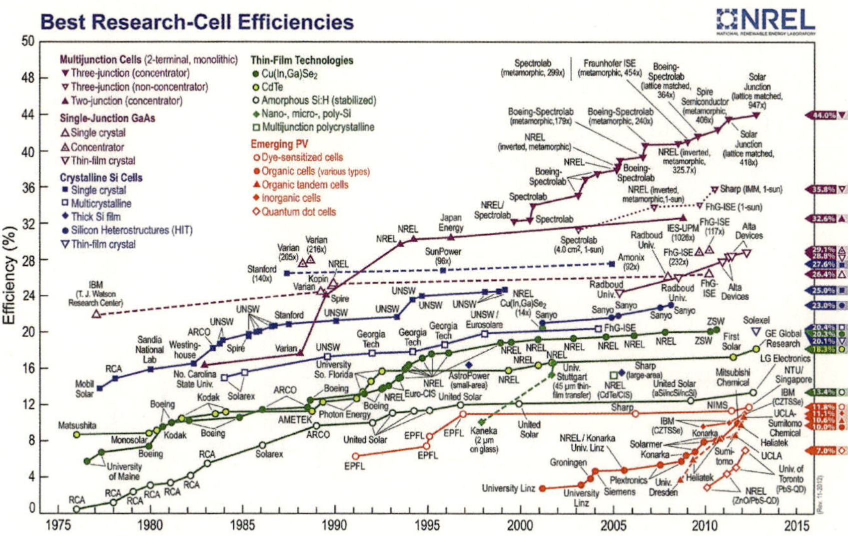

Fig. 1 Power conversion efficiency (PCE) of various solar cells showing their recent performance. Source: National Renewable Energy Laboratory

Sustained development has been achieved for each of the PV strategies over the past decades (Fig. 1). It is important to notice that step changes in efficiency mainly arise with the discovery of a new strategy, but that this is not always upwards. This is because many factors contribute to the cost per Watt, and low efficiency may be counterbalanced by overall cost, which is composed of:

(i)   Energy pay-back time;
(ii) Stability and lifetime;
(iii) Environmentally friendly materials and production;
(iv) Cost and supply of materials;
(v)   Adaptability of shape/form;
(vi) Size/weight.

For PV materials there is a convenient separation of the materials in the broader classification of inorganic and organic. More attention has been devoted to inorganics, which have been known for well over a century and promise very high efficiency. In contrast, the first organic PV (OPV) was crystalline anthracene [H. Kallmann, M. Pope, J. Chem. Phys. **30**, 585 (1959)], this having been first observed in 1959. OPVs have only been only attracting significant attention in the past decade (Fig. 2), at least in part as a result of robust concepts and principles for organic semiconductors set up around 1970 by Alan J. Heeger, Alan G. MacDiarmid, and Hideki Shirakawa, who were awarded the Nobel Prize in 2000 for this contribution. This paved the way for the use of organic materials as PV devices, inducing and stimulating a tremendous interest in research on OPVs for both fundamental and technical purposes.

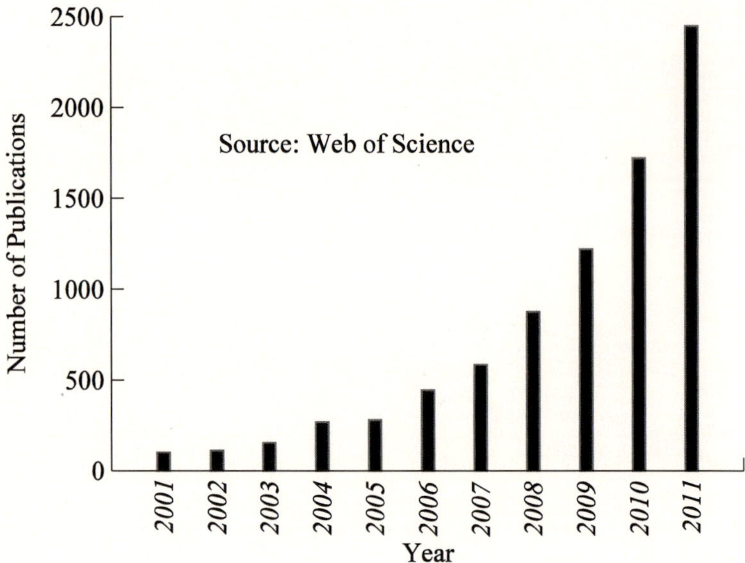

Fig. 2 Approximate number of publications per year for the last 10 years with the topic organic solar cells

In inorganic (thin film, Chap. 5) and organic (Chap. 6) solar cells we focus our attention on case studies of model systems with the aim of showing how neutron techniques of analysis in combination with other techniques contribute to our understanding and resolution of specific challenges in PV materials. The failure of one of the earliest cells, Cu_2S-CdS, due to Cu^+ to Cu^{2+} conversion, illustrates the importance of understanding functional properties of PV materials at a number of levels. PV materials cover a vast range, from soft, almost liquid materials through to hard crystalline materials, with local and large-scale structure, interfaces, and dynamics over a wide range of timescales also being important. The challenge for diffraction is that the required semi-conductors are usually not only non-stoichiometric (due to doping), but also composed of elements with similar atomic numbers. The non-stoichiometry can lead to structural defects that affect the material and electronic properties so it is essential to have a method for identifying and characterising these differences. Chapter 5 shows how the neutron scattering cross-sections enable not only neighbouring elements to be distinguished, but also the study of defect non-stoichiometric structures. Chapter 6 concerns investigating the comparatively weak forces holding the organic and polymeric molecules together, these being both advantageous and disadvantageous for PV applications. Hence, whilst for inorganic PV materials the atoms can be regarded as localized, for organic PV materials molecular dynamics plays at least an equally important role as time-average atomic position. Dynamics on different timescales is responsible for recombination, charge-transfer, processing, and ultimately ageing, all of which are important to the PV's function.

Chapter 6 shows not only the use of neutron diffraction, but also how different neutron spectroscopies can be used to unravel the dynamics that helps or hinders different aspects of the PV process.

Chapter 2
Catalysis

Hervé Jobic

Abstract Catalysis helps to save energy and to produce less waste. Hydrogen will possibly be the energy carrier for the future, but it will not replace oil before several decades so the efficiency of the catalytic processes in petroleum refinery and petrochemistry still has to be improved. Numerous physical techniques are being used to follow catalytic processes. The samples can be subjected to several probes: electrons, photons, ions, neutrons; and various fields can be applied: magnetic, electric, acoustic, etc. Apart from the basic catalyst characterization, the various methods aim to observe surface species (intermediate species are much more tricky), the reaction products, and the influence of diffusion. Coupling of two, three, or more techniques is now common and very powerful. The biggest challenge has always been to perform measurements during the reaction, the term in situ being sometimes replaced by the more recent one operando, when the catalyst is under working conditions of pressure, temperature, flow, and avoiding diffusion limitations.

2.1 Catalysis and Neutron Scattering

The main classes of heterogeneous catalysts are: (i) metals and alloys (supported or not), (ii) metallic oxides (including mixed oxides, heteropolyacids, superacids), (iii) zeolites and molecular sieves in general, and (iv) sulfides. It will take some more years before deciding if metal-organic frameworks (MOFs) become a new member of the catalysts family.

Several neutron techniques are used to study catalytic systems: neutron diffraction (ND), small-angle neutron scattering (SANS), inelastic neutron scattering (INS), and quasi-elastic neutron scattering (QENS). We will limit ourselves here to INS and QENS of hydrogen species and dihydrogen molecules adsorbed on the surface of catalyst particles or inside porous materials.

H. Jobic (✉)
Centre National de la Recherche Scientifique, Institut de Recherches
sur la Catalyse et l'Environnement de Lyon, Lyon, France
e-mail: herve.jobic@ircelyon.univ-lyon1.fr

© Springer International Publishing Switzerland 2015
G.J. Kearley and V.K. Peterson (eds.), *Neutron Applications in Materials
for Energy*, Neutron Scattering Applications and Techniques,
DOI 10.1007/978-3-319-06656-1_2

Hydrogen has recently been associated with the words fuel cells and energy storage, but it is also an essential component in catalytic reactions and the hydrogen produced is mainly used in petroleum refining and ammonia production for fertilizer. Nowadays, about 90 % of the H_2 production comes from catalytic steam reforming of natural gas at high temperatures (subsequent reactions of water–gas shift and preferential oxidation are required to decrease the CO level of the gas mixture to a few ppm before it can be used in a fuel cell). At the time being, one is facing a huge increase of H_2 needs (with related CO_2 emissions), so that until Jules Verne's predictions are realized (water: the coal of the future), we may reach an H_2 deficit, as predicted by some experts.

The applications of INS to catalysis have been mainly focused to systems which are either difficult or impossible to study by other spectroscopies such as transmission or reflection–absorption infrared, and Raman. The kind of catalyst which is studied in INS has generally an inhomogeneous surface, e.g. oxides, sulfides, and metals, although zeolites, which are well-crystallized materials, are well suited. These substrates can be almost transparent to neutrons if they contain a small quantity of hydrogen, in which case the neutron spectrum will be fairly flat and it will be possible to observe all the vibrational modes of the adsorbate.

QENS has been mainly used to measure the diffusion of H or hydrogenated molecules, although the transport of deuterated molecules and of molecules which do not contain hydrogen atoms can now be followed. The dimensionality of diffusion has been studied, even if the samples are in powder form. On zeolites [1], MOFs [2], or clays [3], anisotropic diffusion (one or two dimensional) has been evidenced. The technique allows us to probe diffusion over length scales ranging from an Å to hundreds of Å. The mechanism of diffusion can thus be followed from the elementary jumps between adsorption sites to Fickian diffusion.

2.2 Neutron Spectroscopy of Hydrogen-Containing Materials

There are very few works on atomic-hydrogen diffusion on catalysts by QENS, the H species resulting from H_2 dissociation have been predominantly studied by INS. When hydrogen is bonded to metal atoms, the heavy atoms can be considered as fixed during the local modes of hydrogen. A consequence is that the mean-square amplitudes for the hydrogen atoms will be generally small so that the sample temperature will be of small influence on the INS intensities. Another consequence is that the nondegenerate modes of hydrogen will have nearly the same intensity and an E mode will be twice as intense as an A mode. The INS spectra can thus be fitted and the relative intensities of the bands yield the populations of the various sites.

INS studies of the adsorption of H_2 on various materials, to observe rotational transitions, and of various hydrogen species present on metal and sulfide catalysts were previously reviewed [4, 5]. Only recent work will be considered here.

2.2.1 Sorption of H_2 During the Reduction of Copper Chromite

The activation of a copper–chromium system in H_2 is accompanied by an accumulation of H_2, which can become active for hydrogenation reactions, in the absence of H_2 in the gas phase. This sorption of H_2 is due to a specific mechanism of Cu^{2+} reduction from the $CuCr_2O_4$ structure, a reaction which does not release water. The reduction leads to the formation of metallic copper and to protons substituting for copper cations in the vicinity of O^{2-} anions. INS spectra of copper chromite as prepared and reduced at various temperatures are shown in Fig. 2.1. The initial catalyst shows bands at low frequencies, due to the mass of the atoms, these bands having a low intensity because no proton motion is involved. Upon reduction, a large intensity

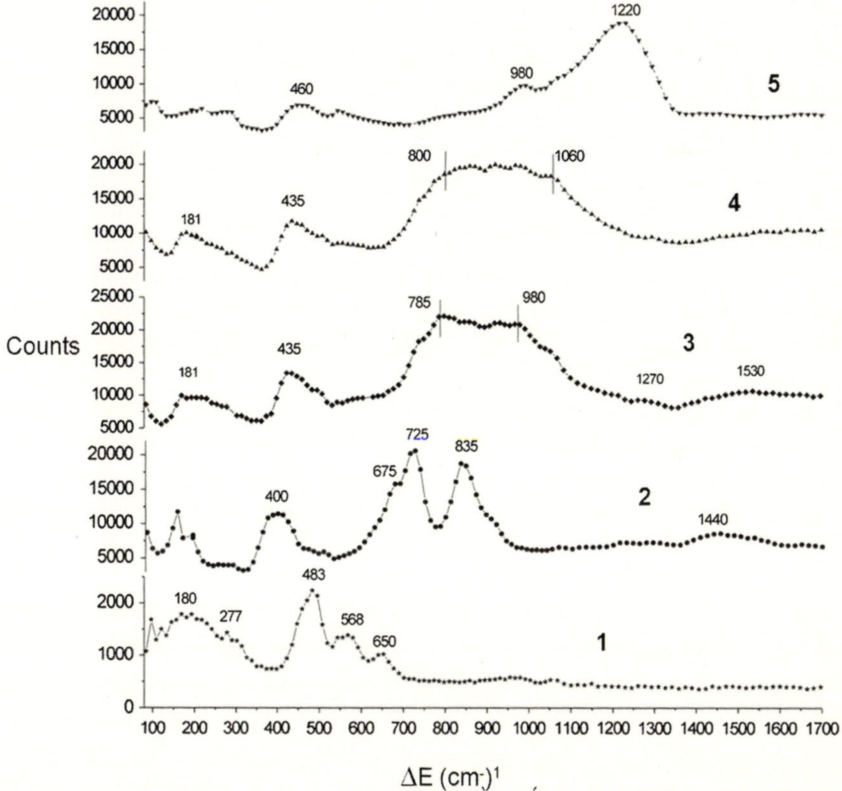

Fig. 2.1 INS spectra of as prepared and reduced samples of copper chromite. *1* Initial spectrum, *2* reduced with hydrogen at 250 °C, *3* at 290 °C, *4* at 320 °C, *5* at 450 °C. The data were obtained on the instrument IN1BeF at the Institut Laue–Langevin (ILL). Adapted with permission from (A.A. Khassin, G.N. Kustova, H. Jobic, T.M. Yurieva, Y.A. Chesalov, G.A. Filonenko, L.M. Plyasova, V.N. Parmon, Phys. Chem. Chem. Phys. **11**, 6090 (2009)) [6]

increase can be measured and OH-groups bending modes are observed in the energy range displayed in Fig. 2.1. Increasing the reduction temperature yields a shift of these bending modes to higher frequencies (from 700–800 to 1,220 cm^{-1}), while the stretching modes shift to lower energy. This indicates that hydrogen bonding strength with neighbouring anions increases with the rise in temperature. The band around 400 cm^{-1} was assigned to librations of two geminal protons (i.e. HOH-groups) [6].

2.2.2 Dissociation of H_2 on Ceria-Supported Gold Nanoparticles

Catalysis by Au nanoparticles has attracted considerable attention since it was discovered that Au particles, with a size less than a few nm, are active. Au nanoparticles deposited on oxides can dissociate H_2 heterolytically on sites involving one Au atom and a nearby surface oxygen atom. This is in contrast with other metals such as M = Ni, Pd, Pt, etc., where H_2 is dissociated homolytically (in a symmetric fashion). The relatively poor activity of Au for hydrogenation reactions was attributed partly to the inability of Au to break the H–H bond and to the instability of Au hydrides (Au–H species), while various M–H species have been clearly evidenced by INS.

The INS spectra obtained after chemisorption of H_2 on 3–4 nm Au particles supported on nanoparticulate ceria (5 nm) show a peak at 400–600 cm^{-1} accompanied by a broad band from 750 to 1,200 cm^{-1}. The first peak was assigned to bridging hydroxyl groups, and the second band to librational modes of water present on the catalyst surface and resulting from the reaction of hydrogen with oxygens of ceria [7]. The lack of observation of Au–H species by INS can be explained by the low percentage of Au on the sample (Au loading: 0.48 wt%). On the other hand, the formation of Au–H could be observed by Fourier-transform infrared (FT-IR) spectroscopy, which seems to indicate a greater sensitivity of IR spectroscopy for this sample.

2.2.3 Hydride Species in Cerium Nickel Mixed Oxides

Bio-ethanol obtained from biomass has been suggested as a promising renewable source of H_2. It is a challenge to find low-cost catalysts (without noble metals) able to break the C–C bond of ethanol at low temperature. A $CeNiH_ZO_Y$ catalyst was recently found to convert ethanol at 60 °C only, by steam reforming coupled with partial oxidation [8]. The distribution of products is similar to what is obtained by steam reforming at high temperatures: H_2 (about 45 %) and mainly CO_2 and CO. The reaction is initiated at 230 °C, but the temperature increases after a short induction period so that a temperature of 60 °C is sufficient to maintain the reaction. This is explained by the occurrence of two exothermic reactions: (i) between hydride species

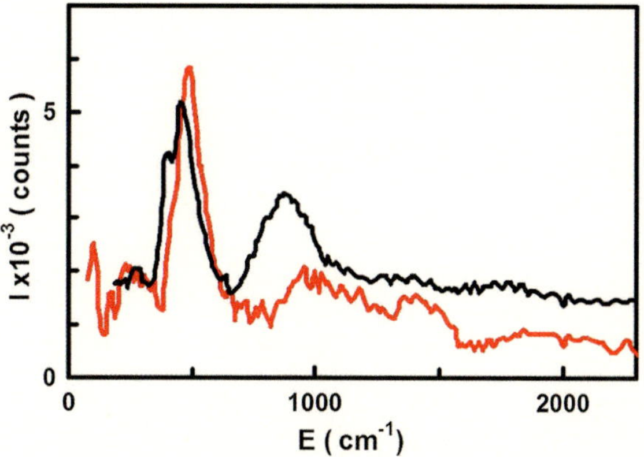

Fig. 2.2 INS spectra of CeNi$_1$O$_Y$ (*black*) and CeNi$_{0.5}$O$_Y$ (*red/grey*) after a treatment at 250 °C under H$_2$ (the INS spectrum of the solid treated in vacuum at 200 °C was subtracted). The data were obtained on the instrument IN1BeF at the ILL

of the catalyst and O$_2$, and (ii) between ethanol and O$_2$ (partial oxidation). The reaction is sustainable because hydride species are replaced and provided by ethanol.

A pre-treatment at 250 °C under H$_2$ is necessary to obtain the active catalyst, which is an oxyhydride. As in the case of copper chromite, large quantities of hydrogen can be stored in CeNi$_X$O$_Y$ mixed oxides. H$_2$ is heterolytically dissociated at an anionic vacancy and an O^{2-} species of the catalyst. The insertion of hydride species in the solid was evidenced by INS. The spectra of CeNi$_1$O$_Y$ and CeNi$_{0.5}$O$_Y$ are shown in Fig. 2.2. The INS spectrum of the solid treated in vacuum at 200 °C, which contains OH groups, has been subtracted. The peak at about 460 cm^{-1} was assigned to hydrides and the band around 870 cm^{-1} to H adsorbed on metallic Ni particles, because the band intensity decreases when the Ni loading is decreased. While the assignment of the first peak appears to be reliable (after re-oxidation, this peak disappears whereas a band corresponding to OH groups emerges around 630 cm^{-1}), the assignment of the higher frequency band to μ_3-H species on Ni^0 particles is less certain, and the contribution from OH groups cannot be excluded.

2.3 Dihydrogen

In the long term, H$_2$ is envisaged as a potential energy carrier. However, one of the issues for portable applications of this energy vector relies on its economic and safe pressure storage under the conditions of transport. Although the targets set for 2015 by the U.S. department of energy (DOE) are difficult to reach, several options are extensively investigated. Compressed or liquefied H$_2$ is not suitable for mobile applications, because of low volumetric energy density and safety problems. A promising way for mobile applications is solid state storage. One can differentiate physisorption in porous materials, including zeolites, MOFs and different types of carbons, and chemisorption resulting in the formation of hydrides. INS has been used to characterize various hydrides, starting from transition-metal hydrides, up to complex hydrides composed with light elements (lithium, boron, sodium or aluminium),

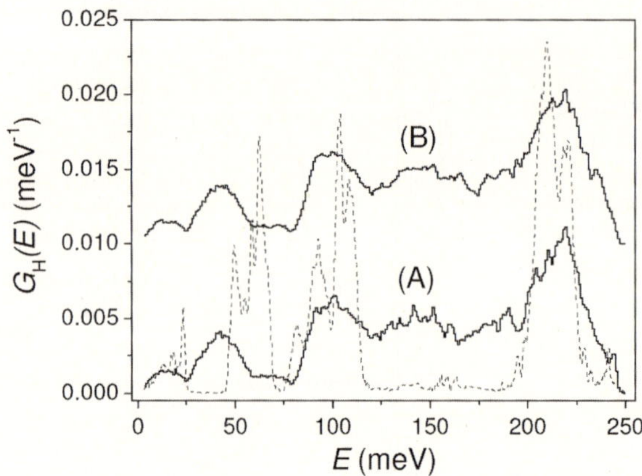

Fig. 2.3 INS spectra from NaAlH$_4$/ACF-25: **A** as prepared, **B** after hydrogen desorption–absorption cycling (*full lines*); the *dashed line* corresponds to bulk NaAlH$_4$. The data were recorded on the instrument TOSCA, at the pulsed neutron and muon source at the Rutherford Appleton Laboratory (ISIS). Reprinted with permission from (D. Colognesi, A. Giannasi, L. Ulivi, M. Zoppi, A.J. Ramirez-Cuesta, A. Roth, M. Fichtner, J. Phys. Chem. A **115**, 7503 (2011)) [11]

which provide high gravimetric H$_2$ density. Catalysts play a role in some cases: for alanates, doping with a titanium catalyst increases the rates of adsorption and desorption of H$_2$ [9], for borohydrides, hydrogen release through a hydrolysis reaction can be controlled catalytically [10].

Since transport can be a limiting step, the use of nanoparticles is an option to improve the kinetics. Raman scattering and INS techniques have been used to find out if the infiltration process of a carbon matrix with NaAlH$_4$ creates new chemical species (e.g., Na$_3$AlH$_6$) and if the nanoparticles of NaAlH$_4$ have a physical state different from the bulk or not [11]. The influence of hydrogen desorption–absorption cycling was also tested. The INS spectra of the melt-infiltrated composite of NaAlH$_4$ and active carbon fibers (ACF-25) are compared to the corresponding spectroscopic data taken from bulk NaAlH$_4$ in Fig. 2.3.

The comparison between spectra (a) and (b) in Fig. 2.3 indicates that INS is not sensitive to the desorption–absorption cycle. On the other hand, the comparison with bulk NaAlH$_4$ shows a broadening of the peaks with some energy shifts, typical of crystal-size effects. The extra intensity observed between 130 and 200 meV was attributed to the presence of a small amount of Na$_3$AlH$_6$, also observed in the Raman spectrum [11].

2.3.1 Hydrogen Diffusion in Nanoporous Materials

Atomic hydrogen diffusion in metal lattices has been studied for a long time by QENS, e.g. [12]. The diffusion of atomic hydrogen on the surface of catalysts has been studied scarcely [13, 14], this could be revisited since the neutron flux and the instrumentation has been tremendously improved since. Dihydrogen diffusion in porous media is a more recent topic. H_2 diffusion is of interest for a number of industrial processes, including membranes for fuel cells or reactions with membrane reactors, where the zeolite acts as a porous membrane. Over the last years, a large number of studies have been conducted on MOFs, to explore the performance of these solids as energy carrier for H_2. In general, for applications involving gas separation, a fast diffusion rate of the gas molecules into the porous system is as important as the adsorption uptake, the selectivity, or the regenerability.

There are few measurements on the diffusion of H_2 in zeolites and MOFs, this is due to the difficulty in measuring fast sorption rates by macroscopic methods. There are also few theoretical studies, a classical approach being not sufficient at low temperatures since the de Broglie thermal wavelength is comparable to the diameter of the confining pores, requiring in this case a quantum mechanical treatment.

The self-diffusivity, D_s, of H_2 in several zeolites was studied by pulsed-field gradient nuclear magnetic resonance (PFG-NMR) and QENS [15]. The self-diffusion coefficients were found to decrease with decreasing pore sizes of the zeolite structures, with one notable exception in silicalite. In this case, the diffusivities were found to be exceptionally small, the explanation being that H_2 molecules are trapped within the pentasil chains forming the structure.

While the self-diffusivity can be determined from incoherent neutron-scattering, the transport diffusivity, D_t, can be derived from coherent neutron-scattering [16]. Normally, D_t is measured under the influence of concentration gradients, i.e. under non-equilibrium conditions. It may seem strange to derive a non-equilibrium quantity from experiments performed at equilibrium. This is because the scattering function corresponding to coherent scattering is related to the full correlation function, $G(\mathbf{r}, t)$, which is itself connected with the evolution of the particle density around equilibrium. This approach was already formulated by Onsager [17] in his regression hypothesis. It was later proven that the response of a system to an external perturbation can be evaluated from correlation functions of the sample at equilibrium.

For a comparison of the different results, the transport diffusivity is often represented in terms of the so-called corrected diffusivity, D_0, which is defined by the relation

$$D_t(c) = D_0(c)\left(\frac{d \ln p}{d \ln c}\right) \tag{2.1}$$

where c denotes the adsorbate concentration in equilibrium with the pressure p. The term $d \ln p / d \ln c$ is the thermodynamic factor. When the adsorption isotherms can be fitted with the Langmuir model, the thermodynamic factor is equal to or larger than 1 [18].

2.3.1.1 Simultaneous Measurement of Self- and Transport Diffusivities

Deuterium is a special case for neutrons since it has comparable coherent and incoherent cross-sections. With D_2 molecules, it is then possible to measure simultaneously D_s and D_t. Such an experiment was performed for different concentrations of D_2 adsorbed at 100 K in the NaX zeolite [19]. The self-diffusivities were checked from the broadenings measured at the same loadings for H_2, after correcting from the mass difference between the two isotopes.

The values of D_s are plotted in Fig. 2.4. The slight increase of the self-diffusivity which is observed when the concentration increases was initially attributed to an interaction with the sodium cations [19]. This was later confirmed by atomistic computer simulations on the basis of the loading dependence of the partial molar configurational internal energy of the sorbate molecules, which indicated the existence of low-energy sites which are preferentially filled at low occupancy [20]. Molecules residing in these sites tend to exhibit lower translational mobility than molecules sorbed elsewhere in the intracrystalline space at higher occupancy.

The values of D_t, obtained at the same loadings, are also reported in Fig. 2.4. The width of the coherent scattering was found to show a minimum corresponding to the maximum of the structure factor. This line narrowing is characteristic of quasi-elastic coherent scattering and was first predicted by de Gennes [21].

It appears that at low D_2 concentration, the self- and transport diffusivities are similar, but for higher loadings the transport diffusivity increases rapidly and exceeds the self-diffusivity, as expected from the contribution of the thermodynamic factor, Eq. (2.1). This means that the lattice gas model, which predicts that the transport diffusivity does not depend on the concentration, does not apply for this system. Only close to the saturation of the zeolite does the transport diffusivity

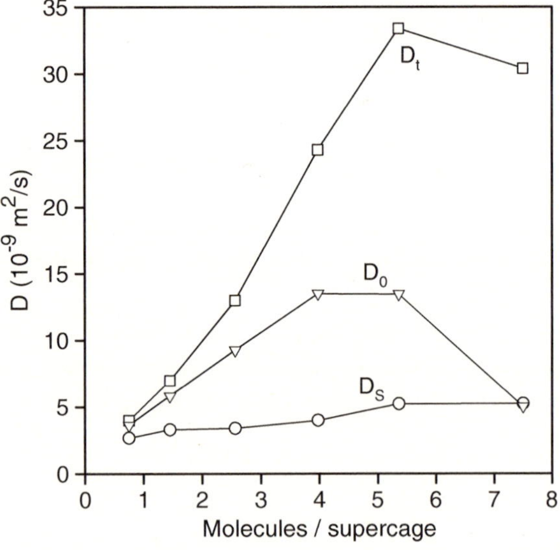

Fig. 2.4 Different diffusion coefficients obtained for D_2 in NaX zeolite at 100 K, as a function of loading: (*squares*) transport diffusivity, (*triangles*) corrected diffusivity, and (*rounds*) self-diffusivity

start to decrease, indicating that collective motions become affected by the packing density. The corrected diffusivity, D_0, was obtained from D_t and from the thermodynamic factor calculated by fitting a Langmuir isotherm to the adsorbed quantities. It is clear from Fig. 2.4 that for D_2 in NaX the corrected diffusivity is not constant, this assumption being often made in the interpretation of macroscopic measurements.

2.3.1.2 Diffusion of Hydrogen in One-Dimensional Metal-Organic Frameworks

MIL-53(Cr), and its isostructural form MIL-47(V), are built up from infinite chains of corner-sharing $Cr^{3+}O_4(OH)_2$ or $V^{4+}O_6$ octahedra interconnected by 1,4-benzenedicarboxylate groups (Fig. 2.5). These three dimensional MOFs contain one-dimensional diamond-shaped channels with pores of nm dimensions. One may note that MIL-53(Cr) exhibits hydroxyl groups located at the metal–oxygen–metal links (μ_2-OH groups) which open up the possibility of additional preferential adsorption sites and thus different adsorption or diffusion mechanisms to that of MIL-47(V) where these specific groups are absent.

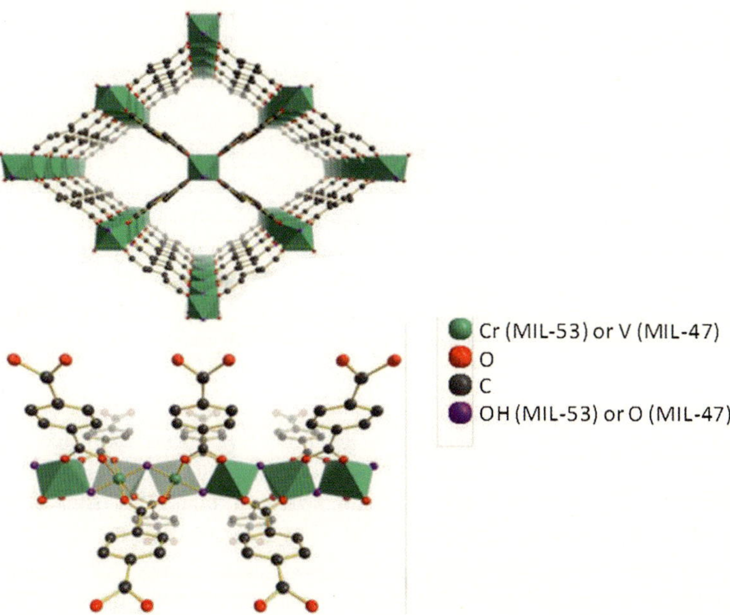

Fig. 2.5 View of the MIL-47(V) and MIL-53(Cr) structures. *Top* view along the chain axis, highlighting the one-dimensional pores system; *bottom* chain of corner sharing $Cr^{3+}O_4(OH)_2$ or $V^{4+}O_6$ octahedra

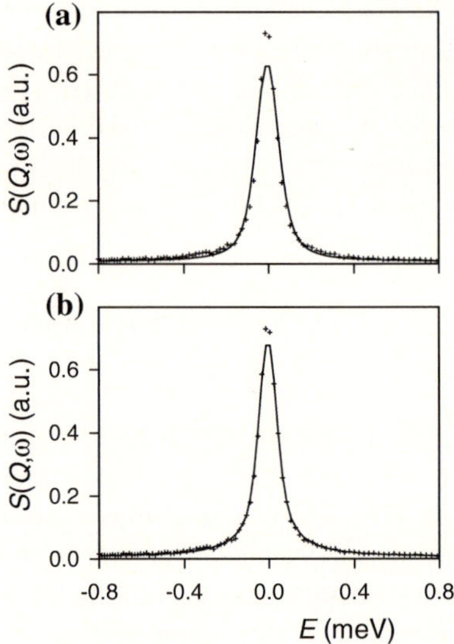

Fig. 2.6 Comparison between experimental (*crosses*) and fitted QENS spectra obtained for H_2 in MIL-53(Cr) at 77 K; the *solid lines* are computed for **a** three-dimensional diffusion, and **b** one-dimensional diffusion (Q = 0.27 Å^{-1}, 3.5 molecules per unit cell on average). The spectra were measured on the IN6 instrument at the ILL. Adapted with permission from (F. Salles, D.I. Kolokolov, H. Jobic, G. Maurin, P.L. Llewellyn, T. Devic, C. Serre, G. Férey, J. Phys. Chem. C **113**, 7802 (2009)) [22]

The self-diffusion of H_2 in these two structures was studied by QENS combined with molecular dynamics (MD) simulations [2, 22]. For the QENS measurements, the frameworks and the μ_2-OH groups in MIL-53(Cr) were deuterated to reduce the signal from the MOF. To illustrate the possibility to obtain information on diffusion anisotropy, one and three dimensional diffusion models are compared with experimental spectra in Figs. 2.6 and 2.7. When the diffusion is isotropic (three-dimensional), the theoretical dynamic structure-factor, $S(Q, \omega)$, corresponding to a translational motion has a Lorentzian profile in energy, but the line shape is more elongated in the case of diffusion in one-dimensional channels, because a powder average has to be made [22].

In MIL-53(Cr), profiles corresponding to three-dimensional diffusion and convoluted with the instrumental resolution do not fit perfectly through the experimental points (Fig. 2.6a and Ref. [22] for spectra obtained at other Q values). A normal one-dimensional diffusion model, with a more waisted shape, fits better the experimental data (Fig. 2.6b and Ref. [22]). This is due to the interaction between H_2 and the μ_2-OD groups, leading to a one-dimensional diffusion along the tunnels via a jump sequence involving these hydroxyl groups. Normal one-dimensional diffusion means that the molecules can cross each other in the tunnels of MIL-53. When the molecules cannot pass each other, the diffusion is called single file [23].

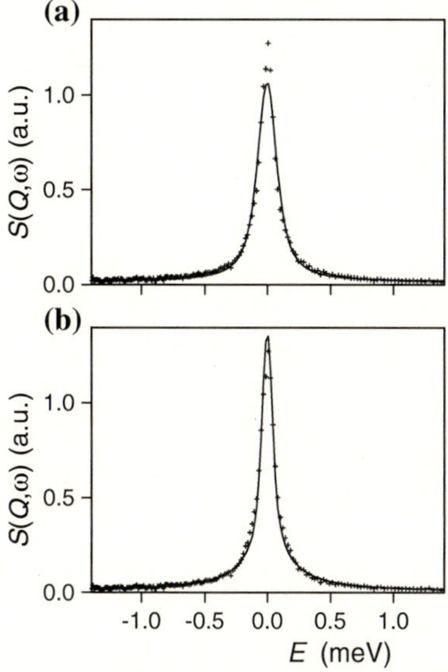

Fig. 2.7 Comparison between experimental (*crosses*) and fitted QENS spectra obtained for H_2 in MIL-47(V) at 77 K. *Solid lines* are computed for **a** three-dimensional diffusion and **b** one-dimensional diffusion ($Q = 0.27$ Å^{-1}, 3.4 molecules per unit cell on average). The spectra were measured on the IN6 instrument at the ILL. Adapted with permission from (F. Salles, D.I. Kolokolov, H. Jobic, G. Maurin, P.L. Llewellyn, T. Devic, C. Serre, G. Férey, J. Phys. Chem. C **113**, 7802 (2009)) [22]

In MIL-47(V), the reverse is found: The three-dimensional diffusion model reproduces better the QENS spectra than one-dimensional diffusion (Fig. 2.7 and Ref. [22]). The motions of H_2 in this MOF are random because there are no specific adsorption sites for hydrogen.

The D_s values of H_2 in both solids are reported in Fig. 2.8 as a function of loading. Contrary to the concentration dependence obtained in NaX (Fig. 2.4), D_s decreases in both MILs when the H_2 loading increases, this is due to steric reasons in these one-dimensional systems, and to the absence of strong adsorption sites. Experiment and simulation find higher diffusivities in MIL-47(V) than in MIL-53 (Cr), whatever the loading. This can be explained by the presence of the μ_2-OD groups in MIL-53(Cr) which act as attractive sites and steric barriers for H_2, leading thus to a slower diffusion process. Further, a high H_2 mobility is observed in both MILs, at low loading, the D_s values are about two orders of magnitude higher than in zeolites (Fig. 2.4 and Ref. [15]). Extrapolating D_s to zero loading in Fig. 2.8 leads to a value of the order of 10^{-6} m^{2}s^{-1} in MIL-47(V). This is comparable to the supermobility predicted in single-walled carbon nanotubes [24].

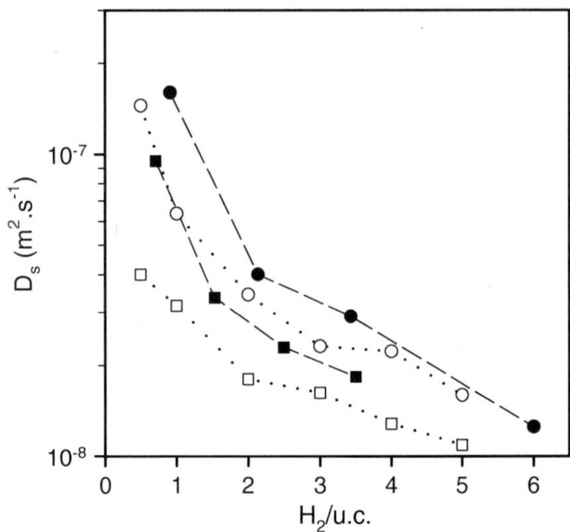

Fig. 2.8 Self-diffusivities of H_2 at 77 K as a function of loading in MIL-47(V) (*rounds*) and MIL-53(Cr) (*squares*): QENS (*full symbols*) and MD (*empty symbols*), where *u.c.* = unit cell

2.3.1.3 Diffusion of Hydrogen in ZIFs

Zeolitic imidazolate frameworks (ZIFs) constitute a sub-family of MOFs. They somewhat bridge the gap between zeolites and MOFs. The structural topologies of ZIFs are similar to zeolite or zeolite-like topologies [25]. ZIFs have tetrahedral frameworks where transition metals (Zn, Co, etc.) are linked by imidazolate ligands, the angle formed by imidazolates when bridging transition metals being close to the Si–O–Si bond angle in zeolites ($\approx 145°$). These new materials exhibit exceptional chemical and thermal stability.

Although ZIFs can have large cavities, the windows apertures whereby mass-transport proceeds have rather small diameters. In ZIF-8, the linker between Zn atoms is 2-methylimidazolate, and the diameter of the window is approximately 3.4 Å [25]. The diffusivities computed from MD, even for H_2, have been found to be quite sensitive to the mobility of the linker [26]. This is illustrated in Fig. 2.9, where a comparison between experimental and calculated self-diffusivities shows that an agreement is only reached for a flexible framework. As detailed in Sect. 2.3.2, quantum corrections were included via the Feynman–Hibbs approach [26]. The influence of flexibility on computed diffusivities in ZIF-3, which has a larger window diameter (4.6 Å), was found to be less pronounced, the agreement with QENS values being not as good as in ZIF-8 [26].

2.3.2 Quantum Effects on the Diffusion of Hydrogen Isotopes

Conventional methods for H_2 isotope separation, such as cryogenic distillation or thermal diffusion, are complex and energy intensive. From a classical viewpoint,

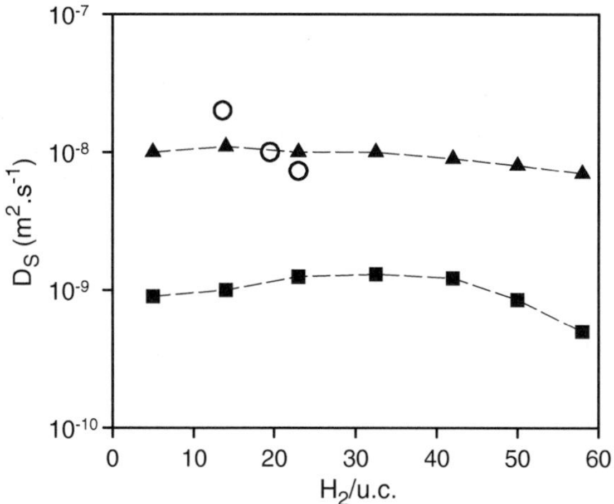

Fig. 2.9 QENS self-diffusivities for H_2 in ZIF-8 at 77 K (*circles*), compared with values computed with a rigid (*squares*) and flexible (*triangles*) framework model

H_2 and D_2 are similar in terms of size and shape as well as energetic considerations. However, H_2 and its isotopes can no longer be treated as classical molecules at low temperatures, as quantum effects become important due to their low mass. Quantum separation should be facilitated by the larger de Broglie wavelength of the lighter isotope. Kinetic molecular sieving, based on differences in diffusivity of the two isotopes, is attractive for porous materials with pore sizes comparable to the de Broglie wavelength.

MD simulations, with quantum corrections incorporated via the Feynman–Hibbs approach, showed that D_2 should diffuse faster than H_2 below 150 K [27], in a hypothetical zeolite rho having a window diameter of 5.43 Å (this corresponding to the distance between the O centres, upon subtraction of the O diameter a free diameter of 2.73 Å is obtained). QENS experiments were performed to test this prediction. However the zeolite rho used for the measurements had a different chemical composition, so that the free diameter of the windows was approximately 3.26 Å [28]. A new set of potential parameters was defined for the MD simulations, which yielded excellent agreement with the QENS diffusion coefficients of H_2 and D_2, but no quantum effect was observed. This was attributed to the larger pore size in the rho sample used experimentally.

Another series of QENS experiments performed with a carbon molecular sieve (CMS) confirmed the predictions made by the simulations [29]. A CMS can be obtained with different pore sizes, and the Takeda 3 Å CMS has pores below 3 Å. In this material, it was indeed found that D_2 diffuses faster than H_2, below 100 K (Fig. 2.10). This shows the extreme sensitivity of the reverse kinetic-selectivity on the window dimension. Transition-state theory and MD calculations indicated that to achieve high flux the cages interconnected by these windows must be small.

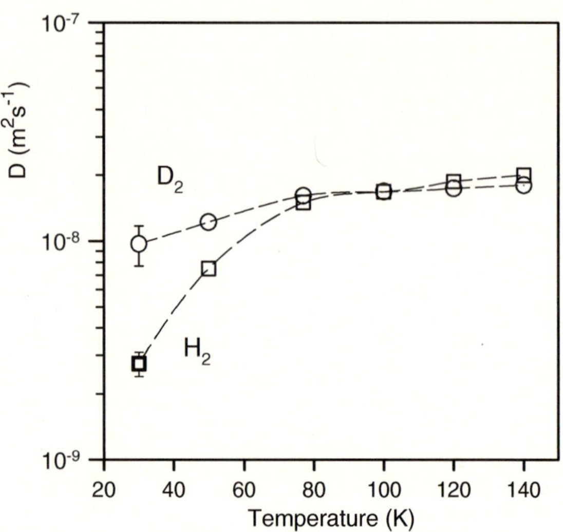

Fig. 2.10 Diffusion coefficients of H_2 (*squares*) and D_2 (*rounds*) in Takeda 3 Å CMS, as a function of temperature (*loading 0.5 mmol/g*)

2.4 Conclusions

Sensitivity is often a problem when studying catalysts with neutrons, only solids which have high surface areas can be studied. One still needs more neutron flux to widen the range of applications.

With INS, the advent of the VISION instrument at the Spallation Neutron Source (SNS) at the Oak Ridge National Laboratory (ORNL), the LAGRANGE instrument being operational at the ILL, and possible improvements on the instrument TOSCA at ISIS should allow us to tackle in the near future grafted catalysts, fuel cells, small supported metal particles (metal loading <1 %), etc. The large Q values which are inherent to these spectrometers at large energy transfers is a serious limitation. The detrimental influence of the Debye–Waller factor on the INS spectra has been known for a long time [30]. This can be partially resolved using direct geometry instruments which allow access of lower Qs [31], but different incident energies have to be selected so that various parts of the spectrum can be combined. On the other hand, the technique is particularly suited to study the different hydrogen species which are formed after dissociation of dihydrogen, during reduction or activation of the catalyst. The problem of characterizing active hydrogen is still a big issue in catalysis.

In the same way as ab initio methods are increasingly being used in INS, molecular simulations are now combined with QENS experiments. Since the space and time scales of the neutron techniques match closely the ones covered by molecular simulations, one expects, and usually finds, good agreement between neutrons and simulations. QENS constitute a benchmark to validate and further develop the modelling work, and the computed trajectories of the sorbate guest molecules within the host matrices are invaluable to understand QENS observables. Discrepancies between experiment and simulation do happen and require in such cases the consideration of a flexible lattice or an improved force field.

References

1. H. Jobic, J. Mol. Catal. A: Chem. **158**, 135 (2000)
2. F. Salles, H. Jobic, G. Maurin, M.M. Koza, P.L. Llewellyn, T. Devic, C. Serre, G. Férey, Phys. Rev. Lett. **100**, 245901 (2008)
3. N. Malikova, S. Longeville, J.-M. Zanotti, E. Dubois, V. Marry, P. Turq, J.M. Ollivier, Phys. Rev. Lett. **101**, 265901 (2008)
4. H. Jobic, in *Catalysis by Metals,* A.J. Renouprez, H. Jobic (eds.) (Les Editions de Physique – Springer, Berlin 1997), p. 181
5. P.W. Albers, S.F. Parker, Adv. Catal. **51**, 99 (2007)
6. A.A. Khassin, G.N. Kustova, H. Jobic, T.M. Yurieva, Y.A. Chesalov, G.A. Filonenko, L.M. Plyasova, V.N. Parmon, Phys. Chem. Chem. Phys. **11**, 6090 (2009)
7. R. Juarez, S.F. Parker, P. Concepcion, A. Corma, H. Garcia, Chem. Sci. **1**, 731 (2010)
8. C. Pirez, M. Capron, H. Jobic, F. Dumeignil, L. Jalowiecki-Duhamel, Angew. Chem. Int. Ed. **50**, 10193 (2011)
9. B. Bogdanovic, M. Schwickardi, J. Alloys Compd. **1**, 253 (1997)
10. C.H. Liu, B.H. Chen, C.L. Hsueh, J.R. Ku, M.S. Jeng, F. Tsau, Int. J. Hydrogen Energy **34**, 2153 (2009)
11. D. Colognesi, A. Giannasi, L. Ulivi, M. Zoppi, A.J. Ramirez-Cuesta, A. Roth, M. Fichtner, J. Phys. Chem. A **115**, 7503 (2011)
12. K. Sköld, G. Nelin, J. Phys. Chem. Solids **28**, 2369 (1967)
13. A. Renouprez, P. Fouilloux, R. Stockmeyer, H.M. Conrad, G. Goeltz, Ber. Bunsenges. Phys. Chem. **81**, 429 (1977)
14. A. Renouprez, R. Stockmeyer, C.J. Wright, J. Chem. Soc. Far. Trans. I **75**, 2473 (1979)
15. N.K. Bär, H. Ernst, H. Jobic, J. Kärger, Magn. Reson. Chem. **37**, S79 (1999)
16. R. Hempelmann, D. Richter, D.A. Faux, D.K. Ross, Z. Phys, Chem. Neue Folge **159**, 175 (1988)
17. L. Onsager, Phys. Rev. **37**, 405 (1931)
18. H. Jobic, D.N. Theodorou, Microporous Mesoporous Mater. **102**, 21 (2007)
19. H. Jobic, J. Kärger, M. Bée, Phys. Rev. Lett. **82**, 4260 (1999)
20. E. Pantatosaki, G.K. Papadopoulos, H. Jobic, D.N. Theodorou, J. Phys. Chem. B **112**, 11708 (2008)
21. P.G. de Gennes, Physica **25**, 825 (1959)
22. F. Salles, D.I. Kolokolov, H. Jobic, G. Maurin, P.L. Llewellyn, T. Devic, C. Serre, G. Férey, J. Phys. Chem. C **113**, 7802 (2009)
23. H. Jobic, K. Hahn, J. Kärger, M. Bée, A. Tuel, M. Noack, I. Girnus, G.J. Kearley, J. Phys. Chem. B **101**, 5834 (1997)
24. A.I. Skoulidas, D.M. Ackerman, J.K. Johnson, D.S. Sholl, Phys. Rev. Lett. **89**, 185901 (2002)
25. R. Banerjee, A. Phan, B. Wang, C. Knobler, H. Furukawa, M. O'Keeffe, O.M. Yaghi, Science **319**, 939 (2008)
26. E. Pantatosaki, H. Jobic, D.I. Kolokolov, S. Karmakar, R. Biniwale, G.K. Papadopoulos, J. Chem. Phys. **138**, 34706 (2013)
27. A.V. Anil Kumar, S.K. Bhatia, Phys. Rev. Lett. **95,** 245901 (2005)
28. A.V. Anil Kumar, H. Jobic, S.K. Bhatia, J. Phys. Chem. B **110,** 16666 (2006)
29. T.X. Nguyen, H. Jobic, S.K. Bhatia, Phys. Rev. Lett. **105**, 085901 (2010)
30. H. Jobic, Chem. Phys. Lett. **106**, 321 (1984)
31. S.F. Parker, D. Lennon, P.W. Albers, Appl. Spectrosc. **65**, 1325 (2011)

Chapter 3
Carbon Dioxide Separation, Capture, and Storage in Porous Materials

Anita Das, Deanna M. D'Alessandro and Vanessa K. Peterson

Abstract Solid porous materials represent one of the most promising technologies for separating and storing gases of importance in the generation and use of energy. Understanding the fundamental interaction of guest molecules such as carbon dioxide in porous hosts is crucial for progressing materials towards industrial use in post and pre combustion carbon-capture processes, as well as in natural-gas sweetening. Neutron scattering has played a significant role already in providing an understanding of the working mechanisms of these materials, which are still in their infancy for such applications. This chapter gives examples of insights into the working mechanisms of porous solid adsorbents gained by neutron scattering, such as the nature of the interaction of carbon dioxide and other guest molecules with the host as well as the host response. The synthesis of many of these porous hosts affords significant molecular-level engineering of solid architectures and chemical functionalities that in turn control gas selectivity. When directed by the insights gained through neutron-scattering measurements, these materials are leading toward ideal gas separation and storage properties.

3.1 The Importance of Carbon Dioxide Capture

As the prime mover of carbon through the atmosphere, carbon dioxide (CO_2), plays a vital role in enabling the cycle of carbon from the Earth's crust (where it is found in elemental graphite and diamond, carbonates, and fossil fuels) to our oceans

A. Das · D.M. D'Alessandro (✉)
School of Chemistry, The University of Sydney, Sydney, NSW, Australia
e-mail: deanna.dalessandro@sydney.edu.au

A. Das
e-mail: anita.das@sydney.edu.au

V.K. Peterson (✉)
Australian Nuclear Science and Technology Organisation, Lucas Heights, NSW, Australia
e-mail: vanessa.peterson@ansto.gov.au

© Springer International Publishing Switzerland 2015
G.J. Kearley and V.K. Peterson (eds.), *Neutron Applications in Materials for Energy*, Neutron Scattering Applications and Techniques,
DOI 10.1007/978-3-319-06656-1_3

(where it occurs in carbonate minerals formed by the action of coral-reef organisms and aqueous CO_2). For hundreds of millions of years, the carbon cycle has maintained a relatively constant amount of CO_2 in the Earth's atmosphere (approximately 400 ppm by volume). While the contribution from human industry is relatively small, its recent growth has shifted this natural balance. Since the start of the Industrial Revolution around 1760, the concentration of CO_2 in the atmosphere has risen dramatically from 280 to 385 ppm today [1, 2]. This significant rise has been attributed to an increasing dependence on the combustion of fossil fuels (coal, petroleum, and natural gas), which account for 86 % of man-made greenhouse-gas emissions, the remainder arising from land use change (primarily deforestation) and chemical processing.

The development of more efficient processes for CO_2 capture from major point sources such as power plants and natural-gas wells is considered a key to the reduction of greenhouse-gas emissions implicated in global warming. Numerous national and international governments and industries have established collaborative initiatives such as the Intergovernmental Panel on Climate Change [3] (IPCC), the United Nations Framework Convention on Climate Change [4], and the Global Climate Change Initiative [5] to achieve this goal. The capture and sequestration of CO_2, the predominant greenhouse gas, is a central strategy in these programmes as it offers the opportunity to meet increasing demands for fossil-fuel energy in the short to medium term, whilst reducing the associated greenhouse-gas emissions in line with global targets. Carbon capture and storage (CCS) will complement other strategies such as improving energy efficiency, switching to less carbon-intensive fuels, and the phasing in of renewable-energy technologies.

Three major technologies are predicted to have the greatest likelihood of reducing man-made emissions to the atmosphere that are implicated in global warming. These processes include postcombustion and precombustion capture from power plants involving $CO_2/N_2/H_2O$ and CO_2/H_2 separations, respectively, and natural-gas sweetening ($CO_2/CH_4/N_2$ separation). The separation processes required for each of these capture applications differs with regard to the nature of the gas mixture and the temperatures and pressures involved, imposing constraints on the materials and processes employed [6, 7].

Conventional CO_2 capture processes employed in power plants world-wide are typically postcombustion 'wet scrubbing' methods, involving the absorption of CO_2 by amine-containing solvents such as methanolamines [8]. Power plant flue-gas streams consist primarily of N_2, H_2O, and CO_2 in a 13:2:2 ratio by weight [9]. Prior to the compression and liquefaction of the captured CO_2 for transportation to storage sites, CCS requires the separation of CO_2 from all other flue-gas components. CO_2 is strongly absorbed by the amine to form a carbamate species [10], however, the high heat of formation associated with the creation of the carbamate leads to a considerable energy penalty for regeneration of the solvent. Since the flue streams from coal-fired power plants contain dilute concentrations of CO_2 (typically 10–15 %) at relatively low pressures and temperatures (1 atm., 40 °C), it is estimated that CO_2 capture and compression will increase the energy requirements of a plant by 25–40 %. Analysis has shown that the thermodynamic minimum energy-penalty for capturing 90 % of the CO_2 from the flue gas of a typical coal-fired power plant is approximately 3.5 % (assuming a flue gas containing 12–15 % CO_2 at 40 °C) [11]. The transportation and storage of CO_2 will necessitate further investment and capital costs. These economic and energy comparisons underscore the immense opportunities and incentives that exist for improved CO_2 capture processes and materials. Despite improvements in conventional postcombustion chemical-absorption methods, wet-scrubbing methods suffer a number of drawbacks and are therefore not cost-effective for large-scale carbon emissions reduction.

While the retrofitting of existing power plants using postcombustion capture methods presents the closest marketable technology, two major alternatives to postcombustion CO_2 capture processes have been proposed, and are currently in the test stages of development [12]. Precombustion processes involve a preliminary fuel-conversion step using a gasification process and subsequent shift-reaction to form a mixture of CO_2 and H_2 prior to combustion. The high pressure of the product gas-stream facilitates the removal of CO_2 from the CO_2/H_2 mixture at pressures of 30–50 bar and temperatures of 50–75 °C [13]. The significant advantage of precombustion capture is that the higher component concentrations and elevated pressures reduce the energy capture penalty of the process to 10–16 %, roughly half that for postcombustion CO_2 capture. A further advantage is that precombustion technology generates an H_2-rich fuel, which can be used as a chemical feedstock in a fuel cell for power generation or in the development of an H_2 economy. In oxyfuel (or denitrogenation) processes, fuel is combusted in O_2 instead of air by the exclusion of N_2, thereby producing a concentrated stream of CO_2 without the need for separation (in high, sequestration-ready concentrations of 80–98 %). Since the separation of interest in this case is air separation (O_2 from mainly N_2 at a pressure of around 100 bar and temperature of 50 °C), reducing the cost of O_2 generation is key to industrial viability. While the emerging technologies associated with precombustion and oxyfuel processes cannot be readily incorporated (via retrofitting) into existing power plants as can postcombustion CO_2 capture processes, the projections from the IPCC indicate that the required extensive capital investments will be compensated by the relatively higher efficiency of the CO_2 separation and capture process [3].

Another important application for CO_2 capture technologies is the 'sweetening' of sour natural-gas wells, where the sweetening refers to the separation of CO_2 from CH_4. Natural gas reserves (mainly CH_4) are typically contaminated with over 40 % CO_2 and N_2 and the use of such fields is only acceptable if the additional CO_2 is separated and sequestered at the source of production. The capture of CO_2 from ambient air has also been suggested, however, the low concentration of CO_2 in air (0.04 %) presents a significantly higher barrier to capture compared with postcombustion methods, and the expense of moving large volumes of air through an absorbing material presents a further challenge in its implementation [6].

The key factor that underscores significant advancements in CCS is materials to perform the capture process [6, 7]. The challenge for gas-separation materials is that the differences in properties between the gases that have to be separated are relatively small. However, differences do exist in the electronic properties of the gases:

CO_2 has a large quadrupole moment (13.4×10^{-40} cm^2 vs. 4.7×10^{-40} cm^2 for N_2 and CH_4 is non-polar) and CH_4 adsorbs preferentially over N_2 due to its higher polarizability (17.6×10^{-25} cm^3 for N_2 and 26.0×10^{-25} cm^3 for CH_4).

A diverse range of promising methods and materials for CO_2 capture applications that could be employed in any one of the abovementioned postcombustion, precombustion, or oxyfuel processes have been proposed as alternatives to conventional chemical absorption. These include the use of physical absorbents, membranes, cryogenic distillation, hydrate formation, chemical-looping combustion using metal oxides, and adsorption on solids using pressure and/or temperature swing adsorption, where the adsorption and desorption temperature/pressures are different and a "swing" is made between them [14]. The key requirements for these new materials are that they exhibit air and water stability, corrosion resistance, high thermal-stability, high selectivity and adsorption capacity for CO_2, as well as adequate robustness and mechanical strength to withstand repeated exposure to high-pressure gas streams. A number of review articles have elaborated the status of a new classes of materials for CO_2 capture [6, 7]. In particular, metal-organic framework (MOF) materials are progressing at a rapid pace.

With respect to new materials, the key scientific challenges are the development of a level of molecular control and modern experimental and computational methods. For crystalline materials, adsorption isotherms and breakthrough–curve measurements under conditions that closely resemble working-condition gas mixtures are essential. In reality, pure gas-adsorption isotherms are often measured, with ideal adsorbed solution theory (IAST) applied in some cases to predict multicomponent adsorption behaviour [15]. A parameter that must be assessed in all cases is the enthalpy of adsorption, since the cost for the regeneration of a capture material is clearly dependent on the energy required to remove the captured CO_2.

Characterization of the molecular transport properties of materials is essential to obtain an understanding of transport processes. Important molecular-level information required includes: how the material structure changes with loading, how adsorbates bind to the material, and how different permeates influence each other's solubility. In situ techniques are particularly powerful as they allow the interactions between gas molecules and the matrix to be probed and determine the material structure under different loading conditions. The most significant information from such measurements is gained from the ability to correlate absorbate uptake with the absorbent structure and the molecular-level absorbate mobility. A comparison between the molecular-level absorbate mobility and its macroscopic diffusion should provide insights into the mechanism of selective transport through these materials.

In parallel with experimental studies, computational modelling methods are being developed, both as a tool to understand further details of the adorbate-absorbent interaction, and as a tool to predict the performance of materials proposed for a given separation process, with the latter enabling large-scale screening of new materials. Ultimately, a clear understanding of the structure- and dynamics-function relations will direct experimental efforts towards a new generation of materials with improved CO_2 capture abilities. Developing force fields for computational work using detailed structures is important for the successful prediction of thermodynamic and transport properties of new materials.

3.1.1 Porous Materials for CO_2 Separation and Storage

Microporous and mesoporous solid-state materials such as activated carbon, carbon-based molecular sieves, mesoporous silicas, and zeolites have been demonstrated to have a significant, and in some cases selective, CO_2 adsorption capacity. Such materials have advantages for CO_2 capture over the amine solvents currently employed in industry as they are endowed with better stabilities and lower energies of regeneration. Zeolites in particular have been widely studied for the purpose of CO_2 capture due to their defined and controllable pore size, insensitivity to moisture, and high uptake at non-extreme conditions (for example, zeolite 13X has a CO_2 uptake of 3.6 mmol.g^{-1} at 25 °C) [16]. At higher, more industrially-relevant temperatures, these zeolites tend to lose adsorption capacity and also suffer low selectivity for CO_2 over other gases (e.g. N_2 and H_2) as a result of the physisorptive nature of the CO_2-adsorbate interaction. To enhance selectivity for CO_2, amine-impregnated or amine-modified materials have been explored, which couple the chemisorption approach used in conventional liquid-amine capture with the physisorption approach traditionally seen in porous solid materials. This technique has also been employed using a number of porous silicas, such as MCM-41 mesoporous molecular sieves impregnated with polyethylenimine [17] and SBA-15 mesoporous silicas covalently tethered with hyperbranched amines [18]. Despite the increase in CO_2 selectivity of such materials achieved using this approach, they often suffer low stabilities over repeated cycles.

For industrial purposes, solid materials with high selectivity and capacity for CO_2 uptake, as well as stability to extreme industrial conditions and a low energy for regeneration, are desired. Metal-organic frameworks (MOFs) are a highly promising class of material for this application due to their structural and chemical versatility, arising from different combinations of metal coordination-spheres as well as multidentate bridging ligands with different lengths, shapes, and directionalities of the coordinating groups. While this versatility sometimes comes at the expense of being able to predict structure accurately, the MOF scaffold provides a unique platform upon which to systematically tune the functionalities of known structures to obtain a desirable property [19]. They may be rationally engineered to have a high surface area and porosity, can be post-synthetically modified to allow for increased selectivity for CO_2, and can possess excellent stability under industrially-relevant conditions [20]. High surface areas and the possibility to possess coordinatively-unsaturated metal sites make MOFs particularly attractive as gas-selective adsorbents. Coordinatively-unsaturated metal centres have been generated in such materials via chelation by post-synthetically modifying bridging ligands or via insertion into open-ligand sites [9]. However, they are most often created through the evacuation of MOFs that have metal-bound solvent molecules. An effective strategy in tuning the selectivity of MOFs for CO_2 is the introduction of functional groups into the pores such as amine [21, 22] and sulfone groups (a SO_2 group attached to two C atoms) [23], known to specifically interact with CO_2 preferentially over other gases of industrial interest.

Topological control is another strategy employed in the design of MOFs for gas separation, however, this approach is frequently serendipitous due to unknown mechanisms of formation of these materials. Despite this, it has been shown that pore size and shape modulation can determine the diffusion dynamics of the molecules to be separated (e.g. metal formates, $M(HCO_2)_2$ (M = Mg, Mn, Co, or Ni), have been shown to selectively adsorb CO_2 over CH_4, suggesting a size-exclusion effect by the small pores) [24]. Generally, attempts to increase pore size through the incorporation of longer ligands results in framework interpenetration. While this is disadvantageous from a gas-storage standpoint, it may be favourable for some guest separations by their kinetic-diameter differences (e.g. CO_2 over CH_4) [25, 26]. This type of "molecular sieving" approach may also be achieved by taking advantage of the structural flexibility in MOFs. For example, the material Cr(OH)(bdc), where bdc = 1, 4-benzenedicarboxylate and also known as MIL-53(Cr), exhibits a two-step CO_2 uptake isotherm compared to a single-step CH_4 uptake isotherm, indicative of a specific "gating" effect [27].

3.2 Neutron Scattering in Studying Porous Materials for CO_2 Separation and Storage

Neutron scattering holds many opportunities for obtaining unique information concerning the porous-solid adsorbent (host) as well as host-adsorbate (host-guest) system. Measurement of structure and dynamics using neutron scattering, across length and time scales pertinent to these systems (also possible at the same time), has been exploited for the better understanding of guest binding in the host and separation mechanisms, as well as the host's response to adsorption, all of which are key to progressing the application of such systems in CCS.

Information about both host and host-guest structure, which yields details of the structural response of the host to adsorption, the location of the guest in the host, and guest-host interaction, are important to determining structure-property relations. As neutron diffraction intensity does not reduce with scattering angle, relatively more fine structural detail is gained than using X-ray diffraction, providing important detail concerning the guest–host and guest–guest interactions. The isotopically-dependent structural information afforded by neutrons allows different contrast between parts of the host framework and/or guest to be gained, providing many advantages for such structural investigations. Examples include distinguishing between guests such as N_2, O_2, and CO_2, and obtaining details of both the host's ligands and metal centres, as well as guests, even within a MOF containing heavy-metal atoms and guests containing light atoms. Additionally, the information obtained can be tuned through isotopic substitution, such as in determining the molecular orientations of CH_4 within a host using the isotopically-substituted CD_4, where D is deuterium (^2H),

The dynamic information obtained through neutron scattering is also isotopically dependent, and spectroscopic neutron techniques allow direct measurement of the local environment and the diffusional transport of the guest within the host. Both structure and dynamics can be measured at the same time, enabling insights into the geometry of the guest motion, in turn allowing the details of the mechanism of diffusion of the guest within the host to be gained.

In situ methods are central in the analysis of MOFs for guest separation and storage applications. Although in situ X-ray single crystal and powder diffraction studies of CO_2 in MOFs facilitate the understanding of the functional mechanism of MOFs for CCS applications [28–30], in situ neutron-scattering methods have significant advantages over X-ray studies of MOF-guest systems, with the penetrating power of neutrons being central to this. Neutrons easily penetrate the often-complex sample environments required for control over temperature of the host at the same time as gas delivery, covering easily the range of temperatures from the relatively cold (about −263 °C) conditions required to "lock in" guests and determine accurate structural details, to the more moderate temperatures required to replicate working post-, pre-, and oxyfuel combustion, as well as natural gas-sweetening conditions (40–75 °C). The relatively high penetrating power of neutrons also allows for the analysis of bulk samples, mg—gram quantities, providing information about the more industrially-relevant "bulk" properties of the material. The bulk-scale analysis also aids in accurately dosing the sample with a known number of guest molecules to determine in detail the nature of their interaction with the host.

3.2.1 Location of CO_2

Neutron powder diffraction (NPD) has been used extensively to determine the location of guest molecules in porous framework materials, and this work extends to CO_2 [31–36]. Two MOFs that have been explored intensively for their selective sorption properties are $M_2(dobdc)$ (M = Mg, Mn, Co, Ni, Zn; dobdc = 2, 5-dioxido-1, 4-benzenedicarboxylate), also known as MOF-74 or CPO-27, and $M_3(btc)_2$ (M = Cu, Cr, Mo; btc = 1, 3, 5-benzenetricarboxylate) with $Cu_3(btc)_2$ also known as HKUST-1. Both materials contain exposed M^{2+} sites, with the $M_2(dobdc)$ material possessing exceptionally large densities of such sites. The location of CO_2 in the two MOF materials $Mg_2(dobdc)$ and $Cu_3(btc)_2$, along with the host–CO_2 structure, was determined using NPD. The nature of the host–CO_2 interaction in both materials was identified to be binding at metal sites via an oxygen with the remainder of the molecule remaining relatively free (see Fig. 3.1), where the adsorbed CO_2 is clearly located above the open Mg ions in $Mg_2(dobdc)$ [32]. Importantly, the presence of coordinatively-unsaturated metal sites in MOFs such as $M_2(dobdc)$ and $Cu_3(btc)_2$ leads to enhanced interactions between adsorbates such

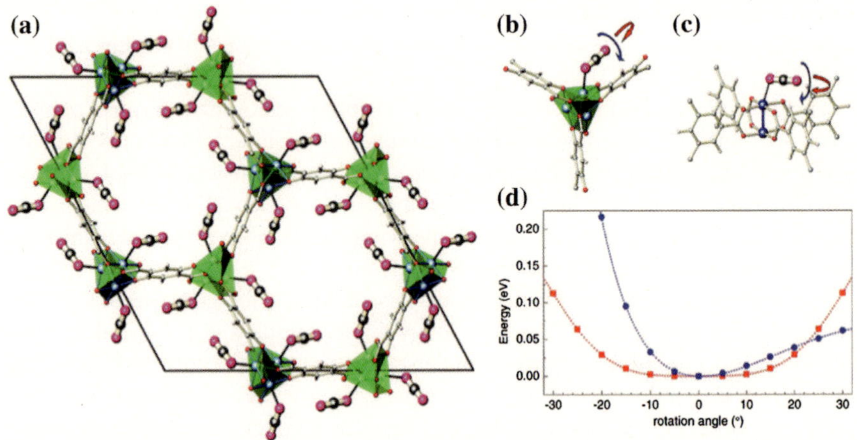

Fig. 3.1 **a** Mg$_2$(dobdc)-CO$_2$ structure determined from NPD showing the strongest CO$_2$ binding-site. (**b–c**) Schematic showing the dominant two motions of the CO$_2$ at the open metal, determined from computational calculations using the NPD-derived structures for Mg$_2$(dobdc) (**b**) and Cu$_3$(btc)$_2$ (**c**). Arrows in (**b–c**) represent CO$_2$ motions occurring about the surface normal largely parallel to the metal—O plane (*red*) and away from this surface normal (*blue*). The potential energy for these two modes occurring in Cu$_3$(btc)$_2$ is shown (**d**) as a function of CO$_2$ rotation angle. Atomic structure is represented with Mg and O forming the central polyhedra (**a**) in Mg$_2$(dobdc) and (axial pair) in Cu$_3$(btc)$_2$ (**c**), with C and H forming the linker, and the pendant CO$_2$ (**a**). Reprinted with permission from (H. Wu, J.M. Simmons, G. Srinivas, W. Zhou, T. Yildirim, J. Phys. Chem. Lett. **1**, 1946 (2010)) [37]. Copyright (2010) American Chemical Society

as CO$_2$ and the host framework, but also guest molecules such as CH$_4$ and H$_2$. Indeed, we will show that the application of NPD to examine competitive binding between CO$_2$ and these other gases represents an area of significant current interest.

Density-functional theory (DFT) calculations performed using the NPD-determined structures allowed evaluation of the representative CO$_2$ motions in Mg$_2$(dobdc) and Cu$_3$(btc)$_2$. These calculations show that the O bound to the open metal can be approximated as the rotational centre. In both materials the open metal and the associated carboxyls from the ligands form a nearly square-planar surface at the CO$_2$ binding-site, such that the metal–CO$_2$ interaction closely represents a surface normal. The CO$_2$ rotations are shown by arrows in Fig. 3.1(b–c), occurring about the surface normal (red arrows) and away from the surface-normal (blue arrows). The mode energies for the motions denoted by the red and blue arrows are 4.3 and 8.5 meV for Mg$_2$(dobdc), respectively, and 0.2 and 3.4 meV for Cu$_3$(btc)$_2$, respectively. To gain more direct information about the CO$_2$-host interaction in Cu$_3$(btc)$_2$ the energy at the open-metal sites (assuming a rigid host framework) was calculated for these two CO$_2$ motions as a function of CO$_2$ rotational angle, and is shown in Fig. 3.1d. As expected, the energy curves are shallow, particularly in the ±10° region, allowing for significant CO$_2$ orientational disorder in the MOF at this site with little effect on the total energy of the MOF–CO$_2$ system. These findings are in excellent agreement with the relatively-large atomic displacement parameters of CO_2 adsorbed at open metal sites obtained from the NPD measurements, in particular for $Cu_3(btc)_2$. These results also point to the presence of disorder (either static or dynamic) in the orientation of the CO_2 molecule, resulting in a relatively large apparent O–C–O bond bend obtained from the NPD data, which is a structural average and strongly biased by the relatively large disorder of the adsorbed CO_2. Since CO_2 is reversibly physisorbed on these open metal sites, a large degree of CO_2 bond activation and bending is unlikely.

Vacancy-containing Prussian blue analogues of the formula $M(1)^{II}_3[M(2)^{III}(CN)_6]^2$ (where M(1) and M(2) are transition metals) are excellent candidate gas adsorbents as 1/3 of their octahedral $M^{III}(CN)_6^{3-}$ units are vacant for charge neutrality, generating both non-vacancy and vacancy pores. Each vacancy pore will possess some of the six bare-metal sites per formula unit (eight per unit cell). The M$(1)_3[Co(CN)_6]_2$ system (M(1) = Mn, Co, Ni, Cu, Zn, Fig. 3.2) displays good selectivity for CO_2 over CH_4 and N_2 [38], with a NPD study revealing two sites for CO_2 binding in the $Fe_3[Co(CN)_6]_2$ material, which has a CO_2 uptake of

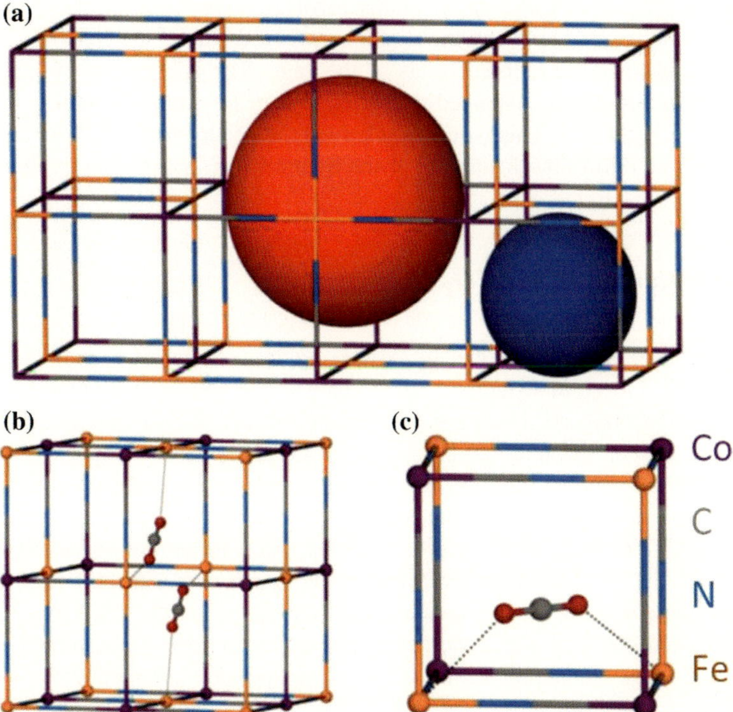

Fig. 3.2 **a** $Fe_3[Co(CN)_6]_2$ structure showing non-vacancy (*right sphere*) and vacancy (*left sphere*) pores. **b** Bridging open-metal CO_2 adsorption site located in a vacancy-type pore. **c** Non open-metal interacting CO_2 adsorption site located above non-vacancy square faces of the framework. Reproduced from (S.H. Ogilvie, S.G. Duyker, P.D. Southon, V.K. Peterson, C.J. Kepert, Chem. Commun. **49**, 9404 (2013)) [39]

2.20 mmol g^{-1} at 35 °C and 1 bar [39]. At one of these sites CO_2 was found to bridge between two open-metal sites, with the quadrupolar CO_2 molecule interacting strongly with the positively-charged Fe sites. The saturation of this site by CO_2 at relatively-low CO_2 concentrations indicated the favourable nature of the interaction, explaining the selectivity of the material.

CO_2 hydrates, consisting of an H_2O-cage encapsulating a CO_2, are another porous material that have great potential for application as CO_2 adsorbents, and these too have been studied using NPD to determine the locations of CO_2 within the cage [40]. This study used a cage in which D was substituted for H, allowing structural details of the cage atoms and their interaction with the CO_2 to be determined. The study also included the temperature-dependence of this CO_2-cage interaction. Data indicate that the CO_2 molecule in the tetrakaidecahedral cage rotates rapidly even at low temperatures and that the interaction between the CO_2 molecule and the D atoms of the cage is strong enough to provide the site dependence of the atomic displacement parameters of the D atoms. Further work on CO_2 hydrates [41] using NPD found CO_2 to have different motions in the small and large cages of this system. In both cages the CO_2 resides at the cage centre, however, in the small cage the O atoms revolved freely around the C atom, in contrast to the large cage where the O atoms revolved around the C atom along the plane parallel to the hexagonal facets of the cage. The analysis of CO_2 hydrates using NPD has also been extended to studies of their formation, including kinetics, using in situ NPD [42]. This work also derived the occupancy of CO_2 in the small and large cage during the formation of the hydrate.

3.2.2 Dynamics of the CO_2-Host System

Crystallographic studies provide the time-averaged position of atoms, and whilst some insight into atom dynamics can be gained through analysis of the average structure and atomic displacement parameters, detailed information regarding the dynamics of the guest and host-guest system are better gained through other neutron-scattering methods. Inelastic neutron scattering (INS) combined with computational calculations permit visualization of the atom dynamics and allow elucidation of the interaction between adsorbed guest molecules and the host.

Measurement of the interaction between CO_2 molecules and the porous host is crucial to understanding the detailed binding mechanism and therefore the observed selectivity and guest-uptake properties of porous hosts. INS cannot directly detect the CO_2 binding interaction within an adsorbent because the incoherent neutron-scattering cross section for these elements (for their naturally-abundant isotopes) is effectively zero, being 0.001 barns each. One approach to overcome this problem is to combine INS and DFT to visualize captured CO_2 molecules within a porous host by investigating the change in the dynamics of the other atoms of the adsorbent structure. INS spectra can be calculated directly using DFT-based computations to obtain the force constants, and then making the harmonic approximation to obtain the eigenvectors and eigenvalues to determine the spectral intensities and frequencies, respectively [43]. INS and DFT-based calculations are a powerful combination in understanding the working mechanism of functionalized materials containing specific gas molecule binding-sites, probing directly the impact of functional groups, and other host features such as topology and pore shape and size on the orientation and type of binding of CO_2 in the host.

An example of this is the application of INS and DFT to study CO_2 in the material $Al_2(OH)_2(bptc)$, where bptc = biphenyl-3, 3′,5, 5′-tetracarboxylate and also known as NOTT-300, where the neutron-scattering signal comes primarily from the H atoms in the $Al_2(OH)_2(bptc)$ hydroxyl groups and benzene rings of the ligand, and the INS signal is perturbed by the binding of CO_2 (Fig. 3.3) [44]. $Al_2(OH)_2(bptc)$ is an Al-hydroxyl functionalized porous-solid exhibiting high chemical and thermal stability as well as high selectivity and uptake capacity for CO_2 and SO_2. The

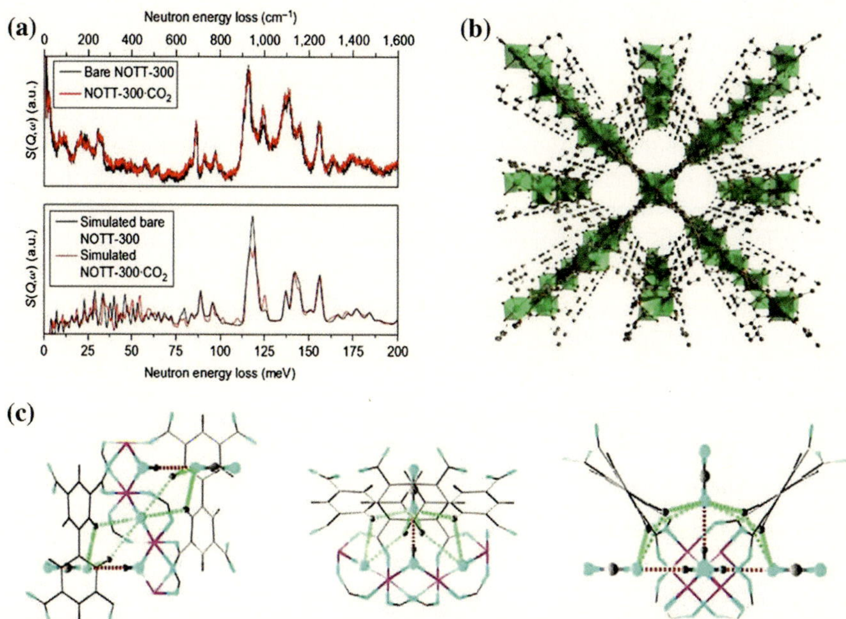

Fig. 3.3 **a** Experimental (*top*) and DFT-simulated (*bottom*) INS spectra for bare and CO_2-loaded $Al_2(OH)_2(bptc)$, also known as NOTT-300. **b** View of the three-dimensional framework structure showing channels along the *c*-axis (into the page). Water molecules in the channel are omitted for clarity. **c** Detailed views of –OH and –CH groups binding CO_2 in the "pocket" cavity of CO_2-loaded $Al_2(OH)_2(bptc)$. Views along the *a*-axis (left), the *b*-axis (*centre*), and the *c*-axis (*right*). The moderate hydrogen bond between O(δ-) of CO_2 and H(δ$^+$) of –OH is dotted *red* (O–H = 2.335 Å). The weak cooperative H bond between the O(δ$^-$) of CO_2 and the H(δ$^+$) from the –CH is dotted green (O—H = 3.029, 3.190 Å with each occurring twice), indicating that each O(δ$^-$) centre interacts with five different H(δ$^+$) centres. Reprinted with permission from (S. Yang, J. Sun, A.J. Ramirez-Cuesta, S.K. Callear, W.I.F. David, D.P. Anderson, R. Newby, A.J. Blake, J.E. Parker, C.C. Tang, M. Schröder, Nat. Chem. **4**, 887 (2012)) [44]. Nature Publishing Group

material exhibits no apparent adsorption of H_2 and N_2, which is attributed to the slow diffusion of these gases through the narrow pore channels. In contrast, unusually high and selective CO_2 and SO_2 uptakes were observed, including at low-pressure. The INS spectra revealed two major increases in peak intensity upon adsorption of 1.0 CO_2 into the formula unit: peak I at lower energy transfers (30 meV) and peak II at higher energy transfer (125 meV). Peaks in the range 100–160 meV were slightly shifted to higher energies in the CO_2-adsorbed material, indicating a hardening of the motion of the $Al_2(OH)_2(bptc)$ host upon CO_2 adsorption.

The INS spectrum derived from DFT calculations show good agreement with the experimental spectrum and confirm that the adsorbed CO_2 molecules are located end-on to the hydroxyl groups. The O–H distance between the CO_2 molecule and the hydroxyl group is 2.335 Å, indicating a moderate to weak hydrogen bond, with the optimized C–O bond distances in CO_2 being 1.183 Å at the hydrogen-bonded end and 1.178 Å at the free end. The CO_2 is linear with a O–C–O bond angle of 180°. Each adsorbed CO_2 molecule is found to be surrounded by four aromatic C–H groups, forming weak cooperative supramolecular interactions between the $O(\delta^-)$ of the CO_2 and the $H(\delta^+)$ of the –CH (where O–H = 3.029 and 3.190 Å and each occurs twice). Peak I in the INS spectrum was assigned to the O–H group wag, occurring perpendicular to the Al–O–Al direction and attributed to the presence of the CO_2. Peak II in the spectrum was assigned to the wag of the four aromatic C–H groups on four benzene rings adjacent to each CO_2, in conjunction with the OH group wag. Hence, in this work the direct visualization of host–guest interactions through INS and DFT calculations was crucial in rationalizing the material's high selectivity for CO_2 and in understanding the detailed binding mechanism of CO_2 in the material. The low H_2 uptake of the material was rationalised in a similar manner, with the contribution from H_2 in the material to the INS data, consistent with that expected for liquid H_2, indicating a weak interaction with the material.

3.2.3 Diffusion and Transport of CO_2

Application of quasielastic neutron scattering (QENS) in tandem with molecular dynamics (MD) simulations brings insights into the dynamics and transport of CO_2 in porous media [45]. Whilst the self-diffusion of molecules containing atoms that have appreciable incoherent neutron-scattering cross sections can be measured directly using QENS, the neutron-scattering cross section of CO_2 is coherent and so the coherent QENS approach must be used to monitor the CO_2 and the transport diffusivity extracted from this. Comparisons between MD simulation and QENS experiments involve the diffusing molecule's self-diffusivity, however, in practical separations and catalytic applications, it is the transport diffusivity that is of greater importance. The transport diffusivity involves the response to a chemical potential gradient and its direct determination calls for non-equilibrium experiments. At the molecular level the dependence of the transport diffusivity and the so-called corrected diffusivity needs to be resolved. Linear response theory allows the transport

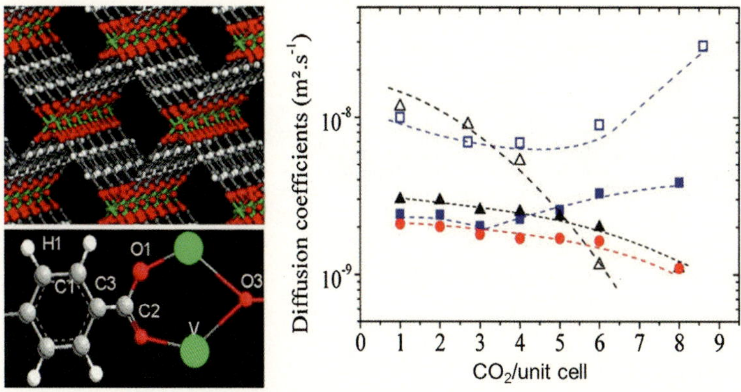

Fig. 3.4 *Left* The orthorhombic (*Pnma*) V(O)(bdc) structure displayed along the z axis, highlighting the one-dimensional pore system (*top*). Labels of the different atoms of the V(O)(bdc) structure (*bottom*). *Right* Evolution of the experimental diffusion coefficients (corrected = △ and transport = □) and simulated diffusion coefficients (self = ●, corrected = ▲, and transport = △) as a function of the CO_2 concentration in V(O)(bdc). The error bars for the simulations are 16, 7, and 12 % for low, intermediate, and high loading, respectively, while the experimental data are defined within an average error bar of 20 %. Reprinted with permission from (F. Salles, H. Jobic, T. Devic, P.L. Llewellyn, C. Serre, G. Ferey, G. Maurin, ACS Nano **4**, 143 (2010)) [48]. Copyright (2010) American Chemical Society

and corrected diffusivity to be determined directly under equilibrium conditions, where chemical potential gradients are absent. Chapter 2 contains details of the relationship between the self-diffusivity, as measured using incoherent QENS, and the transport diffusivity, D_t, as derived from coherent neutron-scattering. Experimentally, this objective can be accomplished by using coherent QENS, which probes the collective motion of guest molecules at equilibrium [46]. The same objective can be accomplished using equilibrium MD simulations.

The joint MD-experimental QENS approach in which simulated CO_2 dynamic properties are validated allows further details of the CO_2 dynamics and transport in the host to be obtained, such as first demonstrated for CO_2 in the NaX and NaY faujasite zeolites [47]. The joint coherent QENS-MD approach was first applied to MOFs in the study of the isostructural Cr(OH)(bdc) and V(O)(bdc) materials and also known as MIL-53(Cr) and MIL-47(V), respectively [48, 49]. This work built on the study of H_2 self-diffusion in these materials as detailed in Chap. 2. V(O)(bdc) contains corner-sharing $V^{4+}O_4O_2$ octahedra connected by bdc linker groups yielding one-dimensional channels (Fig. 3.4). Consequently, V(O)(bdc) has a rel-atively-high working capacity for CO_2 uptake with moderate selectivity for CO_2 in the absence of functional groups within the channel wall. The concentration-dependent self corrected (calculated from MD) and transport diffusivities (measured using QENS) of CO_2 in the rigid V(O)(bdc) material were determined [48], revealing the three-dimensional diffusion of CO_2 through the channels.

Cr(OH)(bdc) differs from V(O)(bdc) by the substitution of μ_2–O groups located at the M–O–M links in V(O)(bdc) by μ_2–OH groups in Cr(OH)(bdc). This

                                                                A. Das et al.

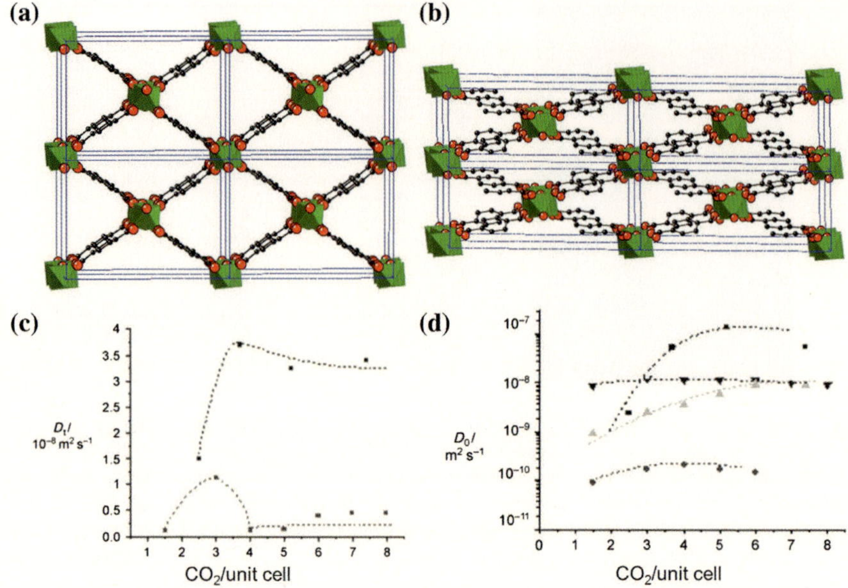

Fig. 3.5 Structural switching of the Cr(OH)(bdc) system induced by CO_2 adsorption between large-pore (*Imcm*) (**a**) and narrow-pore (*C2/c*) (**b**) forms. Experimental (**c**) and simulated (**d**) transport diffusivity (D_t) as a function of CO_2 concentration. Corrected diffusivities (D_0) simulated for the rigid narrow (•) and large pore (▾) forms, along with D_0 simulated using a composite approach (▲) and experimentally determined (■). Top: Reprinted with permission from (A.V. Neimark, F.-X.Coudert, C. Triguero, A. Boutin, A.H. Fuchs, I. Beurroies, R. Denoyel, Langmuir **27**, 4734 (2011)) [50]. Copyright (2011) American Chemical Society. Bottom: Reprinted from (F. Salles, H. Jobic, A. Ghoufi, P.L. Llewellyn, C. Serre, S. Bourrelly, G. Ferey, G. Maurin. Angew. Chem. Int. Edition **48**, 8335 (2009)) [49]

difference makes Cr(OH)(bdc) highly selective for CO_2, in a specific pressure range, which is a consequence of the large-scale breathing modes exhibited by the material. This allows it to switch from a large pore (LP) form to a narrow pore (NP) structure upon CO_2/CH_4 adsorption, with the NP structure able to trap CO_2 and not CH_4. As with the V(O)(bdc) study, the combined coherent QENS-MD approach was used to study the concentration-dependent self- and transport diffusion of CO_2 in Cr(OH)(bdc). This work found a single-file diffusion regime in the material at high CO_2 loading, a phenomenon not previously shown for any MOF (Fig. 3.5).

3.2.4 Evolution of Microstructure of the Host and Adsorption Capacity

Large-scale structure analysis methods such as small and ultra-small angle neutron scattering (SANS and USANS, respectively), yield unique, pore-size-specific insights into the kinetics of CO_2 sorption in a wide range of pores (nano to meso). These methods also provide data that may be used to determine the density of adsorbed CO_2

through the evolution of microstructure and adsorption capacity. This approach has been applied to the analysis of CO_2 in geological samples, including coal. By studying coal exposed to CO_2 at subsurface-like temperature and pressure the phase behaviour of the confined CO_2, particularly the densification occurring on changing from the gaseous to the liquid phase, was found to have significant operational and reservoir capacity ramifications when assessing the suitability of unmineable coal seams for use as CO_2 sequestration reservoirs [51]. The results show that the sorption capacity of coal is sample-dependent and strongly affected by the phase state of the injected fluid (subcritical or supercritical). Subcritical CO_2 densifies in the coal matrix, with details of CO_2 sorption differing greatly between different coals and dependent on the amount of mineral matter dispersed in the coal. A purely organic matrix was found to absorb more CO_2 per unit volume than one containing mineral matter, although the mineral matter markedly accelerated the sorption kinetics [52].

Figure 3.6 shows SANS and USANS data for coal from Seelyville (Indiana, USA) exposed to several pressures of CO_2, which could be described using a power law for the scattered intensity with an exponent of -3, indicating the fractal character of the scattering. The scattering intensity shows Q-dependency as a result of the CO_2 in the pores. After completion of the pressure cycling, the neutron-scattering curves returned to their original shapes within 1 %, implying that the microstructure was not permanently affected by exposure to CO_2 over a period of days. This result indicated that the phenomenon of coal plasticization upon exposure to CO_2 may be less widespread than thought previously. The work also found that the small pores within coal are filled preferentially over larger void-spaces by the invading CO_2, a result echoed by MOFs [32]. Apparent diffusion coefficients for CO_2 in coal are thought to vary in the range 5×10^{-7} to more than 10^{-4} cm^2min^{-1} according to the CO_2 pressure and location. At higher pressures CO_2 is shown to

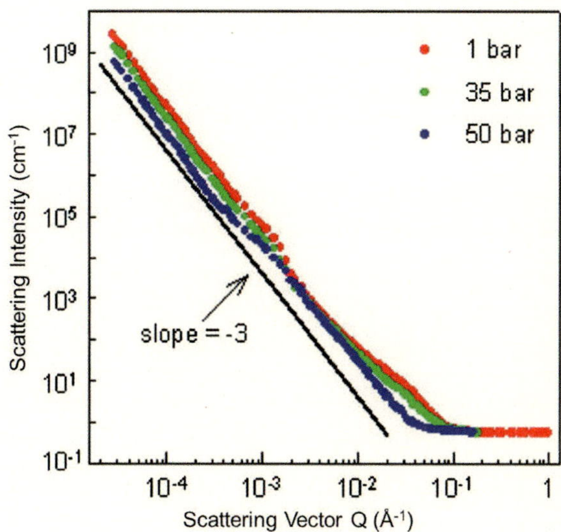

Fig. 3.6 SANS and USANS profiles for Seelyville coal exposed to various pressures of CO_2. Reprinted with permission from (A.P. Radlinski, T.L. Busbridge, E.M. Gray, T.P. Blach, G. Cheng, Y.B. Melnichenko, D.J. Cookson, M. Mastaterz, J. Esterle, Langmuir **25**, 2385 (2009)) [52]. Copyright (2009) American Chemical Society

diffuse immediately into the coal matrix, swelling the coal and changing its macromolecular structure, where it is postulated to create microporosity through the extraction of volatile components [53]. Injection of CO_2 into model subsurface geologic formations has been identified as a key strategy for CO_2 storage. Key to the success of such a strategy is the prevention of leakage from the host by an effective cap with low porosity and permeability characteristics. Shales comprise the majority of caps encountered in subsurface injection sites with pore sizes typically less than 100 nm and whose surface chemistries are dominated by quartz and clays. Analysis of simple, well-characterized fluid-substrate systems can provide details on the thermodynamic, structural, and dynamic properties of CO_2 under conditions relevant to sequestration. In particular, the behaviour of CO_2 interacting with model silica substrates can act as proxies for more complex mineralogical systems. SANS data for CO_2–silica aerogel (95 % porosity; ~ 7 nm pores) indicates the presence of fluid depletion for conditions above the critical density [54].

3.3 Probing Separations for Post and Pre Combustion Capture, as Well as Oxyfuel Combustion

Currently, postcombustion capture methods, which separate CO_2 at low partial pressures from N_2 in flue streams, are the most economically viable CO_2 capture methods in the short- to mid-term as they can be easily retrofitted to existing power plants. Due to the dilute amount of CO_2 present in these flue-gas streams, prospective materials for this purpose are required to have a high selectivity for CO_2 over N_2, which may be achieved by surface or pore functionalisation with strongly-polarising chemical groups. This may be achieved via the incorporation of open-metal cation sites, often exposed upon desolvation or "activation" of the framework, which provide strong, highly-charged binding sites for CO_2 [55, 56]. Another strategy is the introduction of strongly-polarizing organic functional-groups into the pore. Functional groups investigated previously for CO_2 separation from other gases include amines [22, 57], carboxylic acids, nitro, hydroxy, and sulfone groups [23]. Although more strongly polarizing groups enhance CO_2 adsorption manifested through higher isosteric heats of adsorption, this factor must be balanced against the ease of regenerability of the resulting material in an industrial setting.

3.3.1 Diffusion and Transport of CO_2 and N_2

Silicalite membranes exhibit an interesting selectivity for pure CO_2, which flows through the material faster than He or H_2, notwithstanding the larger kinetic diameter of CO_2 [58]. The diffusion of CO_2 and N_2 in silicalite was studied using QENS, with combined coherent QENS-MD used to determine diffusivities for CO_2

and more traditional QENS used to determine the N_2 diffusivity [59, 60]. The work obtained the diffusivity as a function of intra-crystalline occupancy. A direct comparison between computed and measured transport-diffusion allowed a better understanding of the molecular factors governing the occupancy dependence of the corrected diffusivity, a "pure" kinetic parameter that is largely free of the influence of the isotherm. The corrected diffusivity has often been assumed to be independent of sorbate concentration and approximately equal to the self-diffusivity in the limit of zero occupancy, an assumption that was reassessed in light of these results. These measurements pointed to a significant difference in the occupancy dependence of the corrected diffusivity for N_2 and CO_2, where at 300 K the corrected diffusivity for CO_2 was found to decrease with guest loading of the host and for N_2 at 200 K it was fairly constant, exhibiting a weak maximum. At the practical level this work found that interactions are considerably stronger (more attractive) for CO_2 than for N_2 in silicalite, explaining the strong preference of the material for CO_2.

3.3.2 Probing H_2 Separation from CO_2

Precombustion CO_2 capture in natural-gas plants predominantly involves the separation of CO_2 from H_2 at high pressures, resulting in a pure H_2 stream which is used in energy generation [12]. This process has a component with a higher concentration of CO_2, existing at elevated pressures, resulting in the relatively-low energy penalty for carbon capture of 10–16 % [13]. Additionally, due to the large differences in the polarisability and quadrupole moment between CO_2 and H_2, the two gases are more easily separated via chemical methods than other gases such as CO_2 and N_2 [61]. Porous materials with a high density of localized charge, such as achieved through open-metal sites, are particularly promising for this type of separation. Additionally, variances in gas properties such as diffusion rate may also be exploited by adsorbents to increase selectivity. Aside from selectivity for CO_2, the working capacity of the adsorbent is another major factor in determining the effectiveness of candidate materials for precombustion capture processes. However, these factors are generally inversely related as a material with high selectivity will generally suffer low regenerability as the guest molecules are strongly bound to the adsorbent and are difficult to remove via a mild pressure-swing approach.

Whilst little work exploring H_2 separation specifically from H_2/CO_2 mixtures has been performed using neutron scattering, more work using neutron scattering to study H_2 confined in porous materials has been published than for any other guest molecule. This is a direct consequence of the ease of structural characterization of H (as D) using neutron diffraction as well as the unique information that can be gained for H_2 using neutron spectroscopy [62]. This work is extensive and covered in publications concerning H_2 storage [63, 64], where the interaction of H_2 (D_2) with $Zn_4O(bdc)_3$ (also known as MOF-5) [65, 66], $Cu_3(btc)_2$ [67–71], $Mg_2(dobdc)$ and $Fe_2(dobdc)$ as well as its oxidized analogue [72], $Zn_2(dobdc)$ [73], $Al_2(OH)_2(bptc)$

[74], $Zn(mIm)_2$ where mIm = 2-methylimidazolate and also known as ZIF-8 [75], $Cu_3[Co(CN)_6]_2$ [76], as well as many carbonaceous materials and zeolites, have all been elucidated using neutrons. Such work is the subject of Chap. 8.

3.3.3 Probing N_2 Separation from O_2

Oxyfuel combustion involves the combustion of carbon-based fuels in a pure O_2 stream, however, the limiting factor in the industrial implementation of these methods is the large amount of pure O_2 that is required to be generated from air (O_2/N_2 separation). Microporous solids that are able to efficiently perform this separation have the potential to significantly reduce the large energy-costs currently associated with oxyfuel combustion. Small-pore zeolites have been employed for O_2/N_2 separations by exploitation of the difference in the kinetic diameter between the two gases through physical separation involving molecular sieving. The chemical tunability of the pore space of framework materials, however, facilitates the separation of O_2 and N_2 by taking advantage of the electronic differences between the two gases. In particular, MOFs containing electron-rich redox-active sites, such as $Cr_3(btc)_2$ [77] and $Fe_2(dobdc)$ [78], have been shown to reversibly bind O_2 selectively over N_2 via electron transfer from the metal centre to the O_2.

The $Fe_2(dobdc)$ material binds O_2 preferentially over N_2 at 298 K with an irreversible capacity of 9.3 wt%, corresponding to the adsorption of one O_2 per two Fe centres [78]. Remarkably, at 211 K the O_2 uptake is fully reversible and the capacity increases to 18.2 wt%, corresponding to the adsorption of one O_2 per Fe centre. Mossbauer and infrared spectroscopy measurements indicated partial charge-transfer from the Fe^{II} to the O_2 at low temperature and complete charge-transfer to form Fe^{III} and O_2^{2-} at room temperature. NPD data (4 K) confirm this interpretation, revealing O_2 bound to Fe in a symmetric side-on mode with an O_2 intranuclear separation of 1.25(1) Å at low temperature and of 1.6(1) Å in a slipped side-on mode when oxidized at room temperature (Fig. 3.7).

Similar work reported highly selective and reversible O_2 binding in $Cr_3(btc)_2$ [77], with infrared and X-ray absorption spectra suggesting the formation of an O_2 adduct with partial charge-transfer from the Cr^{II} centres exposed on the surface of the framework. NPD data confirm this mechanism of O_2 binding and indicate a lengthening of the Cr–Cr distance within the "paddle-wheel" units of the framework from 2.06(2) to 2.8(1) Å.

Selectivity for O_2 over N_2 was also achieved in polymer/selective-flake nano-composite membranes fabricated with a polyimide and a porous layered aluminophosphate. Using SANS to probe the large-scale structure of the O_2/N_2 host material, the substantially improved selectivities of O_2 over N_2 was shown to occur within only 10 wt% of the AlPO layers [79].

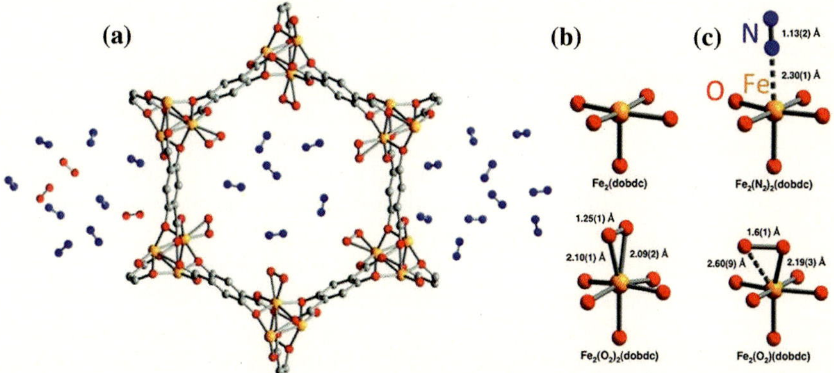

Fig. 3.7 Portion of the crystal structure of Fe_2(dobdc) as viewed approximately along the [001] direction (**a**), where H atoms are omitted for clarity. (**b–e**) First coordination-spheres for the Fe determined from NPD data, where structures are for Fe_2(dobdc) under vacuum (**b**), dosed with N_2 at 100 K (**c**), dosed with O_2 at 100 K (**d**), and dosed with O_2 at 298 K (**e**). Values in parentheses are estimated standard deviations in the final digit. Reprinted from (E.D. Bloch, L.J. Murray, W.L. Queen, S. Chavan, S.N. Maximoff, J.P. Bigi, R. Krishna, V.K. Peterson, F. Grandjean, G.J. Long, B. Smit, S. Bordiga, C.M. Brown, J.R. Long, J. Am. Chem.Soc. **133**, 14814 (2011)) [78]

3.3.4 Probing CO_2/CH_4 Separation for Natural-Gas Sweetening

Natural-gas sweetening (separation of CO_2 from CH_4) is an industrially significant separation process as CO_2 represents a substantial (up to 70 %) impurity in natural-gas wells [80]. The presence of CO_2 reduces the energy content of the natural gas, and its acidity in the presence of water can result in the corrosion of natural-gas lines. Physical solvent-based processes for CO_2 removal from natural-gas are abundant, however, the large amount of water recycling needed makes solvent-based processes highly limited in this application due to solvent degradation and loss during operation [81]. Porous solids present a more efficient and environmentally friendly way to capture CO_2 from natural-gas wells. In this case, separation largely proceeds based on quadrupole moment, due to the similar properties of the two gases in other respects (kinetic diameter, polarizability, dipole moment). Additionally, the flexible structure of some MOFs upon adsorption–desorption (in contrast with "rigid" adsorbents such as carbons and zeolites), may result in dynamic and stepwise adsorption at different pressures. This is generally known as a "gate opening" phenomenon, and arises mainly from the flexibility of the networks and their affinity for particular guests [82]. In MIL-53 [Cr(OH)(bdc)], for example, the selective adsorption of CO_2 over CH_4 is strongly affected by the presence of water which causes dramatic changes in the pore structure [27].

Neutron scattering has also been used extensively to study CH_4 confined in porous materials, in particular to study methane confined in MOFs, commensurate with the increasing work investigating these hosts for application in CO_2/CH_4 separations.

The metal sites, including open-metal sites, in many MOFs also interact with CH_4. NPD studies of the $Mg_2(dobdc)$ material show the binding of one CD_4 molecule per open-metal site, resulting in the large CH_4 storage capacity of $160–174$ cm^3(at standard temperature and pressure, STP)/cm^3, approaching the DOE target of 180 $cm^3(STP)/cm^3$ for solid-based CH_4 storage at room temperature [83]. Direct determination of CD_4 sorption sites in $Zn(mIm)_2$ and $Zn_4O(bdc)_3$ were gained using NPD (Fig. 3.8) [84]. The primary CD_4 adsorption sites are associated with the organic linkers in $Zn(mIm)_2$ and the metal oxide clusters in $Zn_4O(bdc)_3$. In $Zn_4O(bdc)_3$ the first binding sites ("cup" sites) were not found to alter the *Fm* $\bar{3}m$ symmetry of the host–guest system. CD_4 at these primary sites possesses well-defined orientations, implying relatively-strong binding with the framework. With higher CD_4 loading, additional CD_4 molecules populate secondary sites and are confined in the framework. The confined CD_4 at these secondary sites is orientationally disordered and stabilized by the intermolecular interactions. The CD_4 guest is a high symmetry guest whose ordered location (at the primary sites) significantly alters the symmetry of system. The "hex" and "ZnO_2" CD_4 sites caused a symmetry lowering of the system to *I4/mmm* as a result of the symmetry incompatibility of the tetrahedral CD_4 molecules with the local geometry. At higher CD_4 loadings a *P4 mm* structure was found, where CD_4 sites aligned themselves along the *c* axis and further lowered the symmetry.

Using a similar approach, a comprehensive mechanistic study of CD_4 was performed in $Cu_3(btc)_2$, $Cu_2(sbtc)$ where sbtc = *trans*-stilbene-3, 3′, 5, 5′-tetracarboxylate, and $Cu_2(adip)$ where adip = 5, 5′-(9, 10-anthracenediyl)di-isophthalate and also known as PCN-14, allowing a comparison of structures that consist of the same dicopper-paddlewheel secondary-building units (the well-known dicopper acetate unit), but contain different organic linkers, leading to cage-like pores with various sizes and geometries (Fig. 3.9). This work revealed that CD_4 uptake takes place primarily at two types of strong adsorption site: (1) the open Cu sites which exhibit enhanced coulombic attraction toward CD_4, and (2) the van der Waals potential pocket sites in which the total dispersive interactions are enhanced due to the molecule being in contact with multiple "surfaces". Interestingly, the enhanced van der Waals sites are present exclusively in small cages and at the windows to these cages, whereas large cages with relatively flat pore surfaces bind very little CD_4 [85].

The self-diffusion of CH_4 was measured directly using QENS in the isostructural $Cr(OH)(bdc)$ and $V(O)(bdc)$ materials [86]. The hydroxyl groups in $Cr(OH)(bdc)$ were expected to hinder CH_4 mobility, although this work revealed a global one-dimensional diffusion mechanism of CH_4 in both materials, echoing the single-file diffusion regime found for CO_2 [49]. An interesting result of this work was that CH_4 diffusivities are significantly higher in $V(O)(bdc)$ than in $Cr(OH)(bdc)$ over the whole range of investigated CH_4 loadings.

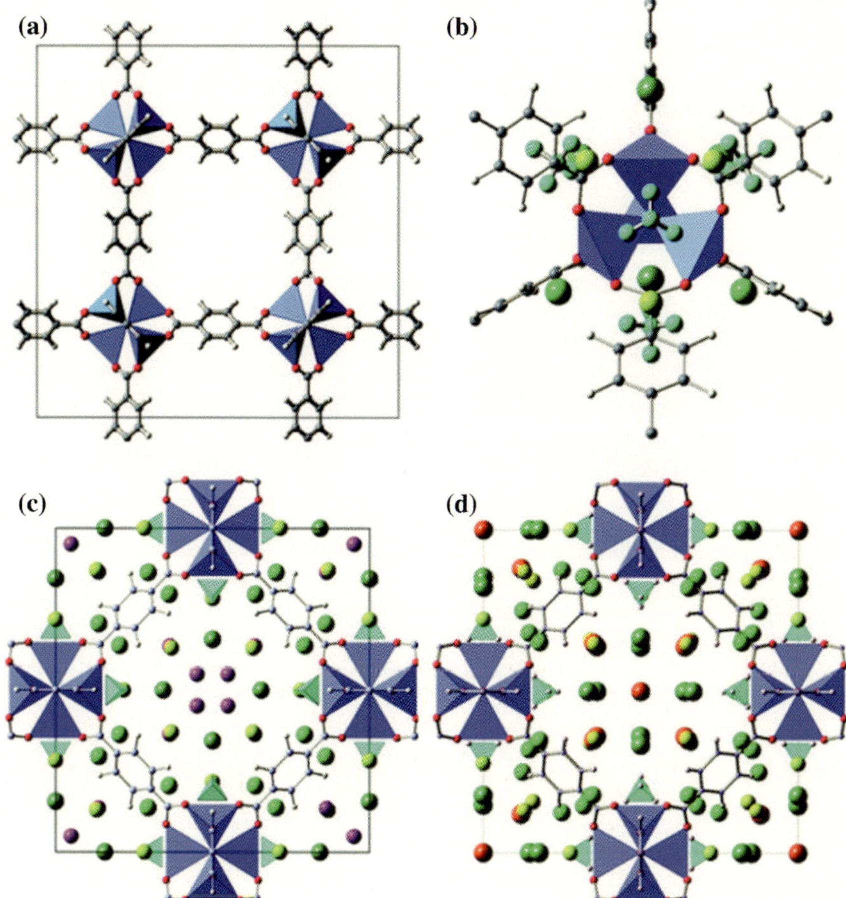

Fig. 3.8 Crystal structure of $Zn_4O(bdc)_3$, with ZnO_4 tetrahedra (*blue*) connected by bdc linkers (**a**). C is grey and D is white. CD_4 adsorption sites in $Zn_4O(bdc)_3$: "cup" sites for the first adsorbed CD_4 with well-defined molecular orientations (*cyan*) and secondary "hex" (*yellow*) and "ZnO_2" sites (*green*) (**b**). [001] view of *I4/mmm* $Zn_4O(bdc)_3$ with CD_4, additional CD_4 (*pink*) were observed near the pore centre (**c**). [001] view of *P4 mm* $Zn_4O(bdc)_3$, where CD_4 sites (*orange*) align along the *c* axis (**d**). Orientationally-disordered CD_4 are shown as spheres for clarity. Reprinted with permission from (H. Wu, W. Zhou, T. Yildirim, J. Phys. Chem. C **113**, 3029 (2009)) [84]. Copyright (2009) American Chemical Society

There have been several neutron-scattering studies targeting the separation mechanism of CO_2 from CH_4. The polymer/selective-flake nanocomposite membranes exhibiting selectivity for O_2 over N_2 (discussed in Sect. 3.3.3) also shows substantial selectivity of CO_2 over CH_4. Again, SANS results revealed that this occurs within only 10 wt% of the AlPO layers. The $Zr_6O_4(OH)(bdc)_6$ material, also known as UiO-66(Zr), is a MOF with encouraging properties for CO_2/CH_4 gas separation, achieved by combining good selectivity with a high working capacity

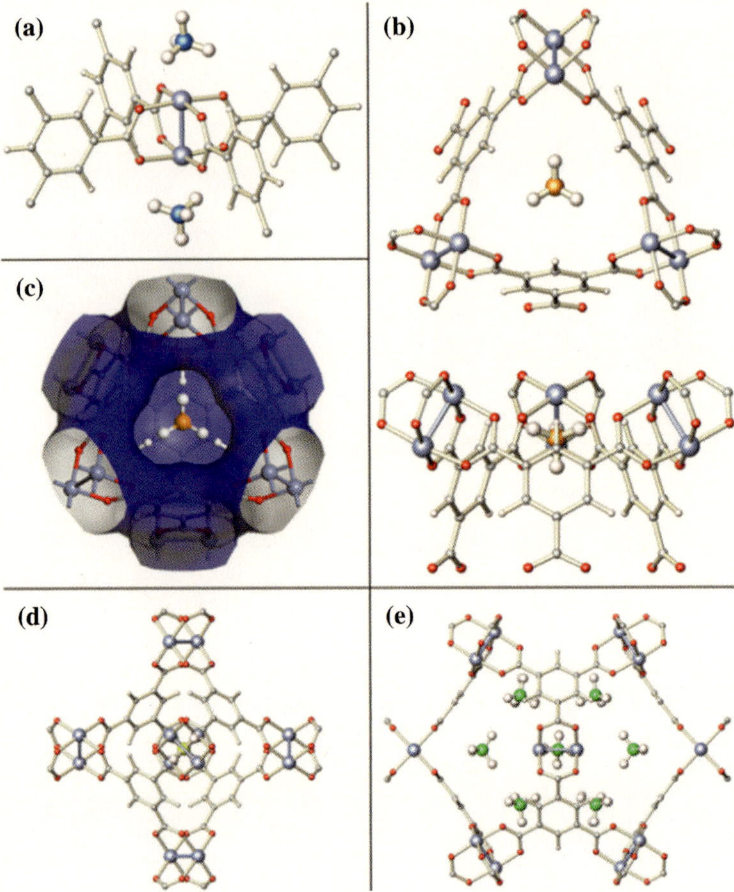

Fig. 3.9 $Cu_3(btc)_2$ with CD_4 molecules adsorbed at the open-Cu sites (**a**) and the small cage window sites (*top and side views*) (**b**). van der Waals surface of the small octahedral-cage, showing the size and geometry of the pore window in excellent match with a methane molecule (**c**). CD_4 molecule adsorbed at the secondary adsorption-site, the centre of the small octahedral-cage (**d**). CD_4 molecule located at the large cage corner-site, also a weak adsorption-site (**e**). C atoms of the CD_4 at different adsorption-sites are colored differently: Open-Cu site is blue, the small cage window-site is orange, the small cage centre site is yellow, and the large cage corner-site is green. Figure adapted from (H. Wu, J.M. Simmons, Y. Liu, C.M. Brown, X.S. Wang, S. Ma, V.K. Peterson, P.D. Southon, C.J. Kepert, H.C. Zhou, T. Yildirim, W. Zhou, Chem. -Eur. J. **16**, 5205 (2010)) [85]

and the ability for regeneration under relatively-mild conditions [87]. $Zr_6O_4(OH)$ $(bdc)_6$ is built from $Zr_6O_4(OH)_4$ octahedra that extend into three-dimensions via bdc ligands, resulting in two types of microporous cages. The dynamics of CO_2 and CH_4 within $Zr_6O_4(OH)(bdc)_6$ was measured using QENS and matched with results from MD simulations [87] (Fig. 3.10). Importantly, this work established the

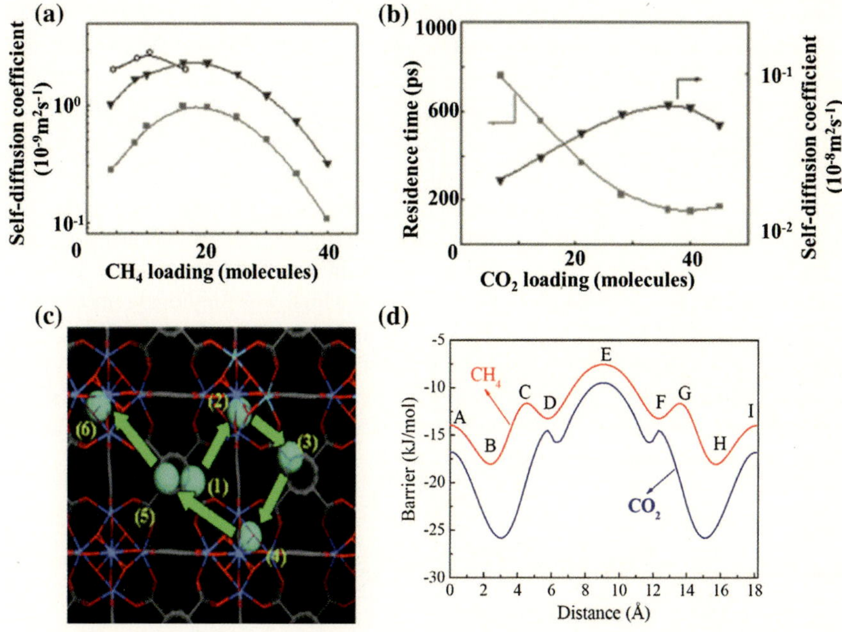

Fig. 3.10 Evolution of self-diffusion coefficients of CH$_4$ in Zr$_6$O$_4$(OH)(bdc)$_6$ at 230 K as a function of concentration (**a**): QENS (*empty circles*), MD simulations using a rigid (*filled squares*) and flexible (*filled triangles*) framework. Simulated self-diffusivity (*triangles*) of CO$_2$ in Zr$_6$O$_4$(OH)(bdc)$_6$ at 230 K as a function of concentration (**b**). The residence times (*squares*) for CO$_2$ molecules in the tetrahedral cages of Zr$_6$O$_4$(OH)(bdc)$_6$ are also shown. Typical illustration of the CH$_4$ diffusion mechanism in Zr$_6$O$_4$(OH)(bdc)$_6$ (**c**). Positions 1–6 correspond to jump sequences of CH$_4$ in the MD calculations. Potential-energy distribution for a CH$_4$ (*upper*) and CO$_2$ (*lower*) in Zr$_6$O$_4$(OH)(bdc)$_6$ as it passes from the centre of one tetrahedral cage to another, via the centre of the octahedral cage (d). Reprinted from (Q.Y. Yang, H. Jobic, F. Salles, D. Kolokolov, V. Guillerm, C. Serre, G. Maurin, Chem. -Eur.J. **17**, 8882 (2011)) [87] with permission

concentration dependence of the diffusivities of CH$_4$ and CO$_2$ (self and transport, respectively) within the material. The flexibility of the framework was found to influence significantly the diffusivity of the two species, and CH$_4$ was found to diffuse faster than CO$_2$ over a broad concentration range, a result that is in contrast to zeolites with narrow windows, for which opposite trends were observed. Further analysis of the MD trajectories for CH$_4$ provided insights into the global microscopic diffusion-mechanism, proposed to occur by a combination of intracage motions and jump sequences between the material's tetrahedral and octahedral cages. The coadsorption of CO$_2$ and CH$_4$ in the material, from both a thermodynamic and a kinetic perspective, was also studied using this approach. It was shown that each type of guest adsorbs preferentially in the two different pores, where CO$_2$ occupies the tetrahedral cages and CH$_4$ the octahedral cages. Further, a very unusual dynamic behaviour was also noted in the study of CH$_4$/CO$_2$ mixtures in Zr$_6$O$_4$(OH)(bdc)$_6$ in that the slower CO$_2$ molecule was found to enhance the mobility of the faster CH_4, again contrasting with the usual observation for CO_2/CH_4 mixtures in narrow-window zeolites, where the molecules diffuse independently or slow the partner species.

The self-diffusion properties of pure CH_4 and its binary mixture with CO_2 within the NaY zeolite have also been investigated by the combined QENS/MD approach. This material combines several favourable features including a good selectivity, high working capacity, and potential easy regenerability that make it a good candidate for the selective adsorption of CO_2 over CH_4 [88]. The QENS measurements at 200 K led to an unexpected self-diffusivity profile for pure CH_4 with the presence of a maximum for a loading of 32 CH_4/unit cell, which was previously unobserved for the diffusion of an apolar species in a zeolite with large windows. The QENS measurements report only a slight decrease of the self-diffusivity of CH_4 in the presence of CO_2 when the CO_2 loading increases. MD calculations successfully reproduce this experimental trend and suggest a microscopic diffusion-mechanism in the case of this binary mixture [89].

3.4 Experimental Challenges and the Importance of In Situ Experimentation

The analysis of porous materials and their interaction with guest molecules using neutron scattering is experimentally challenging. Even with advances in neutron sources and instrumentation, several hundred milligrams of material are usually required for successful neutron-scattering analysis of these systems. Evacuated materials prepared for guest sorption are air-sensitive, mandating their handling in specialist atmospheres such as a helium-filled glove box, where helium is necessary to avoid the heat-transfer medium freezing where the heat-transfer gas is not removed from activated samples prior to low-temperature (<10 K) measurement.

Obtaining a good neutron-scattering signal from the host or guest being studied can involve isotopic substitution, and often with complex ligands that require deuteration. The requirement of neutrons in this work is demonstrated by the recent synthesis of deuterated forms of complex ligands, such as 4, 4′, 4″-benzene-1, 3, 5-triyl-tribenzoic acid, through a technique developed at a specialist deuteration facility associated with a neutron-scattering centre. Such complex chemical synthetic routes are achievements in their own right [90].

The majority of neutron-scattering experiments exploring guest-host interactions in porous adsorbents are in situ in nature. The in situ approach, however, varies in accordance to the experimental need. Most commonly activated materials (porous materials with their pores empty) are analysed at low temperature first, before the introduction of guest molecules to the sample at a temperature where the guest will remain in the gaseous state, and the sample is then cooled slowly to where the guest molecules "lock in" to their equilibrium positions, before the measurement continues. These measurements involve careful control of the temperature of the sample as well as gas-delivery lines through the use of modified cryofurnaces. Advances in neutron instrumentation, particularly large area-detectors and higher-intensity sources, provide the opportunity to resolve in real-time details for such systems [91].

3.5 Perspectives for Neutron Scattering in the Study of Porous Materials for CO_2 Separation, Capture, and Storage

Clearly, the development of more efficient, cost-effective, and industrially-viable CO_2 capture materials is essential for the deployment of large-scale CCS. Novel concepts for porous hosts used for CO_2 capture and separation require a molecular level of control that can take advantage of differences in the chemical reactivity of gas molecules. A challenge in the capture of CO_2 is tuning the selectivity of adsorbents, and coupled with this is the need to examine the adsorption selectivity at the molecular level. Neutron scattering has made important contributions in the understanding of the fundamental separation and storage mechanisms underpinning the functionality of porous materials used in CO_2 capture processes. Great potential exists to develop porous hosts for this purpose using neutron scattering by probing adsorption sites, as well as guest orientation, dynamics, and diffusion in wide range of porous materials. Additionally, the characterization of the hosts themselves and their response to guest adsorption, both on a crystallographic and large-scale structure scale is important.

Postcombustion capture from power-plant flue streams provides one strategy towards reducing CO_2 emissions to the atmosphere, however, there is an urgent need for new methods and materials that perform this separation. In contrast to the low pressure, predominantly CO_2/N_2 separation required for postcombustion capture, materials for precombustion (high pressure, predominantly CO_2/H_2) capture and natural-gas sweetening (predominantly CO_2/CH_4), have distinct requirements. Careful consideration must therefore be afforded to the working conditions of the material at which capture occurs in order to tailor the properties of that material. Commensurate with this requirement is the need for studying materials under relevant working conditions, with an emerging area of particular relevance being the understanding of gas transport in mixed gas and vapour streams. Such co-adsorption experiments, performed for CO_2 and CH_4 mixtures [88, 89], could be extended to study important ternary mixtures such as $CO_2/H_2O/N_2$. This would allow derivation of important competitive gas-sorption mechanisms that are difficult to derive using other methods such as sorption analysis and diffuse-reflectance Fourier-transform infrared spectroscopy. This approach can be expanded further to include mixtures representative of separations that are industrially relevant, and for conversion and catalytic reactions.

References

1. A. Neftel, H. Friedli, E. Moor, H. Lötscher, H. Oeschger, U. Siegenthaler, B. Stauffer, *Trends: A Compendium of Data on Global Change, Carbon Dioxide Information Analysis Center.* (Oak Ridge National Laboratory, US Department of Energy, 1994)
2. C.D. Keeling, T.P. Whorf, M. Wahlen, J. van der Plichtt, Nature **375**, 666 (1995)
3. B. Metz, O. Davidson, H. de Coninck, M. Loos, L. Meyer, IPCC special report on Carbon Dioxide Capture and Storage (2005)
4. http://unfccc.int/2860.php. Accessed 4 Mar 2014
5. http://www.usaid.gov/climate. Accessed 4 Mar 2014
6. K. Sumida, D.L. Rogow, J.A. Mason, T.M. McDonald, E.D. Bloch, Z.R. Herm, T.-H. Bae, J.R. Long, Chem. Rev. **112**, 724 (2011)
7. D.M. D'Alessandro, B. Smit, J.R. Long, Angew. Chem. Int. Ed. **49**, 6058 (2010)
8. J.D. Figueroa, T. Fout, S. Plasynski, H. McIlvried, R.D. Srivastava, Int. J. Greenh. Gas Control **2**, 9 (2008)
9. C.E. Powell, G.G. Qiao, J. Membr. Sci. **279**, 1 (2006)
10. P.D. Vaidya, E.Y. Kenig, Chem. Eng. Technol. **30**, 1467 (2007)
11. E.P.R. Institute, *Program on technology innovation: post-combustion CO_2 capture technology development* (Electric Power Res. Ins, Palo Alto, 2008)
12. P.H.M. Feron, C.A. Hendriks, Oil and gas science and technology—rev. IFP **60**, 451 (2005)
13. L.I. Eide, D.W. Bailey, Oil and gas science and technology—rev. IFP **60**, 475 (2005)
14. M.M. Abu-Khader, Energ. Source. Part A **28**, 1261 (2006)
15. Q. Yang, C. Zhong, J. Phys. Chem. B **110**, 17776 (2006)
16. R.V. Siriwardane, M.-S. Shen, E.P. Fisher, J.A. Poston, Energ. Fuels **15**, 279 (2001)
17. X. Xu, C. Song, B.G. Miller, A.W. Scaroni, Fuel Process. Technol. **86**, 1457 (2005)
18. J.C. Hicks, J.H. Drese, D.J. Fauth, M.L. Gray, G. Qi, C.W. Jones, J. Am. Chem. Soc. **130**, 2902 (2008)
19. B. Moulton, M.J. Zaworotko, Chem. Rev. **101**, 1629 (2001)
20. C. Janiak, Dalton Trans. **14**, 2781 (2003)
21. A. Demessence, D.M. D'Alessandro, M.L. Foo, J.R. Long, J. Am. Chem. Soc. **131**, 8784 (2009)
22. T.M. McDonald, W.R. Lee, J.A. Mason, B.M. Wiers, C.S. Hong, J.R. Long, J. Am. Chem. Soc. **134**, 7056 (2012)
23. E. Neofotistou, C.D. Malliakas, P.N. Trikalitis, Chem.-Eur. J. **15**, 4523 (2009)
24. D.N. Dybtsev, H. Chun, S.H. Yoon, D. Kim, K. Kim, J. Am. Chem. Soc. **126**, 32 (2003)
25. B. Chen, S. Ma, E.J. Hurtado, E.B. Lobkovsky, H.-C. Zhou, Inorg. Chem. **46**, 8490 (2007)
26. B. Chen, S. Ma, F. Zapata, F.R. Fronczek, E.B. Lobkovsky, H.-C. Zhou, Inorg. Chem. **46**, 1233 (2007)
27. P.L. Llewellyn, S. Bourrelly, C. Serre, Y. Filinchuk, G. Férey, Angew. Chem. Int. Ed. **45**, 7751 (2006)
28. C. Serre, S. Bourrelly, A. Vimont, N.A. Ramsahye, G. Maurin, P.L. Llewellyn, M. Daturi, Y. Filinchuk, O. Leynaud, P. Barnes, G. Férey, Adv. Mater. **19**, 2246 (2007)
29. P.D.C. Dietzel, R.E. Johnsen, H. Fjellvag, S. Bordiga, E. Groppo, S. Chavan, R. Blom, Chem. Commun. **41**, 5125 (2008)
30. R. Vaidhyanathan, S.S. Iremonger, G.K.H. Shimizu, P.G. Boyd, S. Alavi, T.K. Woo, Science **330**, 650 (2010)
31. M.R. Hudson, W.L. Queen, J.A. Mason, D.W. Fickel, R.F. Lobo, C.M. Brown, J. Am. Chem. Soc. **134**, 1970 (2012)
32. H. Wu, J.M. Simmons, G. Srinivas, W. Zhou, T. Yildirim, J. Phys. Chem. Lett. **1**, 1946 (2010)
33. W.L. Queen, C.M. Brown, D.K. Britt, P. Zajdel, M.R. Hudson, O.M. Yaghi, J. Phys. Chem. C **115**, 24915 (2011)
34. T.A. Steriotis, K.L. Stefanopoulos, F.K. Katsaros, R. Glaser, A.C. Hannon, J.D.F. Ramsay, Phys. Rev. B **78**, 115424 (2008)

35. K.L. Stefanopoulos, T.A. Steriotis, F.K. Katsaros, N.K. Kanellopoulos, A.C. Hannon, J.D.F. Ramsay, in *5th European Conference on Neutron Scattering*, vol. 340. (Iop Publishing Ltd, Bristol, 2012)

36. T.A. Steriotis, K.L. Stefanopoulos, N.K. Kanellopoulos, A.C. Mitropoulos, A. Hoser, Colloid Surf. A-Physicochem. Eng. Asp. **241**, 239 (2004)

37. H. Wu, J.M. Simmons, G. Srinivas, W. Zhou, T. Yildirim, J. Phys. Chem. Lett. **1**, 1946 (2010)

38. R.K. Motkuri, P.K. Thallapally, B.P. McGraii, S.B. Ghorishi, Cryst. Eng. Comm. **12**, 4003 (2010)

39. S.H. Ogilvie, S.G. Duyker, P.D. Southon, V.K. Peterson, C.J. Kepert, Chem. Commun. **49**, 9404 (2013)

40. T. Ikeda, O. Yamamuro, T. Matsuo, K. Mori, S. Torii, T. Kamiyama, F. Izumi, S. Ikeda, S. Mae, J. Phys. Chem. Solids **60**, 1527 (1999)

41. N. Igawa, T. Taguchi, A. Hoshikawa, H. Fukazawa, H. Yamauchi, W. Utsumi, Y. Ishii, J. Phys. Chem. Solids **71**, 899 (2010)

42. A. Falenty, G. Genov, T.C. Hansen, W.F. Kuhs, A.N. Salamatin, J. Phys. Chem. C **115**, 4022 (2011)

43. A.J. Ramirez-Cuesta, Comput. Phys. Commun. **157**, 226 (2004)

44. S. Yang, J. Sun, A.J. Ramirez-Cuesta, S.K. Callear, W.I.F. David, D.P. Anderson, R. Newby, A.J. Blake, J.E. Parker, C.C. Tang, M. Schröder, Nat. Chem. **4**, 887 (2012)

45. H. Jobic, D.N. Theodorou, Microporous Mesoporous Mat. **102**, 21 (2007)

46. H. Jobic, K. Makzodimitris, G.K. Papadopoulos, H. Schober, D.N. Theodorou, in *Proceedings of the 14th International Zeolite Conference*, Cape Town, 2004

47. D. Plant, H. Jobic, P. Llewellyn, G. Maurin, Eur. Phys. J. -Spec. Top. **141**, 127 (2007)

48. F. Salles, H. Jobic, T. Devic, P.L. Llewellyn, C. Serre, G. Ferey, G. Maurin. ACS Nano **4**, 143 (2010)

49. F. Salles, H. Jobic, A. Ghoufi, P.L. Llewellyn, C. Serre, S. Bourrelly, G. Ferey, G. Maurin. Angew. Chem. Int. Ed. **48**, 8335 (2009)

50. A.V. Neimark, F.-X.Coudert, C. Triguero, A. Boutin, A.H. Fuchs, I. Beurroies, R. Denoyel, Langmuir, **27**, 4734 (2011)

51. Y.B. Melnichenko, A.P. Radlinski, M. Mastalerz, G. Cheng, J. Rupp, Int. J. Coal Geol. **77**, 69 (2009)

52. A.P. Radlinski, T.L. Busbridge, E.M. Gray, T.P. Blach, G. Cheng, Y.B. Melnichenko, D.J. Cookson, M. Mastaterz, J. Esterle, Langmuir, **25**, 2385 (2009)

53. M. Mirzaeian, P.J. Hall, H.F. Jirandehi, J. Mater. Sci. **45**, 5271 (2010)

54. D.R. Cole, A.A. Chialvo, G. Rother, L. Vlcek, P.T. Cummings, Philos. Mag. **90**, 2339 (2010)

55. P.D.C. Dietzel, R.E. Johnsen, R. Blom H. Fjellvåg, Chem.-Eur. J. **14**, 2389 (2008)

56. E.D. Bloch, D. Britt, C. Lee, C.J. Doonan, F.J. Uribe-Romo, H. Furukawa, J.R. Long, O.M. Yaghi, J. Am. Chem. Soc. **132**, 14382 (2010)

57. J. An, S.J. Geib, N.L. Rosi, J. Am. Chem. Soc. **132**, 38 (2009)

58. M.E. Welk, F. Bonhomme, T.M. Nenoff, Prepr. Pap. -Am. Chem. Soc. Div. Fuel Chem., **49**, 245 (2004)

59. H. Jobic, K. Makrodimitris, G.K. Papadopoulos, H. Schober, D.N. Theodorou, in *Recent Advances in the Science and Technology of Zeolites and Related Materials*, Pts a—C, eds. by E. VanSteen, M. Claeys and L. H. Callanan, vol. 154. (Elsevier Science Bv, Amsterdam, 2004) pp. 2056

60. G.K. Papadopoulos, H. Jobic, D.N. Theodorou, J. Phys. Chem. B **108**, 12748 (2004)

61. J.-R. Li, R.J. Kuppler, H.-C. Zhou, Chem. Soc.Rev. **38**, 1477 (2009)

62. A.J. Ramirez-Cuesta, M.O. Jones, W.I.F. David, Mater. Today **12**, 54 (2009)

63. M. Hirscher (ed.), *Handbook of Hydrogen Storage, New Materials for Future Energy Storage* (WILEY-VCH, Weinheim, 2010)

64. P. Earis, ed., *Hydrogen Storage Materials*, (RSC Publishing, 2011)

65. T. Yildirim, M.R. Hartman, Phys. Rev. Lett., 95 (2005)

66. F.M. Mulder, T.J. Dingemans, H.G. Schimmel, A.J. Ramirez-Cuesta, G.J. Kearley, Chem. Phys. **351**, 72 (2008)

67. V.K. Peterson, C.M. Brown, Y. Liu, C.J. Kepert, J. Phys. Chem. C **115**, 8851 (2011)
68. V.K. Peterson, Y. Liu, C.M. Brown, C.J. Kepert, J. Am. Chem. Soc. **128**, 15578 (2006)
69. V.K. Peterson, Y. Liu, C.M. Brown, C.J. Kepert, Mater. Res. Soc. **561–565**, 1601 (2007)
70. C.M. Brown, Y. Liu, T. Yildirim, V.K. Peterson, C.J. Kepert, Nanotechnology, 20 (2009)
71. Y. Liu, C.M. Brown, D.A. Neumann, V.K. Peterson, C.J. Kepert, J. Alloy, Compd **446**, 385 (2007)
72. W.L. Queen, E.D. Bloch, C.M. Brown, M.R. Hudson, J.A. Mason, L.J. Murray, A.J. Ramirez-Cuesta, V.K. Peterson, J.R. Long, Dalton T. **41**, 4180 (2012)
73. Y. Liu, H. Kabbour, C.M. Brown, D.A. Neumann, C.C. Ahn, Langmuir **24**, 4772 (2008)
74. X. Lin, I. Telepeni, A.J. Blake, A. Dailly, C.M. Brown, J.M. Simmons, M. Zoppi, G.S. Walker, K.M. Thomas, T.J. Mays, P. Hubberstey, N.R. Champness, M. Schroder, J. Am. Chem. Soc. **131**, 2159 (2009)
75. H. Wu, W. Zhou, T. Yildirim, J. Am. Chem. Soc. **129**, 5314 (2007)
76. M.R. Hartman, V.K. Peterson, Y. Liu, S.S. Kaye, J.R. Long, Chem. Mater. **18**, 3221 (2006)
77. L.J. Murray, M. Dinca, J. Yano, S. Chavan, S. Bordiga, C.M. Brown, J.R. Long, J. Am. Chem. Soc. **132**, 7856 (2010)
78. E.D. Bloch, L.J. Murray, W.L. Queen, S. Chavan, S.N. Maximoff, J.P. Bigi, R. Krishna, V.K. Peterson, F. Grandjean, G.J. Long, B. Smit, S. Bordiga, C.M. Brown, J.R. Long, J. Am. Chem. Soc. **133**, 14814 (2011)
79. H.K. Jeong, W. Krych, H. Ramanan, S. Nair, E. Marand, M. Tsapatsis, Chem. Mater. **16**, 3838 (2004)
80. H. Lin, E. Van Wagner, R. Raharjo, B.D. Freeman, I. Roman, Adv. Mater. **18**, 39 (2006)
81. P.D.C. Dietzel, V. Besikiotis, R. Blom, J. Mater. Chem. **19**, 7362 (2009)
82. M. Tagliabue, D. Farrusseng, S. Valencia, S. Aguado, U. Ravon, C. Rizzo, A. Corma, C. Mirodatos, Chem. Eng. J. **155**, 553 (2009)
83. H. Wu, W. Zhou, T. Yildirim, J. Am. Chem. Soc. **131**, 4995 (2009)
84. H. Wu, W. Zhou, T. Yildirim, J. Phys. Chem. C **113**, 3029 (2009)
85. H. Wu, J.M. Simmons, Y. Liu, C.M. Brown, X.S. Wang, S. Ma, V.K. Peterson, P.D. Southon, C.J. Kepert, H.C. Zhou, T. Yildirim, W. Zhou, Chem. -Eur. J. **16**, 5205 (2010)
86. N. Rosenbach, H. Jobic, A. Ghoufi, F. Salles, G. Maurin, S. Bourrelly, P.L. Llewellyn, T. Devic, C. Serre, G. Ferey, Angew. Chem. Int. Ed. **47**, 6611 (2008)
87. Q.Y. Yang, H. Jobic, F. Salles, D. Kolokolov, V. Guillerm, C. Serre, G. Maurin, Chem. -Eur. J. **17**, 8882 (2011)
88. Q.Y. Yang, A.D. Wiersum, H. Jobic, V. Guillerm, C. Serre, P.L. Llewellyn, G. Maurin, J. Phys. Chem. C **115**, 13768 (2011)
89. I. Deroche, G. Maurin, B.J. Borah, S. Yashonath, H. Jobic, J. Phys. Chem. C **114**, 5027 (2010)
90. T.A. Darwish, A.R.G. Smith, I.R. Gentle, P.L. Burn, E. Luks, G. Moraes, M. Gillon, P.J. Holden, M. James, Tetrahedron Lett. **53**, 931 (2012)
91. F.M. Mulder, B. Assfour, J. Huot, T.J. Dingemans, M. Wagemaker, A.J. Ramirez-Cuesta, J. Phys. Chem. C **114**, 10648 (2010)

Chapter 4
Materials for the Nuclear Energy Sector

Michael Law, David G. Carr and Sven C. Vogel

Abstract Current and future nuclear-technologies such as fission and fusion reactor-systems depend on well-characterized structural materials, underpinned by reliable material-models. The response of the material must be understood with science-based models, under operating and accident conditions which include irradiation, high temperature and stress, corrosive environments, and magnetic fields. Neutron beams offer methods of characterizing and understanding the effects of radiation on material behaviour such as yield and tensile strength, toughness, embrittlement, fatigue and corrosion resistance. Neutron-analysis techniques improve our understanding of radiation damage, which is essential in guiding the development of new materials.

4.1 Introduction

Radiation damage changes structural materials; the role of microstructure, stress, and radiation flux on swelling, creep, embrittlement, and phase transformations must all be understood. This knowledge will allow development of materials with superior resistance to fast-neutron fluence and high temperatures.

The ability to model and predict the performance and life of materials in nuclear power plants is essential for the reliability and safety of these technologies. These systems may include new environments such as high-pressure water, molten salt, molten metal, and helium which all increase the potential for material degradation

M. Law (✉) · D.G. Carr
Australian Nuclear Science and Technology Organisation, Lucas Heights, NSW, Australia
e-mail: michael.law@ansto.gov.au

D.G. Carr
e-mail: david.carr@ansto.gov.au

S.C. Vogel
Los Alamos National Laboratories, Los Alamos, MN, United States
e-mail: sven@lanl.gov

© Springer International Publishing Switzerland 2015
G.J. Kearley and V.K. Peterson (eds.), *Neutron Applications in Materials for Energy*, Neutron Scattering Applications and Techniques,
DOI 10.1007/978-3-319-06656-1_4

and corrosion. Significant material-development challenges must be met as components of Generation IV reactor systems will experience higher fluxes, temperatures, and sometimes stresses, than conventional light-water reactor systems. The same applies for fusion reactors which in the current developmental phase pose significant challenges to available structural materials. Only by improved characterization can we move to science-based models of material behaviour. Understanding material behaviour from the atomic level up to the full-component scale is essential in developing new materials for these applications.

Creep and creep-fatigue of reactor materials is poorly understood. When the effects of irradiation are added, it is obvious that a better understanding is required. These same effects are intensified in welds due to texture, material inhomogeneity, residual stress, and thermal-expansion mismatch.

Irradiation can cause significant microstructural changes including atomic displacement, helium bubble formation, irradiation-induced swelling and irradiation-induced creep, crystalline-to-amorphous phase transitions, and the generation of point defects or solute aggregates in crystalline lattices. Irradiation also creates defects resulting from atomic displacement or from transmutation products. These defects increase the yield and tensile strengths while reducing ductility and causing embrittlement.

The neutron-beam techniques relevant to the nuclear-energy sector are residual stress and texture measurements, crystallographic phase analysis to establish phase diagrams and reaction kinetics, neutron radiography and tomography, prompt-gamma activation analysis, and small-angle neutron scattering. As exposure to neutron beams activates many materials, neutron facilities generally have the infrastructure to accommodate radioactive materials which allows post-irradiation examination of samples.

The ability to characterize materials in situ is essential, at the appropriate temperature and environment, rather than bringing the sample back to ambient conditions. This also allows the evolution of material behaviour to be studied rather than just the properties at the start and endpoint.

4.2 Steels

Steels are important structural materials in modern nuclear-power systems. Common alloy systems include ferritic materials which have high resistance to radiation induced swelling, nickel-based alloys for high-temperature applications, and austenitic (stainless) steels which typically offer superior corrosion resistance. Typical applications are piping, pressure vessels, heat exchangers, steam generators, and general structural components. Some reactor vessels are made from a ferritic shell which is then clad with stainless steel for corrosion resistance. Structural integrity issues which have been investigated by neutron methods include stress corrosion cracking, weld and cladding cracking, and loss of creep strength and embrittlement due to exposure to temperature and radiation.

Hydrogen can be absorbed into metals, particularly at high temperatures. This interstitial hydrogen can significantly affect the strength and ductility of the material, leading to reduced structural integrity. Hydrogen flows along gradients of interstitial spacing, particularly the variations caused by temperature and stress. Thus it will flow to hot areas associated with welding, and to stress concentrations associated with crack tips. As a strong incoherent scatterer of neutrons, the hydrogen distribution in metals can be investigated with radiography and prompt-gamma neutron activation analysis.

A new class of steels, the oxide-dispersion strengthened steels, have recently been developed specifically for improved radiation tolerance. This material contains a fine dispersion of nano-sized, precipitate-like features which improve high-temperature creep properties and act as sinks for transmutation-produced helium, provide better void-swelling resistance and promote recombination of irradiation-induced point defects. Small-angle neutron scattering has proved to be a particularly useful technique for bulk characterisation of the nano-cluster distribution, volume fraction, shape, interface and size.

4.2.1 Residual Stress

Residual stresses are those that remain in a component after external forces are removed; they are self-equilibrating in nature and are often caused by deformation or uneven heating during manufacture; particularly casting, forging, forming, or welding operations. Residual stresses are significant in the failure of components as they contribute to fracture, fatigue, stress corrosion cracking (SCC), hydrogen-assisted cold cracking, hydride formation, or lead to unacceptable deformation during manufacture. An understanding of residual stress is essential in developing new components, materials, and joining techniques for nuclear-energy systems.

As most metal-forming operations involve heating and or deformation, residual stresses are almost always present due to differential thermal-strains, phase transformations, or plastic mismatch. These mismatches cause elastic strains, which result in residual stresses. The structural issues that arise from residual stresses are of two types, those that are conventional structural-integrity issues (fracture, fatigue, stress corrosion cracking susceptibility, creep crack growth, hydride formation, and distortion) and issues that come about due to the interaction of residual stress and radiation with service exposure (stress relaxation, creep, and swelling; all induced by radiation).

Structural integrity assessments of nuclear components rely on accurate values of the residual stresses; in the absence of better information these must be conservatively assumed to be equal to the yield strength, leading to small critical-defect sizes and loads. The regular use of residual stress measurements by neutron diffraction has been able to safely reduce the conservatism of these estimates by providing accurate, validated measurements. Stresses cannot be measured directly, only the elastic strains locked into the material. Broadly speaking, there are two methods of measuring residual stresses; compliance methods and methods that measure lattice strains in crystalline materials (typically metals) [1–3]. Compliance methods assess deformations that occur during cutting or other methods of material removal. Lattice strains are typically measured by diffraction and comparing lattice spacings in the strained and unstrained condition, this can be done by X-ray (including synchrotron) and neutron diffraction. Neutron diffraction has many advantages over other methods of measuring residual stresses due to its good penetrating power and spatial resolution within the bulk of the component.

A difficulty with diffraction measurements is that a stress-free sample is often required for reference. Furthermore, the compositional strain-variation that may occur across welds leads to unavoidable sources of error due to so-called chemical strains. There may be significant variation in the weld position and composition between the stress-free sample and the measured component. Where one stress component is known to be near zero measurement at a range of angles normal to the surface using the $\sin^2\theta$ technique [3] can obviate the need for a stress-free sample.

4.2.1.1 Welding and Joining

Welds are often located at mechanical-stress concentrations, and welds can be regarded as a form of metallurgical notch due to degraded local material properties, defects, and residual stresses. The combination of high stresses, reduced material properties, and probable defects often leads to failure by fracture, fatigue, stress corrosion cracking, or even creep cavitation.

Although developing materials for demanding applications is essential, joining these materials will be a major difficulty. Welds and other methods of joining are inevitably the weak point due to metallurgical inhomogeneity, defects, residual stresses, and dissimilar mechanical properties. Well-characterized residual stresses are essential for assessing the structural integrity of welds. In a similar manner to the stress redistribution that may occur with high temperature, irradiation can change the residual stress distribution in the weld (Fig. 4.1).

Residual stresses may occur between layers in composites such as the commonly-used austenitic (stainless steel) cladding on ferritic pressure vessels (Fig. 4.2). These layered composites have two forms of stress, any residual stress due to the bonding process, and a thermal mismatch which is a function of the different coefficients of thermal expansion, and the difference between the bonding temperature and the current (measurement or operational) temperature. In some cases the cladding is too thin for residual stress measurements and only the residual stresses in the base material can be measured [5].

The welds and heat-affected zones (HAZs) are areas of concern for SCC because of the presence of as-fabricated flaws, high residual stresses, elevated plastic strains, chemical heterogeneity, and microstructural differences relative to base metals. Dissimilar metal welds are critical areas in nuclear-power systems due to higher residual stresses than for similar-metal welds, and additional thermal stresses during operation due the different coefficients of thermal expansion.

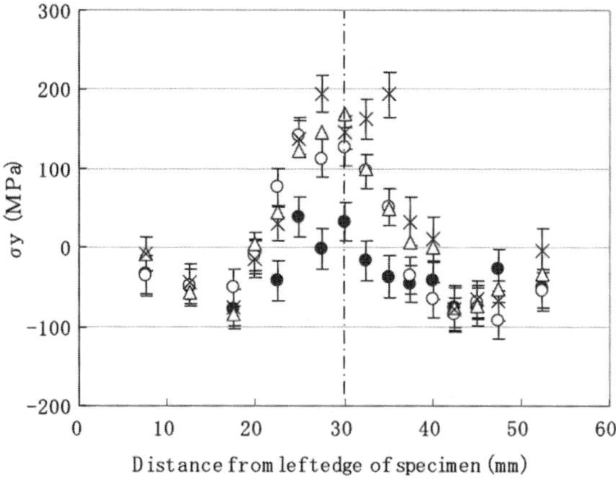

Fig. 4.1 Comparison of residual stress distributions in stages of irradiation (Reprinted with permission from (Y. Ishiyama, R.B. Rogge, M. Obata, J. Nucl. Mater. **408**, 153 (2011)) [4]. Copyright (2011) Elsevier

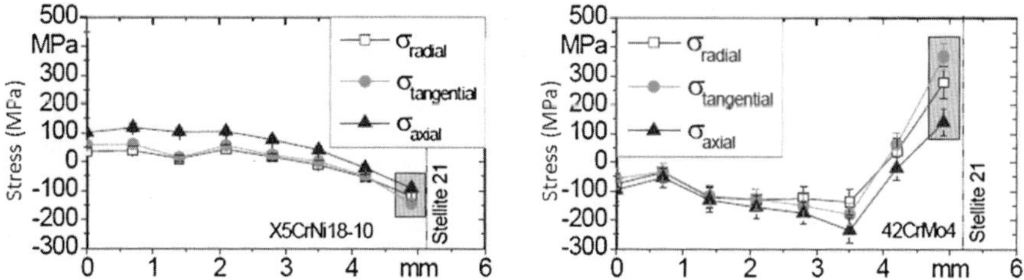

Fig. 4.2 Residual stresses in base material only in Stellite-clad steel specimens. Reprinted with permission from (H. Köhler, K. Partes, J.R. Kornmeier, F. Vollertsen, Phys. Procedia **39**, 354 (2012)) [5]. Copyright (2012) Elsevier

Many of the materials of interest for nuclear-power systems are difficult for neutron diffraction due to large grain size in the weld (stainless steels, U, Zr), low scattering (Zr, Ti), or strong attenuation (W). Hexagonal and orthorhombic crystal structures (Ur, Zr) can complicate residual stress measurement due to type II (inter-granular) stresses from elastic, thermal, and plastic anisotropy which are superimposed on the type I macroscopic stresses.

Dissimilar metal welds (austenitic to ferritic) or bonding system (e.g. copper-tungsten composites for the plasma facing component in fusion systems) requires measurement of different reflections necessitating a different instrument configuration for each material.

Inevitably components will need to be modified or repaired and weld repairs complicate an already complex residual stress field. Repair welds are of special interest as they are often made without post-weld heat treatment, producing welds with higher levels of residual stress (and sometimes hydrogen) than conventional welds. Nearly

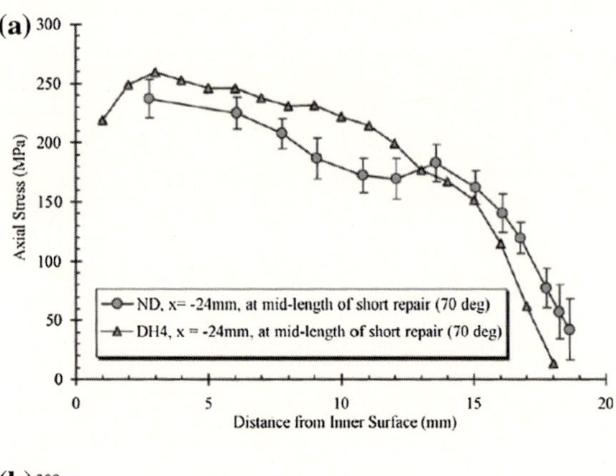

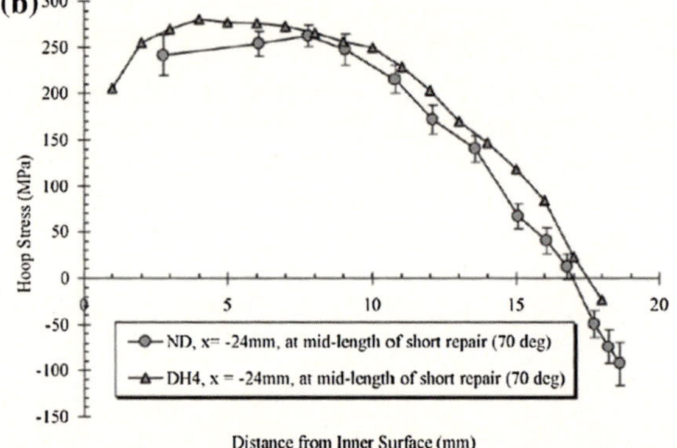

Fig. 4.3 Comparison of measured residual stresses in the HAZ of a short repair weld: **a** axial, **b** hoop. Reprinted with permission from (P. J. Bouchard, D. George, J.R. Santisteban, G. Bruno, M. Dutta, L. Edwards, E. Kingston, M. Smith, Int. J. Press. Vessels Pip. **82**, 299 (2005)) [7]. Copyright (2005) Elsevier

half of repair welds made on high-energy components in the power-generation industry subsequently fail. Repair welds are more complex than normal fabrication welds as the repair may have significant stop/start thermal fields, may possess further transformation stresses, and overlay an existing residual stress field. Edwards et al. [6] and Bouchard et al. [7] made neutron-diffraction measurements of residual stresses in typical repair welds for the nuclear industry. Figure 4.3 shows a good comparison between residual stresses measured by deep hole drilling and by neutron diffraction for a short weld-repair in a 20 mm thick 316 stainless vessel.

4.2.1.2 Measurement Validation

Due to experimental uncertainties, there is considerable variation in residual stress measurements; a common strategy is to validate measurements by using two or more methods. Some neutron residual stress measurements have been validated (Figs. 4.4 and 4.5) by modelling [8], deep hole drilling [7], or contour methods [8, 9].

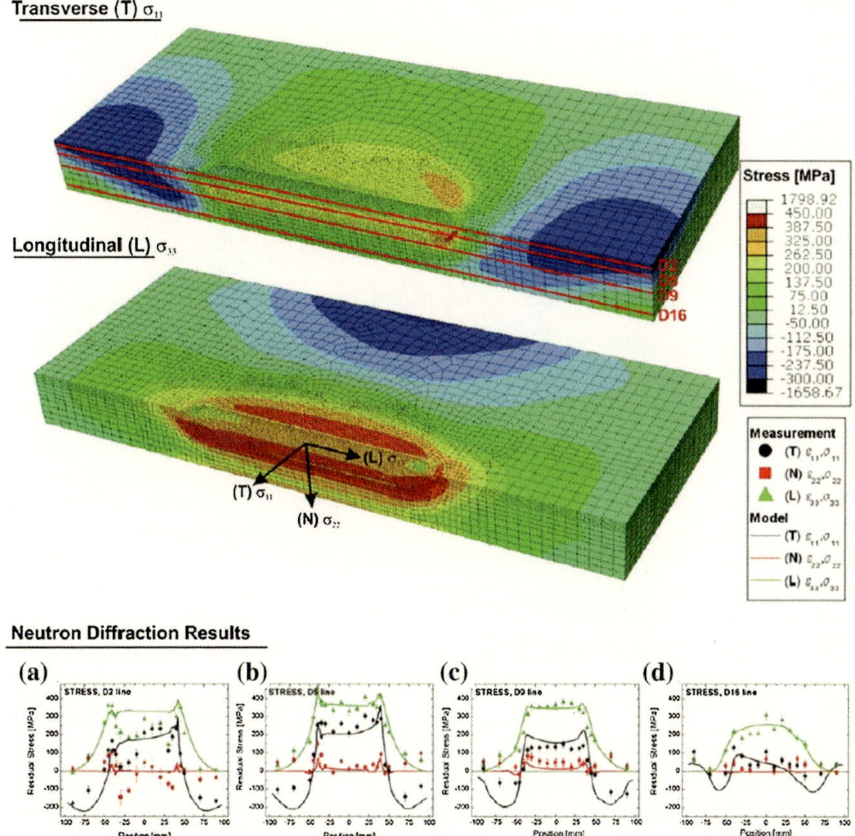

Fig. 4.4 Comparison of model and neutron diffraction measurements. Contour plots of predicted transverse ($\sigma11$) and longitudinal ($\sigma33$) residual stresses. Predicted stresses compared with (**a–d**) neutron diffraction-measured residual stresses along the *D lines*. Reprinted with permission from (O. Muránsky, M.C. Smith, P.J. Bendeich, T.M. Holden, V. Luzin, R.V. Martins, L. Edwards, Int. J. Solids Struct. **49**, 1045 (2012)) [9]. Copyright (2012) Elsevier

Due to the time and complexity of residual stress measurements, computer modelling of residual stresses has been pursued by many groups. The development and validation of models relies heavily on residual stress measurements. As there are significant uncertainties in both modeling and in residual stress measurements, the two activities inform each other's results. Agreement between modelling and measurements, or between different measurement techniques, improves confidence in both sets of results.

The Versailles Project on Advanced Materials and Standards (VAMAS) Technical Working Area (TWA) 20 ring-and-plug strain round-robin specimen has been used to validate neutron diffraction measurements with good correspondence in results [11].

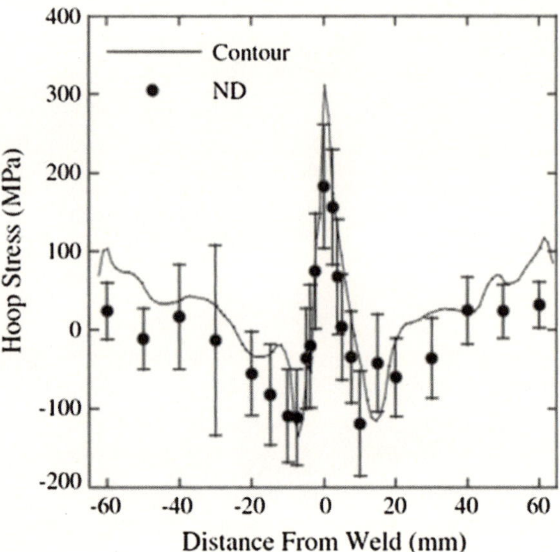

Fig. 4.5 Hoop residual stresses in E-beam welded uranium cylinder measured by *neutron diffraction* and by the *contour method*. Reprinted with permission from (D.W. Brown, T.M. Holden, B. Clausen, M.B. Prime, T.A. Sisneros, H. Swenson, J. Vaja, Acta Mater **59**, 864 (2011)) [10]. Copyright (2011) Elsevier

The European network on neutron techniques standardization for structural integrity (NeT) round robin had a number of samples for both residual stress measurement (by neutron diffraction, with some deep hole drilling and also using the contour method) and modelling [12–15]. The results show good correspondence (Fig. 4.6), although there were some systematic shifts in modelling and contour-method results, when compared to other methods. This led to changes in material descriptions used in modelling, and an appreciation of the effects of localized yielding on contour-method results during cutting.

There are difficulties comparing results as the neutron method averages stresses within the gauge volume. Finite-element analysis (FEA) results are discrete so the results should be volume-averaged to produce values over similar gauge volumes to neutron diffraction results. As real welds often have significant distortion, the spatial position of neutron results should be considered, if they were taken in a straight line they will often be at different distances from the surface while FEA results made on a 'straight' weld will be from different areas.

4.2.2 Nano-Particle Strengthening

Modern steels are strengthened by finely dispersed nano-particulates, particularly in high-temperature structural materials. Radiation damage and temperature can cause these to change their shape, size, and distribution, leading to embrittlement. Oxide dispersion strengthened (ODS) steels are designed for high-temperature operation; they contain oxide nano-clusters (e.g. yttrium-titanium oxide) in a ferritic-steel matrix.

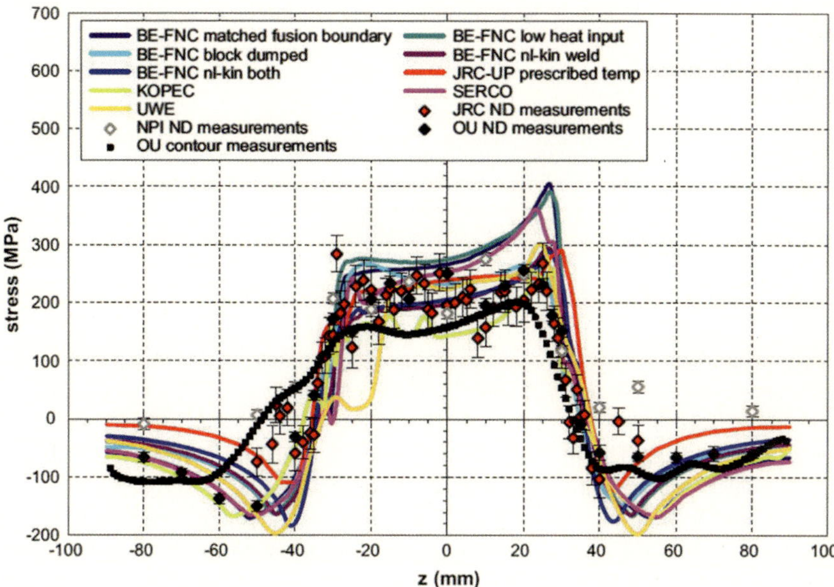

Fig. 4.6 Comparison of modelling predictions and neutron diffraction and contour method measurements for the NeT task group 1 single weld-bead on plate. Reprinted with permission from (M.C. Smith, A.C. Smith, Int. J. Press. Vessels Pip. **86**, 79 (2009)) [13]. Copyright (2009) Elsevier

The size and morphology of these particles may change with exposure to stress, temperature, or radiation; all conditions which occur in nuclear-power systems, typically increasing the size and decreasing the number of strengthening particles. Changes in the volume fraction, surface interface and size distribution of these particles can cause significant changes to the strength, ductility, and toughness of these materials. Advanced reactor-materials use nano-particulates to act as sinks for radiation-induced phenomena such as vacancies, self-interstitials, and as trapping sites for He bubbles from nuclear reactions in the material. Small-angle neutron scattering (SANS) can non-destructively analyse these nanoscale features over a large sample volume.

The use of a strong magnetic field can differentiate between magnetic and non-magnetic scattering at right angles to each other, and the strengthening nano-particulate oxides are typically non-magnetic. With ODS steel samples the SANS curves for no magnetic field (nuclear scattering only) and with field (nuclear + magnetic scattering) are compared. If they have no difference then the scattering is entirely non-magnetic and it can be assumed that the scattering is from the non-magnetic oxide nano-clusters. If there is a difference then the contribution from iron-based magnetic scatterers can be removed. This experimental technique should be used in conjunction with an oxide-free reference sample of the same manufacturing route (if possible), for correct de-convolution of the SANS from oxide nano-clusters.

The size of strengthening nano-particles is in the range 1–20 nm, and the few techniques available for studying these are transmission-electron microscopy (TEM), atom-probe tomography (APT), and SANS, however TEM and APT only sample a small volume. As a result, SANS has become a critically important technique for the development, understanding and characterisation of these irradiation-resistant materials [16, 17].

4.3 Zirconium and Its Alloys

Zirconium and its alloys are extensively used in nuclear applications due to the combination of a low neutron-absorption cross section, good mechanical properties under high stress and temperature conditions, low hydrogen/deuterium uptake and corrosion resistance. The main uses are for some structural components (particularly the calandria in CANada Deuterium Uranium (CANDU)-type reactors) and fuel cladding. It is anticipated to be used in the primary containment-vessel of high temperature D_2O inside the core of the fourth generation supercritical-water-cooled reactor (SCWR). Its behaviour during manufacturing and in service as well as under accident scenarios is therefore of great importance and a topic of extensive research.

Pure zirconium has a hexagonal closed-packed (hcp) crystal structure up to 866 °C and transforms to a body-centred cubic (bcc) crystal structure at higher temperatures. Their complex deformation-mechanisms have been investigated with neutron diffraction where different crystallographic planes have different elastic constants and post-yield behaviour, leading to strain partitioning. The main alloying elements for nuclear applications are niobium and tin, the latter forming together with other minor elements the so-called Zircaloys (*Zircaloy is a trademark of Westinghouse Electric Company, Pittsburgh, PA.).

Neutron diffraction has been applied to zirconium and its alloys to characterize welds, as an in situ mechanical test technique to characterize deformation modes to allow predictive modelling of deformation, to investigate the development of texture under temperature and stress, and to characterize the phase transformations including texture-variant selection during the hcp/bcc phase transformation. The following sections describe some of these experiments.

Zircaloys are also prone to brittle-hydride formation, particularly in welds. Hydrides can also form in parent plate if unfavourable textures are present due to a particular manufacturing route. Thus, many investigations have been performed on hydrides, such as imaging their location (radiography and prompt gamma), and assessing their susceptibility to hydride formation (residual stress and texture). Hydrogen accumulation or "pick-up" can also occur in the Zircaloy cladding of nuclear fuel and can cause embrittlement. In this case, the hydrogen content can be spatially visualized and quantified by neutron radiography. Blistering of Zircaloy fuel-cladding has also been investigated using neutron radiography, as X-rays are not effective in practice due to the high background, which includes gamma radiation from decay products.

4.3.1 Deformation

Due to the complex deformation behaviour of highly-anisotropic Zirconium alloys, these materials have been studied extensively by neutron diffraction, in combination with polycrystalline deformation models, the elasto-plastic and visco-plastic self-consistent models [18], to understand the deformation and texture development.

The difference in tensile and compressive behaviour in Zircaloy-2 was demonstrated by MacEwen et al. [19]. The strains were measured for each lattice plane on a time-of-flight (TOF) instrument (the General Purpose Powder Diffractometer at the intense-pulsed neutron source (IPNS), which is no longer operational). Neutron time-of-flight instruments can measure multiple lattice planes without reorienting the sample, as is required on a constant-wavelength instrument.

The residual stresses, stress tensor, and inter-granular stresses were characterized after 5 % strain by Pang et al. [20]. The measured lattice strains were in good agreement with those predicted by elasto-plastic self-consistent models, which predict deformation modes such as slip and twinning.

Zirconium and its alloys have a hcp structure, and there are too few slip systems for standard plasticity so twinning is an important contributor to plastic deformation. Rangaswamy et al. [21] compared changes in texture and twin volume-fractions to predictions from a visco-plastic self-consistent polycrystal model, which described both slip and twinning.

Balogh et al. [22] examined the deformation behaviour of Zr-2.5Nb samples by full-pattern diffraction line-profile analysis (DLPA) to determine the evolution of the density and type of the dislocation-structure induced by irradiation and plasticity. Control samples were compared to samples removed from a CANDU nuclear reactor pressure-tube to determine the evolution of microstructure and plasticity characteristics during deformation (27 % cold work during manufacture). The pressure tube was in service for 7 years at ~ 250 °C with a neutron fluence of $1.6 \times 1024 \text{ m}^{-2}$ (E > 1 MeV). Results show that fast-neutron irradiation significantly increases the overall dislocation density, accomplished entirely by an increase in the $\langle a \rangle$ Burgers vector dislocations.

4.3.2 Residual Stress

Welding is commonly used to fabricate zirconium-alloy components. Welding commonly results in residual stresses, and these can be complex due to the anisotropy of the material. Residual stresses can lead to hydride formation, so are significant to the structural integrity of the material. Using time-of-flight neutron diffraction, Carr et al. [23, 24] measured the residual stresses (Fig. 4.7) in a Zircaloy-4 gas tungsten arc weld. They also determined the lattice strains, texture, and the strain evolution in the weld during loading.

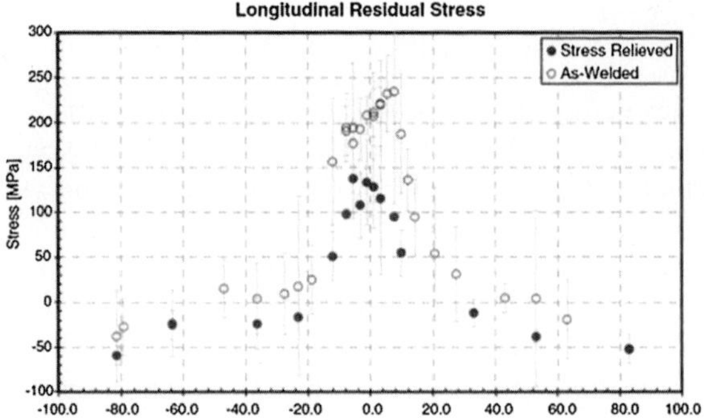

Fig. 4.7 Variation of the macroscopic residual-stresses in a Zircaloy-4 weld after a stress-relieving heat-treatment (*filled circles*) and as-welded (*open circles*). The as-welded stresses are reduced by about 40 % by the heat treatment. Measurements were taken across the plate at the mid-thickness position. Reprinted with permission from (D.G. Carr, M.I. Ripley, D.W. Brown, S.C. Vogel, T.M. Holden, J. Nucl. Mat. **359**, 202 (2006)) [24]. Copyright (2006) Elsevier

Fig. 4.8 Residual stresses in Zircaloy weld. Reprinted with permission from (P. Bendeich, V. Luzin, M. Law, Australian Nuclear Science and Technology Organisation report (2012)) [25]

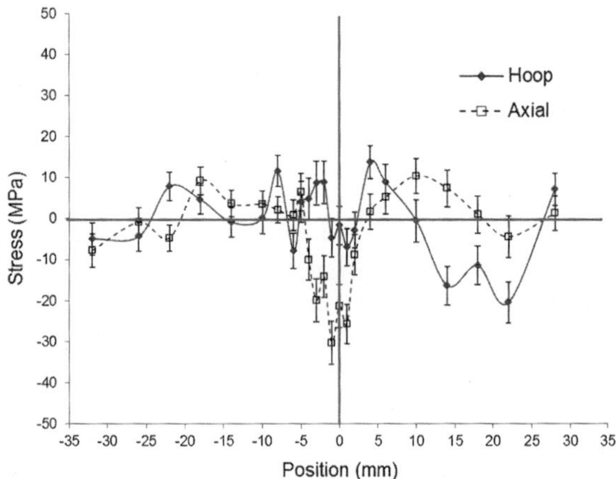

Welding often leads to grain growth and crystallographic texture changes, which may result in limited numbers of crystals which satisfy the diffraction condition, leading to poor statistics and errors. Methods of overcoming this include increasing the sampled area by increasing the gauge volume, and rocking or sweeping the sample through a larger volume. The residual stresses using such techniques were measured in a girth weld in the Zircaloy containment vessel for a cold-neutron source in a research reactor [25]. The rotational symmetry of the vessel was taken advantage of and the weld was rotated back and forth around the cylinder axis to increase the number of grains sampled. The residual stresses were low (Fig. 4.8) as the weld was subject to post-weld heat treatment to reduce residual stresses and drive off hydrogen trapped in the weld, in an effort to reduce hydrogen cracking and hydride formation.

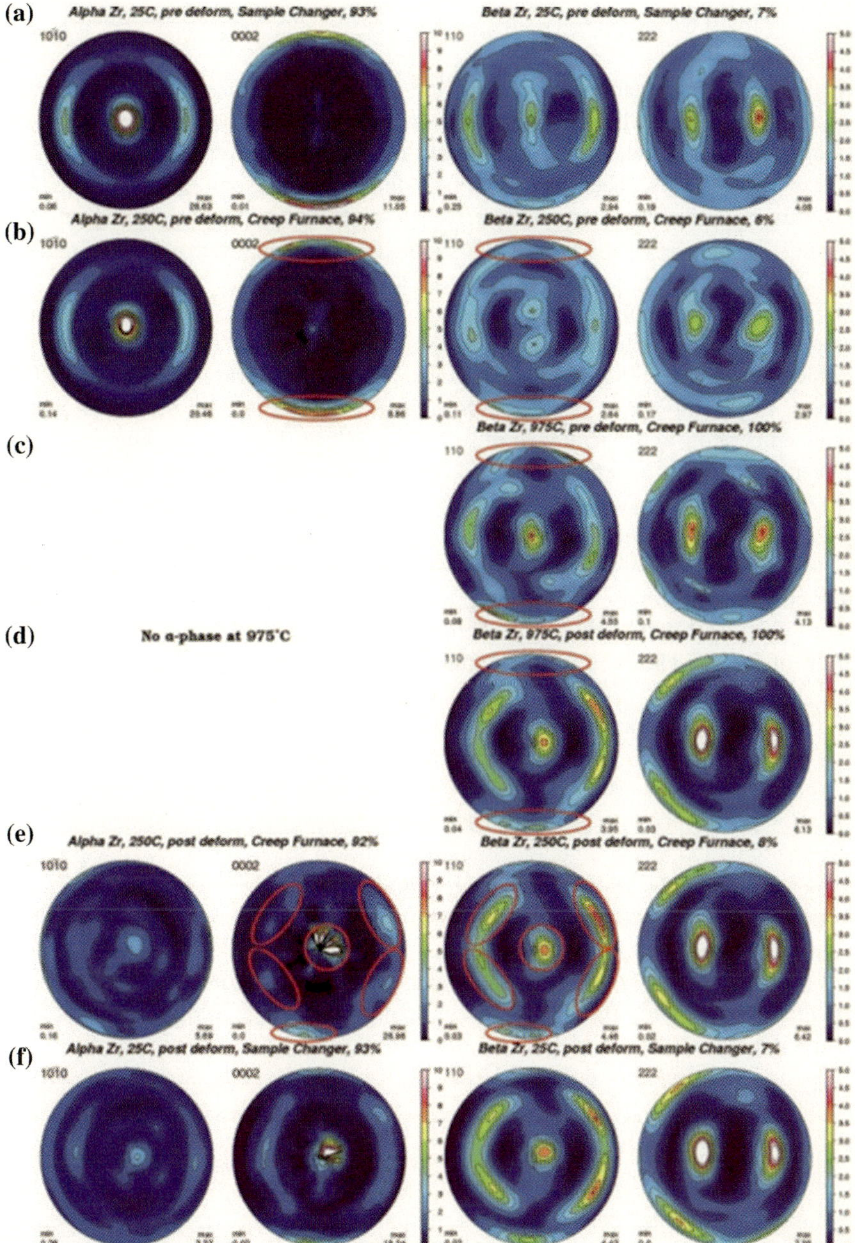

Fig. 4.9 Texture of the α-phase (*left column*) and β-phase (*right column*) of Zr-2.5Nb during heating, with 20 % strain in the axial direction (original extrusion direction), and cooling. The radial direction is *horizontal* on the pole figure and the hoop direction *vertical*. Reprinted from (S.C. Vogel, ISRN Materials Science **2013**, 24 (2013)) [32]

4.3.3 Texture

Texture is the orientation distribution of the single crystals forming the poly-crystalline aggregate. Due to the anisotropy of the single crystals, particularly the hexagonal alpha phase, texture is of great importance in the deformation behaviour of zirconium alloys. Therefore, understanding texture evolution during thermo-mechanical processing steps and service is a necessary precursor to predicting texture. Texture is affected by temperature [18, 26] and strain rate [27]. In situ diffraction is essential in understanding texture evolution.

Figure 4.9 shows an example of texture evolution in Zr-2.5Nb, with both temperature changes and deformation. This measurement was made on the HIPPO diffractometer [28, 29] at the Los Alamos Neutron Scattering Center (LANSCE) using a high-temperature deformation furnace [30]. The texture was measured at room temperature and at a number of steps up to 975 °C [31]. At this temperature the sample experienced a compressive strain of 20 %. Texture measurements were made at 975 °C, then at the same temperature steps during cooling to room temperature.

At room temperature the bcc beta phase is meta-stable and increases during heating at locations on the rim of the pole figure where the 0001 alpha pole-figure previously showed maxima, showing that alpha grains transform directly to beta grains [33]. At 975 °C only the bcc beta phase exists, at this point the sample had a strain of 20 %. The beta grains transform to alpha during cooling and (by the Burgers orientation relationship) the {0002} planes of the alpha phase become {110} planes of the beta phase, which is reflected in the texture evolution.

Pre and post treatment conditions of the sample are confirmed in Fig. 4.6a/b and e/f, respectively. The Burger orientation-relationship determines the transition of (0002) hcp/α to maxima in (110) bcc/β during a temperature increase to 975 °C. The resulting textures at high temperature are shown in (c). The (222) bcc/β planes align with the applied compression direction (d). To study the evolution of many properties such as texture, it is necessary to measure at intermediate steps, rather than just the start and end of processing.

Using Bragg-edge transmission [34] and neutron imaging in combination makes simultaneous mapping of the strains and texture (Fig. 4.10) [35] of the crystallites within the entire sample possible.

4.3.4 Zirconium Hydride

Hydride formation in Zircaloys reduces strength and ductility significantly. Hydrides form preferentially in areas of higher stress [36] such as near welds or at crack tips. In loss of coolant accidents (LOCAs), overheated Zircaloy cladding may react with cooling water in a complex manner with different phase-transformation temperatures depending on other species such as oxygen or hydrogen. The solubility of hydrogen in zirconium and Zircaloy differs by almost an order of magnitude between the alpha and beta phases [37], and excess hydrogen may form embrittling hydrides, particularly at

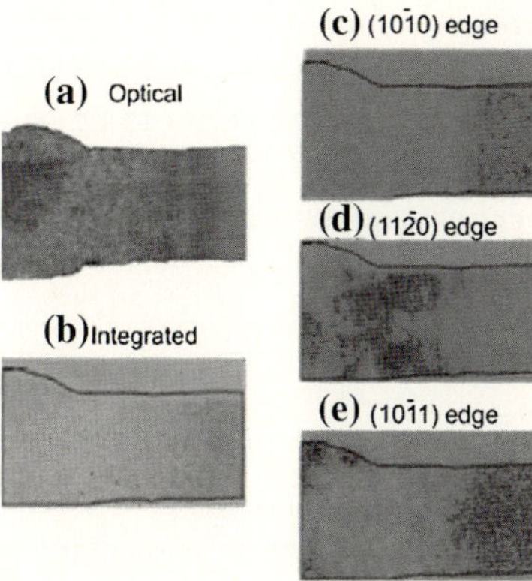

Fig. 4.10 Optical (**a**) and neutron-radiography images (**b, c, d, e**) of the welded Zircaloy-4 plates. Figure **b** was produced using neutrons with wavelengths between 1.4 and 2 Å, whilst Figures **c**, **d** and **e** are images of the height of selected Bragg edges. In figure **c** the start of the HAZ clearly shows the disappearance of the (10–10) Bragg edges, whilst the (11–20) edge in figure **d** reveals the differences between the *outer* and *inner layers* of the plate. Reprinted with permission from (J.R. Santisteban, M.A. Vicente-Alvarez, P. Vizcaino, A.D. Banchik, S.C. Vogel, A.S. Tremsin, J.V. Vallerga, J.B. McPhate, W. Lehmann Kockelmann, J. Nucl. Mater. **425**, 218 (2012)) [35]. Copyright (2012) Elsevier

Fig. 4.11 Steel target-rod with Zr cladding after exposure to protons in the target of the Swiss Spallation Neutron Source, SINQ. The *black points* are attributed to the agglomeration of hydrogen on the *inner side* of the Zr cladding. Because the sample is highly activated, a well-shielded facility was used for the investigations. Reprinted with permission from (E.H. Lehmann, P. Vontobel, G. Frei, C. Bronnimann, Nucl. Instrum. Meth. Phys. Res. A **531**, 228 (2004)) [43]. Copyright (2004) Elsevier

crack tips etc. Despite decades of work, the mechanisms of corrosion of zirconium-based alloys, particularly in reactor conditions, are still an active research field.

As hydrogen is a strong neutron-scatterer, the hydrogen concentration (Fig. 4.11) can be measured by neutron radiography [38] with sensitivities of $\sim$1,000 wt. ppm [39]. Neutron diffraction and radiography can be combined [40, 41] to identify and map hydride phases. The distribution [42] and kinetics [37] of hydrogen in materials can be identified by neutron radiography.

The hydrogen concentration in metals (particularly in zirconium-based alloys) can be also be measured by cold neutron prompt-gamma activation analysis PGAA [44]. However, these techniques are not suitable for imaging and rather provide the bulk average from the volume illuminated by the incident neutron beam. Cold neutron PGAA is based on measuring prompt gamma rays following the absorption of cold neutrons by hydrogen. This method has to be performed in a close proximity to a neutron source. The technique does not change the sample and so is well suited to assessing hydrogen uptake during interrupted corrosion testing.

4.4 Uranium

In this section we describe several applications of neutrons to investigate the crystallography and phase composition of uranium, particularly in fuels. Metallic uranium has an orthorhombic crystal structure, leading to anisotropic single-crystal properties such as a negative thermal-expansion along the crystallographic b axis. This leads to substantial integrity problems when the material is heated. Therefore, the vast majority of nuclear fuels in power reactors consist of cubic uranium-oxide whereas in research reactors metallic uranium-molybdenum alloys are also used, with the molybdenum stabilizing the cubic gamma-structure. During operation, i.e. during heating and irradiation, atoms rearrange and phase transformations may occur, with the phases in a spent fuel having different material properties to those of a fresh fuel. For example, thermal gradients of several hundred degrees exist between the centre and outside of a fuel rod, leading to a spatial distribution of crystallographic phases over a distance on the order of a centimeter. The identification of the new phases, determination of their formation conditions and kinetics, as well as establishing their properties, are of paramount importance for new and existing fuel types. Similar considerations apply to structural materials, e.g. cladding or pressure-tubing materials in accident scenarios, and actinide-bearing minerals for mining and waste deposition [45].

In fuels, the elements of interest are high Z-number (uranium and other actinides) and low Z-number (oxygen, nitrogen, carbon). Neutron diffraction is better at determining these structures while X-ray diffraction is biased towards the heavy atoms.

There are three structure models proposed for cubic UC_2 [46–48] a non-quenchable phase existing between $\sim 1,823$ and $\sim 2,104$ °C [49]. The three crystal structures differ in the arrangement of the carbon atoms. Simulated X-ray diffraction patterns are similar due to the bias towards the uranium lattice. Simulated neutron diffraction patterns can discriminate between the three different structures (Fig. 4.12) due to the sensitivity to the carbon atoms. Experimental neutron-diffraction data matches well with the structure proposed by Bowman [48].

This example illustrates the great advantages neutron diffraction offers over X-ray diffraction for crystal-structure investigations of nuclear materials and in particular nuclear fuels.

The smaller low Z-number elements are typically the mobile species and their rearrangement as a function of temperature leads to phase transitions, and neutron diffraction may be sensitive to these while X-ray diffraction will not. As many phases are non-quenchable, in situ techniques offer great advantages. Classical methods to study phase transitions, such as dilatometry or calorimetry, do not identify the

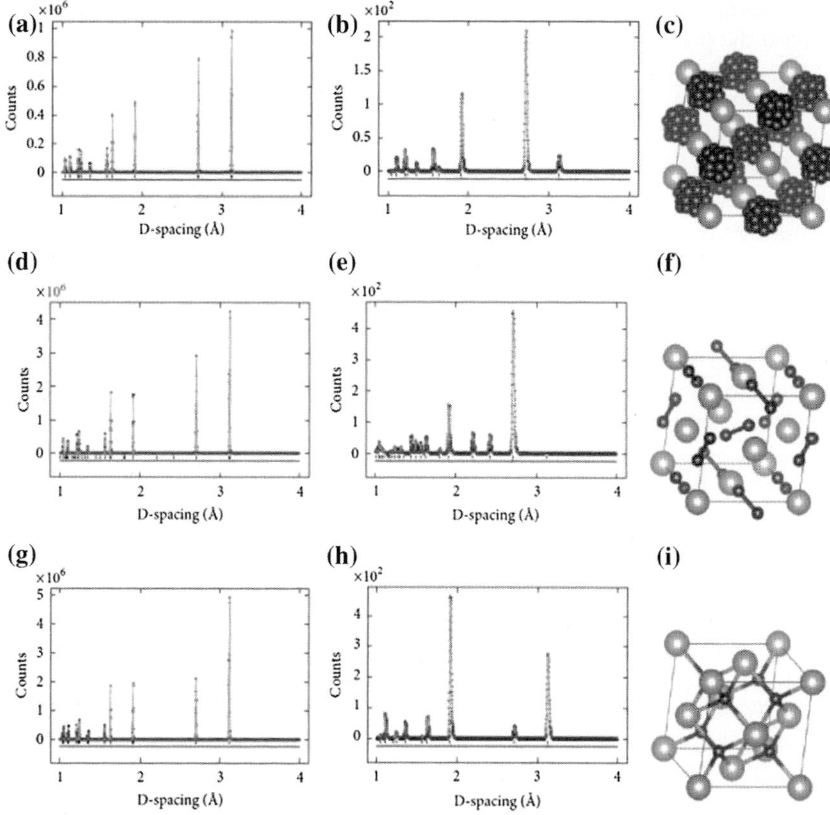

Fig. 4.12 Simulated diffraction patterns for X-rays (*left column*) and neutrons (*right column*) of the three structures for cubic UC$_2$ proposed by Bowman [48] (*top row*), Bredig [46] (*middle row*), and Wilson [47] (*bottom row*) showing the superiority of neutron diffraction in determining the structure of the cubic UC$_2$ phase. Reprinted from (S.C Vogel, ISRN Materials Science **2013**, 24 (2013)) [32]

newly-formed phases and are also sensitive to changes in chemical composition, e.g. rearrangement of the oxygen atoms, without formation of a new, distinct phase. Not surprisingly, neutrons have played a vital role over the past decades in elucidating the properties of nuclear fuels.

4.4.1 Structure

Structures of fuels with both light and heavy elements are best determined by neutron diffraction. Urania, with a large range of hyper-stoichiometric oxygen (up to UO$_{2.25}$), has been investigated by neutron diffraction. Many of the phases in the U-O phase diagram were determined by neutron diffraction.

As examples, Willis [50] utilized diffraction from a $UO_{2.13}$ single crystal to establish the positions of the excess oxygen atoms in the CaF_2-type structure; the crystal structure of beta-U_4O_{9-y} was solved by Bevan et al. [51] and later refined by Cooper [52].

Besides Bragg diffraction, where the crystallographic information is derived from the integrated peak intensities via the structure factor, diffuse scattering can be used to study disorder and oxygen diffusion, e.g. Clausen et al. [53], Goff et al. [54]. As the energy of thermal neutrons is similar to the generation or annihilation energies for phonons, neutrons can also be used to probe the lattice dynamics of nuclear materials, e.g. the first phonon–dispersion curve for UO_2 at room temperature measured by Dolling et al. [55]. See also the review article by Hutchings 1987 [56] for a review of earlier neutron scattering work on UO_2 and ThO_2. It is important to stress that the phases and their behaviour has to be studied in situ as the phenomena relevant to operation and accident conditions are temperature controlled.

4.4.2 Kinetics of Phase Transformations

High neutron-flux and improved detectors allow faster data acquisition; if sufficiently fast compared to the reaction rate, in situ kinetic studies are possible. The sensitivity of neutron diffraction to the crystallographic phases and the rate of data acquisition can allow not only the equilibrium state, but any intermediate phases and reaction rates to be determined.

To study the kinetics of phase transformations, hyper-stoichiometric uranium oxide was cycled across a phase boundary [57]. Desgranges et al. [58] performed in situ studies on transitions between four different uranium-oxide phases which depended on both temperature and oxygen partial-pressure. Knowledge of intermediate phases led to a better understanding of the phase-transition process and growth kinetics.

4.4.3 Radiography and Tomography

Neutrons are particularly important in the imaging of nuclear-fuel rods which are strong γ-sources (X-ray background) and are made from heavy metal elements i.e. uranium or lead with a high attenuation of X-rays. A resolution of 50 μm or better can be accomplished and this has been utilized to characterize nuclear materials, e.g. fuel rods [43] or cladding materials [42], see the section on hydrides above. Recent detector developments allowing for spatially and time-resolved neutron detection [59] are likely to open new avenues of characterization of nuclear materials and nuclear fuels in particular as they allow for isotope-sensitive imaging via neutron resonance absorption [60].

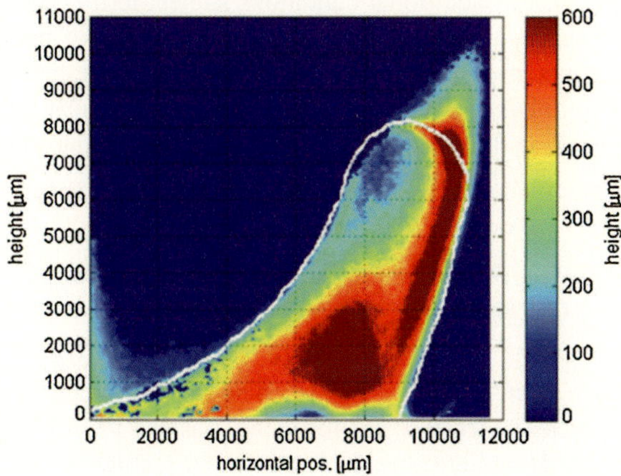

Fig. 4.13 Liquid film thickness on the surface of a vane in a coolant channel. The *white line* is the surface of the vane. Reprinted with permission from (R. Zboray, J. Kickhofel, M. Damsohn, H.M. Prasser, Prasser, Nucl. Eng. Des. **241**, 3201 (2011)) [62]. Copyright (2011) Elsevier

4.5 Coolant

As hydrogen is such a strong scatterer, the presence of hydrogen, water, and hydrides are easily detected in neutron radiography and tomography [61]. Coolant behaviour in the fuel bundle of a boiling-water reactor (BWR) was examined by Zboray et al. [62] with neutron tomography (Fig. 4.13), allowing the coolant flow and channels to be optimized.

4.6 Measurement of Radioactive Samples

Active samples pose additional difficulties; safe handling and sample preparation, transport, shielding, and disposal may all add to the experimental complexity. The standard methods of reducing operator exposure include limiting the amount or size of the sample, increasing the distance from the sample to the operator, limiting exposure time, and the use of shielding [63].

For SANS and texture measurements, the sample size can be favourably small (~ 10 mm^3 in some cases). The sample size in radiography is generally defined by the project scope and the scale of the component to be imaged. When measuring residual stresses, reducing the sample size changes the constraint conditions which can significantly alter the measured residual stress; there are simplified methods of assessing the allowable sample size [64]. An additional issue is that reducing sample size by cutting active samples increases operator exposure and leads to active waste. All these factors must be considered for the safe conduct of the experiment and as well as additional conditions subject to the safety policy of the laboratory. However, in order to understand the real effects of neutron irradiation damage to reactor materials the benefits can sometimes outweigh the complexities involved. By nature, neutron facilities have detailed radiological safety procedures already in place, making the study of radioactive samples perhaps more viable than at other facilities.

4.7 Outlook and Perspectives

Neutron characterization techniques are particularly important in understanding materials for the nuclear-energy sector for many reasons. In structural components the combination of radiography, residual stress, and texture measurements allows assessment of structural integrity and optimization of manufacturing, while SANS provides a unique tool to track the effects of radiation damage. In fuels and waste forms, particularly when heavy and light elements are combined, neutron diffraction has inherent advantages over X-rays.

Neutron characterization techniques will be increasingly important for understanding materials for the nuclear sector. The increased functionality comes partly from new techniques, and partly from improvements in existing techniques.

New techniques such as energy-dispersive neutron radiography and Bragg-edge strain imaging offer new insights into materials. As the technologies are new, large improvements are to be expected in technique and analysis.

Well established techniques will benefit from improvements in neutron flux, detectors, and analysis, enabling in situ or kinetic studies, and smaller samples. Sample environments will increase in complexity to better mimic the studied operating conditions. Shielding or dedicated beamlines will allow characterization of active materials.

Combinations of techniques, such as diffraction and radiography, will provide detailed crystallographic information in combination with a spatially-resolved distribution of the properties of interest.

References

1. P.J. Withers, H.K. Bhadeshia, Mater. Sci. Technol. **17**, 355 (2001)
2. P.J. Withers, H.K. Bhadeshia, Mater. Sci. Technol. **17**, 366 (2001)
3. M.T. Hutchings, P.J. Withers, T.M. Holden, T. Lorentzen, *Introduction to the characterization of residual stress by neutron diffraction* (Taylor & Francis, Boca Raton, 2005)
4. Y. Ishiyama, R.B. Rogge, M. Obata, J Nucl Mater **408**, 153 (2011)
5. H. Köhler, K. Partes, J.R. Kornmeier, F. Vollertsen, Phys. Procedia **39**, 354 (2012)
6. L. Edwards, P.J. Bouchard, M. Dutta, D.Q. Wang, J.R. Santisteban, S. Hiller, M.E. Fitzpatrick, Int. J. Press. Vessels Pip. **82**, 288 (2005)

7. P.J. Bouchard, D. George, J.R. Santisteban, G. Bruno, M. Dutta, L. Edwards, E. Kingston, M. Smith, Int. J. Press. Vessels Pip. **82**, 299 (2005)
8. O. Muránsky, M.C. Smith, P.J. Bendeich, L. Edwards, Comput. Mater. Sci. **50**, 2203 (2011)
9. O. Muránsky, M.C. Smith, P.J. Bendeich, T.M. Holden, V. Luzin, R.V. Martins, L. Edwards, Int. J. Solids Struct. **49**, 1045 (2012)
10. D.W. Brown, T.M. Holden, B. Clausen, M.B. Prime, T.A. Sisneros, H. Swenson, J. Vaja, Acta Mater **59**, 864 (2011)
11. G.A. Webster, J. Wimpory, Mater. Process. Technol. **117**, 395 (2001)
12. C. Ohms, R.C. Wimpory et al., *Residual stress measurements in a three-bead slot weld in a 20 mm carbon steel plate*. PVP2009-77496 ASME Pressure Vessels and Piping Conference, Prague, Czech Republic (2009)
13. M.C. Smith, A.C. Smith, Int. J. Press. Vessels Pip. **86**, 79 (2009)
14. S. Pratihar, M. Turski, L. Edwards, P.J. Bouchard, Int. J. Press. Vessels Pip. **86**, 13 (2009)
15. P.J. Bouchard, Int. J. Press. Vessels Pip. **86**, 31 (2009)
16. G. Albertini, R. Coppola, Nucl. Instrum. Meth. Phys. Res. **A314**, 352 (1992)
17. R. Coppola, F. Fiori, E.A. Little, M. Magnani, J. Nucl. Mater. **245**, 131 (1997)
18. C. Tomé, P. Maudlin, R. Lebensohn, G. Kaschner, Acta Mater. **49**, 3085 (2001)
19. S.R. MacEwen, J. Faber, A.P.L. Turner, Acta Metall. **31**, 657 (1983)
20. J.W.L. Pang, T.M. Holden, P.A. Turner, T.E. Mason, Acta Mater. **47**, 373 (1999)
21. P. Rangaswamy, M. Bourke, D.W. Brown, G.C. Kaschner, R.B. Rogge, M.G. Stout, C.N. Tome, Metall. Mater. Trans. A **33**, 757 (2002)
22. L. Balogh, F. Long, M.R. Daymond, R.A. Holt, D.W. Brown, C.D. Judge, The influence of fast neutron irradiation on the plasticity induced evolution of dislocation densities and operating deformation modes in Zr-2.5Nb, Mecasens, Sydney (2013)
23. D.G. Carr, M.I. Ripley, T.M. Holden, D.W. Brown, S.C. Vogel, Acta Mater. **52**, 4083 (2004)
24. D.G. Carr, M.I. Ripley, D.W. Brown, S.C. Vogel, T.M. Holden, J. Nucl. Mat. **359**, 202 (2006)
25. P. Bendeich, V. Luzin, M. Law, Australian Nuclear Science and Technology Organisation report (2012)
26. G.C. Kaschner, J.F. Bingert, C. Liu, M.L. Lovato, P.J. Maudlin, C.N. Tome, Acta Mater. **49**, 3097 (2001)
27. T.A. Sisneros, D.W. Brown, B. Clausen, D.C. Donati, S. Kabra, W.R. Blumenthal, S.C. Vogel, Mater. Sci. Eng. A **527**, 5181 (2010)
28. H.R. Wenk, L. Lutterotti, S.C. Vogel, Nucl. Instrum. Meth. Phys. Res. A **515**, 575 (2003)
29. S.C. Vogel, C. Hartig, L. Lutterotti, R.B. Von Dreele, H.R. Wenk, D.J. Williams, Powder Diffr. **19**, 65 (2004)
30. H.M. Reiche, Advanced sample environments for in situ neutron diffraction studies of nuclear materials. Ph.D. thesis, New Mexico State University, Las Cruces, New Mexico, U.S.A. (2012)
31. H.M. Reiche, S.C. Vogel, Rev. Sci. Instrum. **81**, 1 (2010)
32. S.C Vogel, ISRN Materials Science **2013**, 24 (2013)
33. D. Bhattacharyya, G.B. Viswanathan, S.C. Vogel, D.J. Williams, V. Venkatesh, H.L. Fraser, Scripta Mater. **54**, 231 (2006)
34. S. Vogel, Ph.D. thesis, Christian-Albrecht-University, Kiel (2000), http://e-diss.uni-kiel.de
35. J.R. Santisteban, M.A. Vicente-Alvarez, P. Vizcaino, A.D. Banchik, S.C. Vogel, A.S. Tremsin, J.V. Vallerga, J.B. McPhate, W. Lehmann Kockelmann, J. Nucl. Mater. **425**, 218 (2012)
36. E. Garlea, H. Choo, G.Y. Wang, P.K. Liaw, B. Clausen, D.W. Brown, J. Park, P.D. Rack, E. Kenik, Metall. Mater. Trans. A **41**, 2816 (2010)
37. M. Grosse, M. Steinbrueck, E. Lehmann, P. Vontobel, Oxid. Met. **70**, 149 (2008)
38. E.H. Lehmann, P. Vontobel, N. Kardjilov, Appl. Radiat. Isot. **61**, 503 (2004)
39. R. Yasuda, M. Nakata, M. Matsubayashi, K. Harada, Y. Hatakeyama, H. Amano, J. Nucl. Mater. **320**, 223 (2003)
40. E. Sváb, G. Mészáros, Z. Somogyvári, M. Balaskó, F. Körösi, Appl. Radiat. Isot. **61**, 471 (2004)

41. A.M. Shaikh, P.R. Vaidya, B.K. Shah, S. Gangotra, K.C. Sahoo, BARC Newsl. **273**, 104 (2006)
42. M. Grosse, G. Kuehne, M. Steinbrueck, E. Lehmann, J. Stuckert, P. Vontobel, J. Phys. Condens. Matter **20**, 104263 (2008)
43. E.H. Lehmann, P. Vontobel, G. Frei, C. Bronnimann, Nucl. Instrum. Meth. Phys. Res. A **531**, 228 (2004)
44. A. Couet, A.T. Motta, R.J. Comstock, R.L. Paul, J. Nucl. Mater. **425**, 211 (2012)
45. P.C. Burns, R.C. Ewing, A. Navrotsky, Science **335**, 1184 (2012)
46. M.A. Bredig, J. Am. Ceram. Soc. **43**, 493 (1960)
47. W.B. Wilson, J. Am. Ceram. Soc. **43**, 77 (1960)
48. A.L. Bowman, G.P. Arnold, W.G. Wittman, T.G. Wallace, N.G. Nereson, Acta. Crystallogr. **21**, 670 (1966)
49. T.B. Massalski, *Binary Alloy Phase Diagrams* (William W. Scott, Jr, Materials Park, Ohio, 1990)
50. B.T.M. Willis, Nature **197**, 755 (1963)
51. D.J.M. Bevan, I.E. Grey, B.T.M. Willis, J. Solid State Chem. **61**, 1 (1986)
52. R.I. Cooper, B.T.M. Willis, Acta Cryst. A **60**, 322 (2004)
53. K. Clausen, W. Hayes, J.E. Macdonald, R. Osborn, M.T. Hutchings, Phys. Rev. Lett. **52**, 1238 (1984)
54. J.P. Goff, B. Fak, W. Hayes, M.T. Hutchings, J. Nucl. Mater. **188**, 210 (1992)
55. G. Dolling, R.A. Cowley, A.D.B. Woods, Can. J. Phys. **43**, 1397 (1965)
56. M.T. Hutchings, J. Chem. Soc. Faraday Trans. **2**, 1083 (1987)
57. J.D. Higgs, W.T. Thompson, B.J. Lewis, S.C. Vogel, J. Nucl. Mater. **366**, 297 (2007)
58. L. Desgranges, G. Baldinozzi, G. Rousseau, J.C. Niepce, G. Calvarin, Inorg. Chem. **48**, 7585 (2009)
59. A.S. Tremsin, J.B. McPhate, J.V. Vallerga, O. Siegmund, J.S. Hull, W.B. Feller, E. Lehmann, Nucl. Instrum. Meth. A **604**, 140 (2009)
60. W.E. Lamb, Phys. Rev. **55**, 190 (1939)
61. Ross AM (1976) Practical applications of neutron radiography and gaging, ASTM STP 586, pp 195–209
62. R. Zboray, J. Kickhofel, M. Damsohn, H.M. Prasser, Nucl. Eng. Des. **241**, 3201 (2011)
63. J.H. Root, C.E. Coleman, J.W. Bowden, M. Hayashi, J. Press Vessel Technol. **119**, 137 (1997)
64. M. Law, O. Kirstein, V. Luzin, J. Strain Anal. **45**, 567 (2010)

Chapter 5
Chalcopyrite Thin-Film Solar-Cell Devices

Susan Schorr, Christiane Stephan and Christian A. Kaufmann

Abstract In order to understand the importance of the structural properties of compound semiconductors for the operation of a thin-film solar cell, this section aims to explain the operation principle using the example of a $Cu(In,Ga)Se_2$ (CIGSe) thin-film solar cell. For detailed information the reader is kindly referred to the literature for a more extensive overview of the recent developments [1], device operation [2] and material preparation [3].

5.1 Introduction

CIGSe thin-film solar cells are made of a stack of metal and semiconductor thin films in the following sequence: a molybdenum back contact (metal), a polycrystalline CIGSe absorber layer (p-type semiconductor), CdS buffer layer (*n*-type semiconductor), and ZnO front contact (n-type semiconductor). Together, the CdS and the ZnO are often referred to as the 'window' of the device. The core of the device is the p-n-heterojunction between the p-type absorber layer and the n-type window layers. The resulting energy band line-up and a cross sectional scanning electron microscope view of a complete device are shown in Fig. 5.1. Due to the high doping of the ZnO front contact layer the field, which develops upon contact of the n- and p-type materials in the interface region, is located almost entirely inside the absorber layer. In comparison to homojunctions, the heterojunction has the advantage that the n-type component can be chosen such that its band gap is

S. Schorr (✉) · C. Stephan
Institut Für Geologische Wissenschaften, Freie Universität Berlin,
Malteserstr. 100, 12249 Berlin, Germany
e-mail: susan.schorr@fu-berlin.de

C. Stephan
e-mail: christiane.stephan@fu-berlin.de

S. Schorr · C.A. Kaufmann
Helmholtz-Zentrum Berlin Für Materialien Und Energie, Berlin, Germany

© Springer International Publishing Switzerland 2015
G.J. Kearley and V.K. Peterson (eds.), *Neutron Applications in Materials for Energy*, Neutron Scattering Applications and Techniques,
DOI 10.1007/978-3-319-06656-1_5

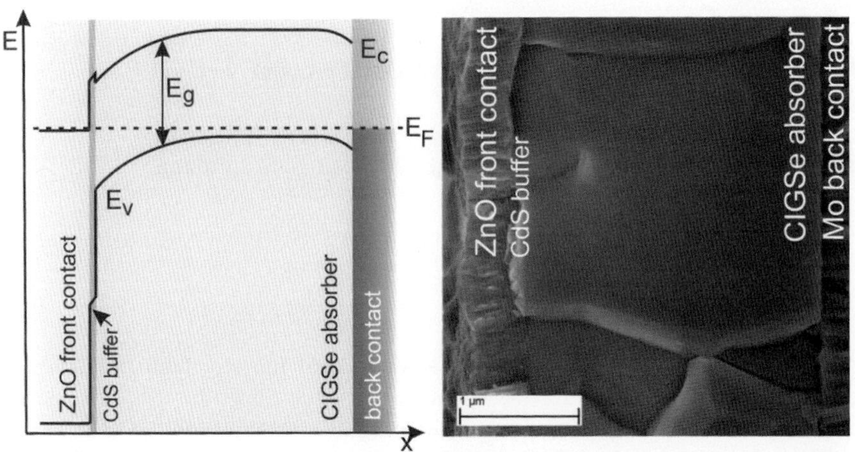

Fig. 5.1 Energy band diagram of a standard Cu(In,Ga)Se$_2$ thin film solar cell (*left*) in correlation to the cross sectional view of a complete solar cell device as seen in the scanning electron microscope (*right*)

large and the photoactive band-gap E$_g$ of the solar-cell device, determined by the band gap of the absorber-layer material, is optimized to reach high conversion-efficiencies.

Figure 5.2 illustrates the basic working principle of a CIGSe thin-film solar cell. Electron-hole pairs are generated by light absorption within the absorber thin-film. Absorption of photons with an energy higher than E$_g$ results in the loss of excess energy via thermalization. If an electron-hole pair is excited within the depletion

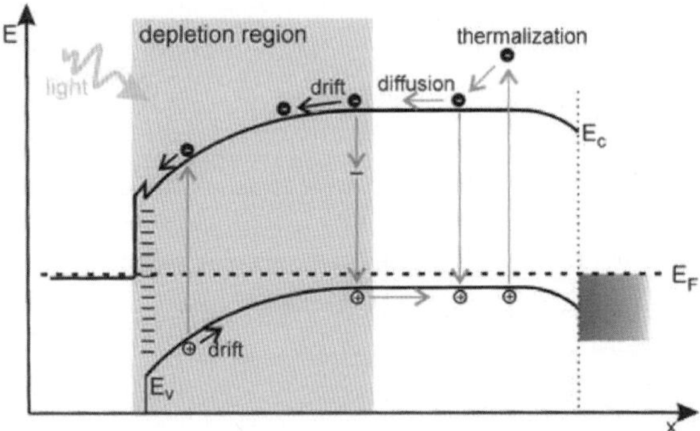

Fig. 5.2 Working principle of a CIGSe thin film solar cell: electron-hole pairs are excited by the absorption of incident light; within the depletion region of the p-n-heterojunction they are immediately spatially separated; charge carriers, which are generated in the bulk of the thin film, can diffuse into the depletion region. When $h\nu >$ E$_g$ excess energy is lost by thermalization. A number of recombination channels are present in the bulk, in the depletion region and at the interfaces

region near the font interface of the device, it is immediately spatially separated by the electric field that is present. If generation happens outside the depletion region, the minority charge carrier has to diffuse into the depletion region in order to contribute to the photo current. Recombination of the electron-hole pairs can occur by a number of processes: radiative and non-radiative, band-to-band or via one or several defect levels located in the band gap of the absorber material, within the bulk of the thin-film material, within the depletion region, and at the front interface of the device or possibly at the grain boundaries of the polycrystalline CIGSe thin-film. The nature of grain boundaries in CIGSe thin-films however, has proven to be considerably more benign than in other semiconductor materials such as for example silicon, the exact reasons for which are still under discussion [4].

CIGSe absorbers for photovoltaic application are mostly fabricated slightly Cu-poor and with an overall Ga content of [Ga]/([Ga] + [In]) $\sim$ 0.3. The resulting CIGSe material is a p-type semiconductor material ($E_g \sim$ 1.15 eV), which is highly compensated. This means that there are acceptor- and donor-type defects present within the material and p-type conductivity is established due to the exceptionally-low formation energy of Cu vacancies. In addition, the formation of defect complexes, such as $2V_{Cu}^{-} + In_{Cu}^{2+}$, seems to play an important role in terms of the electronic properties of the material, and also regarding phase formation and crystal structure [5] (Figs. 5.2 and 5.3).

For most applications rigid soda-lime glass is used as a substrate material, but flexible metal or polyimide foils have been used successfully. Working on soda lime glass, it has been established that Na plays an important role in increasing the carrier concentration within the CIGSe absorber [6]. It diffuses at elevated process temperatures from the glass substrate through the Mo back contact into the growing CIGSe thin-film [7], and also has an effect on the morphology and material inter-diffusion in the growing layer [8, 9]. On samples, which do not intrinsically contain sodium, it has to be externally supplied.

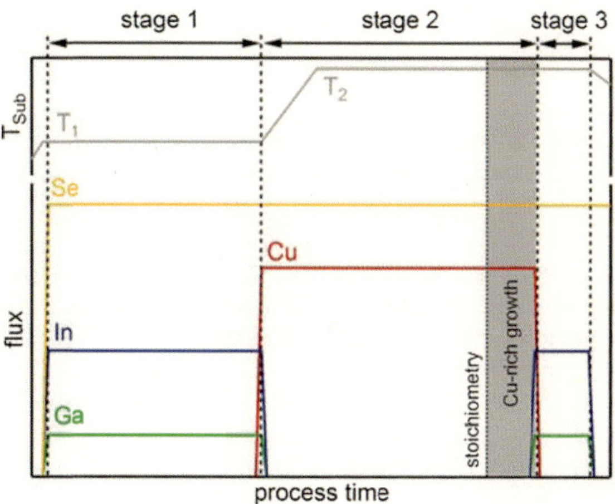

Fig. 5.3 Elemental flux and substrate-temperature profiles for a typical 3-stage co-evaporation process

While ZnO and Mo are usually sputtered, CdS is most widely deposited via chemical-bath deposition. As the use of Cd-containing components is viewed critically, Cd-free alternatives are in development and, in some cases, already incorporated in the commercially-available product [10]. The methods for CIGSe thin-film fabrication can be grouped into two main categories: co-evaporation and sequential processing [10]. The method of co-evaporation relies on the more or less simultaneous evaporation of the elements Cu, In, Ga, and Se to form a CIGSe thin-film on a heated, Mo-coated substrate in vacuum. For sequential processing on the other hand, a so-called precursor layer in either metallic, binary, or nanoparticle form is treated in a reactive atmosphere in order to make the CIGSe thin-film absorber. The current record conversion-efficiency of 20.3 % [11] is reached by a CIGSe thin-film device that was deposited via 3-stage co-evaporation [12].

The 3-stage co-evaporation process for CIGSe deposition leads to a thin film, which has a characteristic compositional in-depth profile, shown in Fig. 5.4. While Ga accumulates near the Mo/CIGSe interface during the inter-diffusion of Cu-Se in stage 2, the last stage of the process also provides for a slightly increased Ga content near the surface. Of course the latter depends on the Ga flux during stage 3. The Ga profile is of relevance to the resulting solar-cell device, as the Ga content determines the energy band-gap of the CIGSe material. A graded compositional profile is therefore equivalent with an in-depth band-gap grading within the device. Compositional gradients, as shown in Fig. 5.4, can be observed even within single grains. Not only does the Ga content have an impact on the resulting energy band-gap of CIGSe, but also on the Cu-deficiency, in particular near the absorber surface. This is possibly caused by the nature of the last stage of the 3-stage process, which can affect its band gap. Cu-deficient CIGSe phases, such as $Cu(In,Ga)_3Se_5$, show slightly-larger band gaps than the stoichiometric compound. The presence of such Cu-poor phases at the front interface has been argued to be of relevance for efficient CIGSe thin-film devices [13]. Other evidence points towards entirely Cu-free film surfaces, which ensure favourable interface formation between the CIGSe and the buffer layer [14]. It is most likely that the exact process-parameters determine which

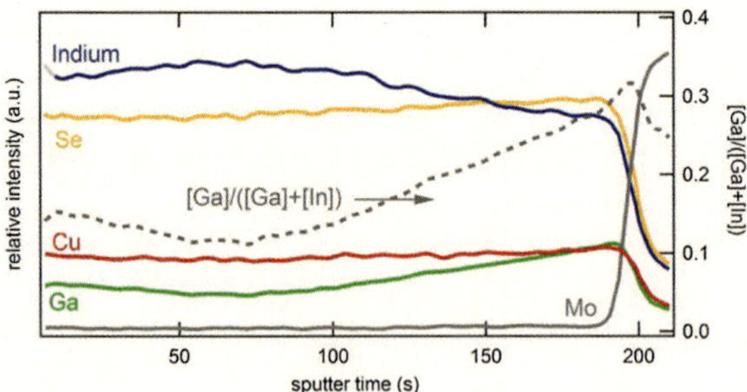

Fig. 5.4 Compositional depth-profile of a solar cell grade $Cu(In,Ga)Se_2$ thin-film, recorded by glow discharge optical-emission spectroscopy [1, 7]. The profile, that is displayed here, is typical for a CIGSe thin-film, that is deposited via 3-stage coevaporation

of the different scenarios holds in each case. CIGSe thin-films from other manu-facturing routines, such as the sequential processing, will have grown under very different thermodynamic and kinetic growth conditions and, so far, only little is known of the extent to which the resulting thin films can be considered identical.

In order to understand the correlation between growth conditions, material properties and final device quality, much has to be understood regarding the basic material properties of the absorber material in question. Topics such as the occu-pation density of the different sites within the crystal lattice, i.e. defect formation under certain growth conditions, at certain material compositions, or in the presence of foreign elements are areas where neutron diffraction can provide valuable input for the design of growth models, analytical material science, and also for compu-tational methods.

5.2 Material Properties of Chalcopyrite-Type Compound Semiconductors

$CuB^{III}C_2^{VI} \left(C^{VI} = Se, \ S\right)$ compound semiconductors are part of the chalcopyrite (ch) family and are located in the middle of the ternary system $Cu–B^{III}–C^{VI}$, on the pseudo-binary section $Cu_2C^{VI}–B_3^{III}C_2^{IV}$ (see Fig. 5.5). The band gap (E_g) ranges from 1.0 to 1.5 eV for a single junction thin-film solar cell. The $CuB^{III}C_2^{VI}$ compounds crystallize in the chalcopyrite-type crystal structure, named after the mineral $CuFeS_2$. This tetragonal crystal structure (space group $I\bar{4}2d$) consists of two specific cation-

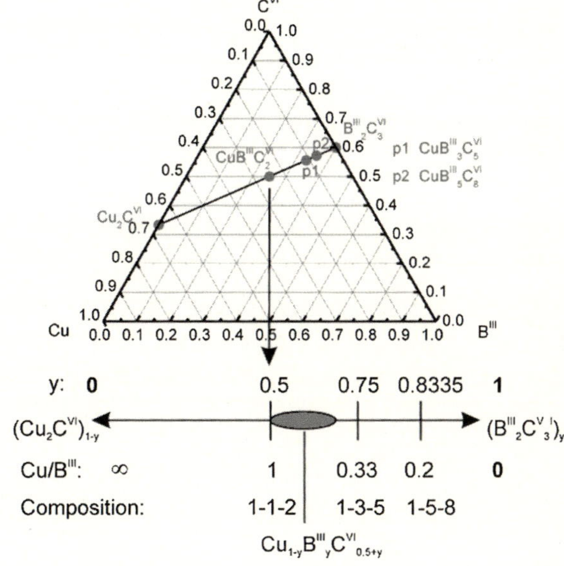

Fig. 5.5 The ternary system $Cu-B^{III}-C^{IV}$ and the pseudo-binary tie line $Cu_2C^{VI}-B^{III}_2C^{VI}_3$

sites. The monovalent cations are sited on the $4a$ (0 0 0) and the trivalent cations (In^{3+}, Ga^{3+}) on $4b$ (0 0 ½) position. All cations are tetrahedrally coordinated by the anions ($8d$ (x ¼⅛)) and vice versa.

A closer examination of the pseudo-binary tie-line reveals a stability of the ch-phase over a defined compositional-range. That means, the ch-phase accepts a deviation from ideal stoichiometry ($CuB^{III}C_2^{VI}$) by maintaining the crystal structure, and without the formation of any secondary phase. The compound $Cu_{1-y}In_ySe_{0.5+y}$ is single phase in the region of $0.513 \leq y \leq 0.543$ and contains within this area only the chalcopyrite-type phase [15]. The common highly-efficient $Cu(In,Ga)Se_2$ thin-film solar devices all exhibit an overall off-stoichiometric composition, due to the multi-stage process applied to grow these absorbers. Such deviations from stoichiometry always cause structural inhomogeneities and charge mismatches, which influence the material properties. One effect is the generation of point defects, which influences the electronic and optical properties of the compound semiconductor. In general 12 intrinsic point-defects can exist within the ch-type crystal structure.

- 3 vacancies: on the two cation and one anion sites (V_{Cu}, V_B^{III}, V_C^{VI})
- 6 anti-site defects: B_{Cu}^{III}, Cu_B^{III}, C_{Cu}^{VI}, Cu_C^{VI}, $B_C^{III\ VI}$, $C_B^{VI\ III}$
- 3 interstitial defects: Cu_i, B_i^{III}, C_i^{VI}

These intrinsic point-defects cause different defect levels in the energy gap of the semiconductor (see Table 5.1) and therefore influence the electronic and optical properties, sensitively. Consequently, it is of great importance to know where the atoms are.

In addition to the generation of point defects, the anion position ($x(C^{VI})$) of chalcopyrite crystallites is also affected by off stoichiometry. A change of the anion position is proposed to be directly correlated with a change in E_g. The x-parameter controls the position of the valence-band maximum and conduction-band

Table 5.1 Cation point-defect formation energies and defect types in off-stoichiometric $Cu_{1-y}In_ySe_{0.5+y}$ [16]

Defect	Formation energy (eV) Cu-poor/In-rich	Formation energy (eV) Cu-rich/In-poor	Defect type
V_{Cu}^0	−1.4	0.5	Acceptor
V_{Cu}^-	−2.4	−0.5	Acceptor
V_{In}^0	3.0	1.0	Acceptor
V_{In}^{3-}	1.2	−1.7	Acceptor
Cu_{In}^0	3.5	−0.5	Acceptor
Cu_{In}^{2-}	2.3	1.7	Acceptor
In_{Cu}^0	1.3	5.3	Donor
In_{Cu}^{2+}	−0.1	3.9	Donor
Cu_i^0	5	3	Donor
Cu_i^{+-}	4	2	Donor

minimum, and therefore E_g. Current studies have shown that E_g decreases for the Cu-poor composition in $CuInSe_2$ caused by a change in x(Se), which is weakly dependent on the concentration of copper vacancies (V_{Cu}) [17]. The interplay between the crystal structure and the optical and electronic properties is a fundamental problem, which has to be understood when tailoring high efficiency thin-film devices with a compound semiconductor as absorber layer. For instance, an uncontrolled change in E_g within an absorber layer is undesirable because it is less optimal for absorption of the incoming sunlight.

It is difficult to identify and quantify very small changes in the crystal structure, such as point defects or changes in atomic positions by imaging techniques. Therefore, it is preferred to study such effects by diffraction methods. The method of Rietveld refinement is applied to refine the crystal structure using an X-ray or neutron powder diffraction pattern of the off-stoichiometric compound in detail. This method provides information about the cation distribution and the position of the atoms within the structure with high accuracy. In the following section we discuss the reasons for the preferred use of neutrons in the description of structural changes in detail, and how point defects in compound semiconductors can be identified.

5.3 Structural Analysis of Off-Stoichiometric Chalcopyrites

Among various experimental possibilities, structural analysis of compound semiconductors by diffraction techniques using X-rays or neutrons has become a technique of choice. The reasons are various, and here we place the focus on the use of neutrons as a radiation source.

The highest efficiency thin-film devices consist of a Cu-poor $Cu(In,Ga)Se_2$ absorber layer. The interplay of structural with electronic and optical properties is therefore interesting to study for this quaternary compound and the corresponding ternaries ($CuGaSe_2$ and $CuInSe_2$). In the case of $CuGaSe_2$ or $Cu(In,Ga)Se_2$ the cations Cu^+ and Ga^{3+} have an identical number of electrons (28). This is a problem for the differentiation of these two cations by conventional X-ray diffraction, where the diffraction is at the electron shell of the atoms. Since atomic scattering form factors f for X-rays are proportional to the atomic number Z, the positions of the unit cell atoms of similar atomic number and the fractional occupation of the Wyckoff sites are difficult to distinguish. Hence, a differentiation of these cations in the atomic structure by conventional X-ray diffraction is impossible. In the case of neutron diffraction, the scattering is at the nucleus and the neutron-scattering lengths of copper and gallium are different ($b_{Cu} = 7.718(4)$ fm, $b_{Ga} = 7.288(2)$ fm [18]). Moreover, in the case of X-rays, destructive interference effects lead to a decrease of the scattering amplitude with angle. In contrast to diffraction at the electron shell, the atomic nuclei cross-sections are very small and the interference effects are also very small. Therefore, neutron-scattering amplitudes do not decrease rapidly with angle, resulting in the advantage of high intensities of Bragg-reflections observed even at high Q-values. This is especially important for the determination of atomic positions and atomic site occupations. For the analysis of point defects in compound semi-conductors and position parameters, for instance of the anion, neutron powder diffraction is advantageous.

5.3.1 Rietveld Refinement

To get detailed information about the cation and anion distribution within the materials, the neutron powder diffraction data has to be analysed by the Rietveld refinement method [19, 20]. To obtain reliable results it is important to perform the Rietveld procedure in a physically-reasonable sequence. In the case of the chalcopyrite-type compound semiconductors the chalcopyrite-type crystal structure was used as basis model. The free structural parameters of the fit were the lattice constants, the anion position parameter x, the cation site occupancy factors (SOF) of the $4a$ and $4b$ site (SOF_{4a}, SOF_{4b}) and atomic displacement parameters (ADP). The following example sequence of free parameters can be applied for the analysis of $Cu(In,Ga)(Se,S)_2$ compounds:

1. Refined parameters in the first step were the global parameters, zero shift and the scaling factor. Profile parameters like the u, v, w, x and y parameters, defining the full width at half maximum (FWHM), were fixed at the values of a previously-measured standard sample such as Y_2O_3. Also the background values were fixed in the beginning of the refinement. Other fixed parameters were the structure parameters: lattice constants, ADPs, SOFs, and the atomic position parameters. The SOFs were fixed at values according to the chemical composition as known from chemical analysis. The isotropic ADPs (b_{iso}^A; $A = 4a, 4b; 8d$) were kept at 1.0
2. Subsequently, lattice constants were refined, the u, v, w, x and y parameters remained fixed, as well as ADPs, SOFs, and position parameters.
3. The b_{iso}^A of $4a$, $4b$, and $8d$ positions are refined, keeping the SOFs of the respective positions fixed.
4. The cation site occupancy factors (SOF_{4a}, SOF_{4b}) are then refined whilst b_{iso} were fixed to their previously-refined values.
5. Step 3 and 4 are repeated until no change of parameters occurs. This procedure has been expanded with the use of anisotropic ADPs, b_{ii}^A, ($A = 4a$, $4b$; i = 1–3), since the tetragonal system leads to the need for anisotropic atomic displacement.
6. The b_{ii}^A and the SOFs are refined simultaneously until convergence.
7. If necessary, background and profile parameters: u, v, w, x and y as well as asymmetry values are refined.

This refinement strategy results in reliable values for SOFs of the respective species. Using the following method of average neutron-scattering length [22], it is possible to determine the point defect concentration in complex off-stoichiometric compounds over a large sample volume.

5.3.2 Point-Defect Analysis by the Method of Average Neutron-Scattering Length

The real crystal structure of a compound is defined by a deviation of the average structure caused by different types of defects. Extended defects, such as grain boundaries and dislocations influence the structure on a grand scale, without a variation on the level of individual unit cells. Small defects like single, isolated point defects, or correlated defects forming small defect-clusters, generally cause systematic variations of unit cells, which can be detected by diffraction methods. The method of average neutron-scattering length is an approach that gives a possible solution of point defects. By this method it is possible to determine the cation distribution within the material for a wide range of structure types, and will be explained here using $CuInSe_2$ and $CuGaSe_2$ as examples.

The approach is based on the fact, that vacancies (V_{Cu} and V_{In}), as well as anti-site defects (In_{Cu} and Cu_{In}), will change the neutron-scattering length of the cation sites $4a$ (copper site) and $4b$ (indium site) in the chalcopyrite-type structure significantly [15]. In a neutron diffraction experiment, the neutron interacts with the atomic nucleus and the neutron-scattering lengths of copper and indium are different ($b_{Cu} = 7.718$ (4) fm, $b_{In} = 4.065$ (2) fm, $b_V = 0$ [18]). Thus, the distribution of copper and indium on both cation sites of the structure can be revealed from the SOFs determined by Rietveld analysis of the neutron diffraction data.

If different species like Cu, In, and vacancies occupy the same structural site j, the average neutron-scattering length of this site is defined by:

$$\bar{b}_j = N_{Cu_j} \cdot b_{Cu} + N_{In_j} \cdot b_{In} + N_{V_j} \cdot b_V \qquad (5.1)$$

Here, N is the fraction of the species on the corresponding site and b are the neutron-scattering lengths. As an additional requirement the full occupation of the site j has to be taken into account, which is achieved when $\sum N_i = 1 (i = Cu, In, V)$. In an example case the chalcopyrite-type crystal structure is used as basis model for the Rietveld refinement. Thereby, Cu occupies the $4a$ and In the $4b$ site. An experimental average neutron-scattering length for the two sites using the SOFs can be calculated by:

$$\bar{b}_{4a}^{exp} = SOF_{4a} \cdot b_{Cu} \qquad\qquad \bar{b}_{4a}^{exp} = SOF_{4b} \cdot b_{In} \qquad (5.2)$$

where SOF_{4a} and SOF_{4b} are the cation site occupancy factors of the chalcopyrite-type crystal structure from the Rietveld analysis. The evaluation of the experimental average neutron-scattering length of a series of different Cu/In ratios, and therefore different degrees of off-stoichiometry, reveals a decrease of $\bar{b}_{4a}^{exp}$ (see Fig. 5.6) with

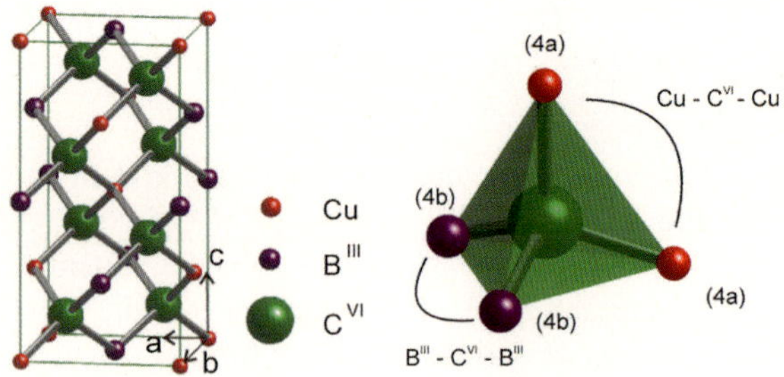

Fig. 5.6 Experimental average neutron-scattering length as a function of Cu/In ratio

decreasing Cu/In. Taking into account $b_V < b_{In} < b_{Cu}$, a decrease of $\bar{b}_{4a}^{exp}$ can only be caused by the presence of vacancies (V_{Cu}) or anti-site defects of type In_{Cu}. Since the decrease of $\bar{b}_{4a}^{exp}$ becomes stronger with decreasing Cu/In, an increase of the particular defect concentration can be assumed. Clearly, the value for $\bar{b}_{4b}^{exp}$ increases with decreasing Cu/In ratio. This can only be due to the presence of anti-site defects of type In_{Cu} ($b_{Cu} > b_{In}$).

The experimental average neutron-scattering lengths have to be compared with theoretical values. In a first step, these values are derived from a cation distribution model built on the basis of the known chemical composition of the material. The general formula for the calculation of the cation distribution is given by Eq. (5.1). A simultaneous comparison of $\bar{b}_j$ with $\bar{b}_j^{exp}$ during variation of the cation-distribution model leads to the corresponding amounts of isolated point-defects.

Using this approach it was possible to determine the cationic point-defect concentration for various chalcopyrite-type compounds. It was clearly shown that $CuInSe_2$ tends, when being Cu-poor, to form V_{Cu}, In_{Cu}, and Cu_{In} defects, resulting in a partially disordered chalcopyrite-type crystal structure. In contrast to $CuInSe_2$, $CuGaSe_2$ exhibits the same crystal structure, but due to the small ionic radii of Ga^{3+} [23] the material tends to form interstitial defects of type Ga_i [24]. These differences sensitively influence the properties of a final solar device. The kind of point defects present in $CuInSe_2$ allow a neutralization of isolated point-defects by clustering together forming a neutral defect complex of type ($2V_{Cu} + In_{Cu}$), which has a considerable binding-energy [16]. Such a neutralization of point defects cannot be assumed for $CuGaSe_2$. This crucial difference is one reason why it is possible to design a thin-film solar cell with a very off-stoichiometric absorber and high defect-concentration, like $CuInSe_2$, but maintain considerable efficiency. Tailoring high efficiency devices with a $CuGaSe_2$ absorber layer is still a challenging task due to different problems. One aspect is the difference in the presence of intrinsic point-defects in Cu-poor $CuGaSe_2$ discovered by neutron powder diffraction.

5.4 Low-Temperature Thermal Expansion in Chalcopyrite-Type Compound Semiconductors

Solids whose crystal structures are based on tetrahedrally-coordinated ions may show the intriguing property of negative thermal-expansion.

The ternary $A^I B^{III} C_2^{VI}$ semiconductors (A = Ag, Cu; B = Al, Ga, In; C = S, Se, Te), exhibit such a tetrahedral coordination (see Fig. 5.7). The coordination tetrahedron around an anion (sulfur or selenium) consists of two monovalent and two trivalent cations. The chemical bonds within such a tetrahedron are of mixed covalent and ionic character, whereby the ionicity of the bonds is different for the A^I–C^{VI} and B^{III}–C^{VI} bonds. These different interactions result in different bond lengths ($R_{AC} \neq R_{BC}$) as well as bond angles and lead to a displacement of the anions from the ideal tetrahedral site by a quantity u = |x − ¼| (where x is the anion x coordinate).

The linear thermal-expansion coefficients are closely related to the Grüneisen parameters γ of lattice vibrations [25]. The occurrence of a negative thermal-expansion can be understood using the notation of a balance between acoustic shear and compression modes of the observed crystal structure. The Grüneisen parameters of the shear modes show a tendency to negative values, while those of the compression modes are positive [25–27]. Hence, the temperature dependence of the thermal expansion is determined by the degree of excitations of the various modes and can change its sign when the relative thermal-population of the modes varies.

In $A^I B^{III} C_2^{VI}$ chalcopyrite-type semiconductors the thermal-expansion behaviour is described by the independent linear thermal-expansion coefficients α_a and α_c with

$$\alpha_a(T) = \frac{1}{a(T)}\frac{da(T)}{dT} \quad \text{and} \quad \alpha_c(T) = \frac{1}{c}\frac{da(T)}{dT}. \tag{5.3}$$

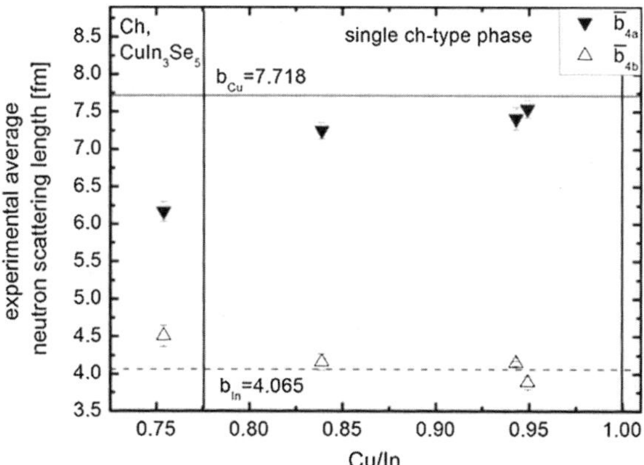

Fig. 5.7 The tetragonal chalcopyrite-type crystal structure (*left*). This structure consists of a periodic sequence of tetrahedra, each one built of a central anion surrounded by 4 cations (*right*)

The uniaxial chalcopyrite-type structure comes with two independent Grüneisen parameters γ_a and γ_c, which are related to α_a and α_c according to [28]:

$$\gamma_a = \frac{V_m}{C_p}\left[(c_{11}^S + c_{13}^S)\alpha_a + c_{31}^S\alpha_c\right] \quad \text{and} \quad \gamma_c = \frac{V_m}{C_p}\left[2c_{13}^S\alpha_a + c_{33}^S\alpha_c\right]. \tag{5.4}$$

Here V_m is the molar volume, C_p the molar specific-heat at constant pressure and c_{ij} are the adiabatic elastic-stiffnesses. With increasing ionicity the Grüneisen parameter should become more negative [29]. Thus, the covalent character of the chemical bond is expected to strongly affect the Grüneisen parameter.

The determination of linear thermal-expansion coefficients by dilatometry or X-ray diffraction [30–33] has shown that α_a and α_c vary independently with temperature. This is caused by the axial symmetry of the chalcopyrite-type crystal structure and the difference in strength of the Cu-C^{VI} and B^{III}-C^{VI} cation-anion bonds.

The investigation of the negative thermal-expansion is conveniently achieved by neutron powder diffraction. One aspect for the use of neutrons is again the high intensity in the diffraction pattern at high Q-values, important for an exact determination of the chalcogen position. It is important to monitor the change of this position at low temperatures to describe the bond stretching during cooling. The negative thermal-expansion has been studied for several chalcopyrite-type compounds, whereby the focus now lies on Cu(In$_x$Ga$_{1-x}$)Se$_2$ once with high (x = 0.918) and once with low indium content (x = 0.096), to show the effect of different bond ionicities on the negative thermal-expansion. Neutron powder diffraction patterns were collected for temperatures between 1.5 K > T > 300 K and structures refined by the Rietveld method according to the previously-described sequence. The ionicity can be calculated following Phillip's definition [34]:

$$f_i = 1 - e^{\frac{-(X_A - X_B)^2}{4}} \tag{5.5}$$

with X_A and X_B the electronegativity of the elements A and B (X_{Cu} = 1.9; X_{Ga} = 1.81; X_{In} = 1.78). According to Phillip's definition the bond ionicities increase from Cu–Se (f_i = 0.1002) to In–Se (f_i = 0.115) and Ga–Se (f_i = 0.128). Thus the ionicity of the B^{III}–Se cation-anion bond is increasing with increasing substitution of indium by gallium. From this it follows that with a high amount of gallium the difference in bond ionicity between the Cu–Se and B^{III}–Se cation-anion bond increases, which lead to an increased anisotropy.

The higher anisotropy affects the change of lattice parameters with decreasing temperature, which is stronger for the gallium-rich Cu(In,Ga)Se$_2$ and pure CuGaSe$_2$ than for indium-rich Cu(In,Ga)Se$_2$ (see Fig. 5.8). Applying a third-order polynomial fit to the lattice parameters the thermal-expansion coefficients α_a and α_c, can be derived. The temperature at which the linear thermal-expansion becomes negative (T_0), is seen to vary with the chemical composition (see Table 5.2).

Fig. 5.8 Lattice constants a and c as a function of temperature for **a** the In-rich sample with In/(In + Ga) = 0.918 **b** the Ga-rich sample with In/(In + Ga) = 0.096 and **c** pure CuGaSe$_2$

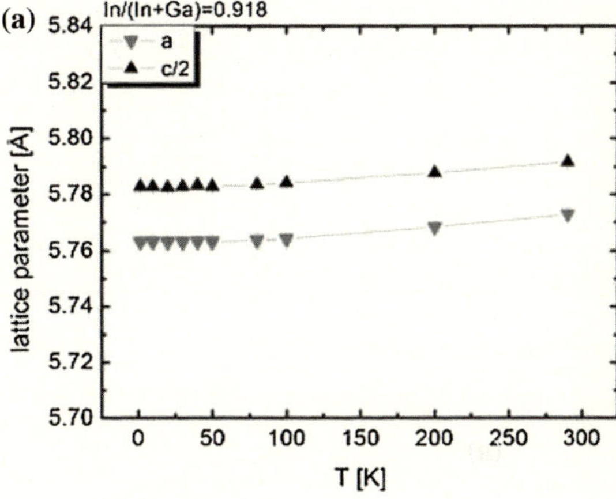

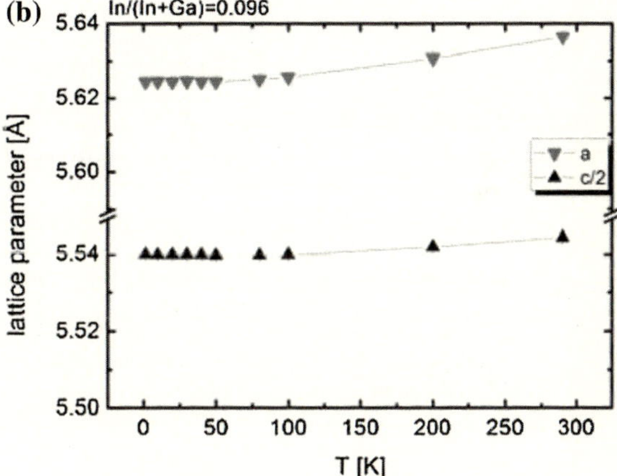

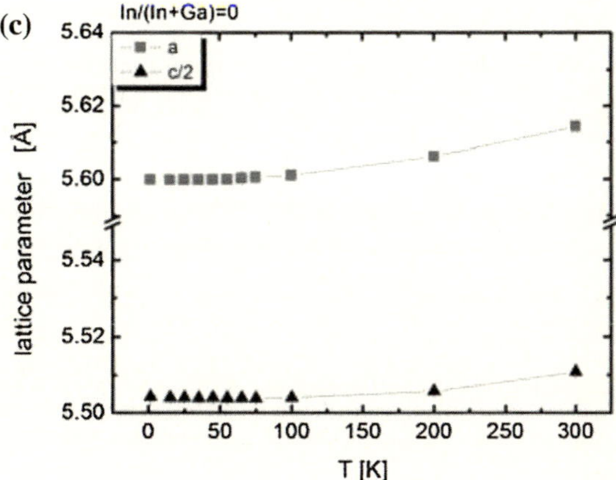

Table 5.2 Comparison of bond ionicity and the respective critical temperatures of α_a and α_c as well as of the average thermal-expansion coefficient

x	f_i (BIII–Se)	$T_0\ \alpha_a$ (K)	$T_0\ \alpha_c$ (K)
1	0.115	48.10	97.8
0.918	0.116	24.05	10.94
0.096	0.127	26.65	52.27
0	0.128	24.60	78.30

Fig. 5.9 The tetragonal deformation u as a function of temperature for the indium-rich and gallium-rich Cu $(In_xGa_{1-x})Se_2$ compounds

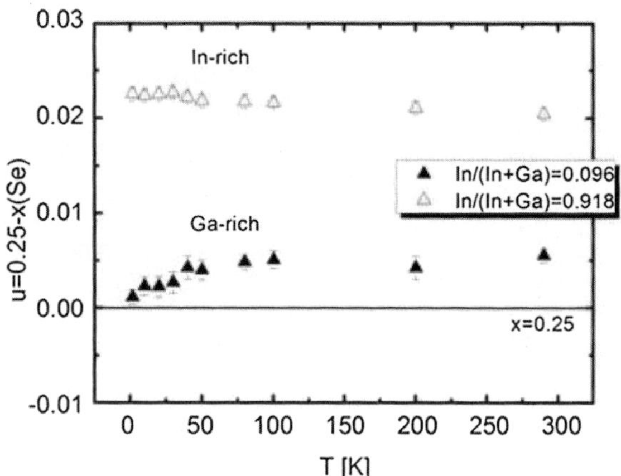

The variation of T_0 with chemical composition should be discussed in context with the bond ionicity of the B^{III}-Se bonds, which increase with increasing substitution of indium by gallium. Thus, with increasing gallium content the ionicity increases and the temperature, for which the linear thermal-expansion coefficient changes its sign, increases. This is observed for $Cu(In_xGa_{1-x})Se_2$ with different x-values as summarized in Table 5.2.

The In/(In + Ga) ratio strongly influences the character of the covalent-ionic B^{III}–Se cation-anion bond, and therefore the behaviour of the linear thermal-expansion coefficients of the two lattice constants α_a and α_c.

Also, the x-parameter of the selenium anion as a function of temperature is strongly affected by the different bond ionicities. In the Ga-rich sample the tetragonal deformation $u = 0.25–x(Se)$ strongly tends to zero with decreasing temperature, whereas it stays almost constant for the In-rich sample (Fig. 5.9). This effect is explained by the higher bond-ionicity for the Ga–Se bond compared to the In–Se bond.

The change of the tetragonal distortion and the anion position parameter x(Se) is reflected by the change in the average cation-anion bond distances and angles, which change markedly for the Ga-rich sample compared to In-rich sample (see Fig. 5.10).

Fig. 5.10 The cation-anion-cation bond angles as a function of temperature for **a** Ga-rich and **b** In-rich Cu $(In_xGa_{1-x})Se_2$

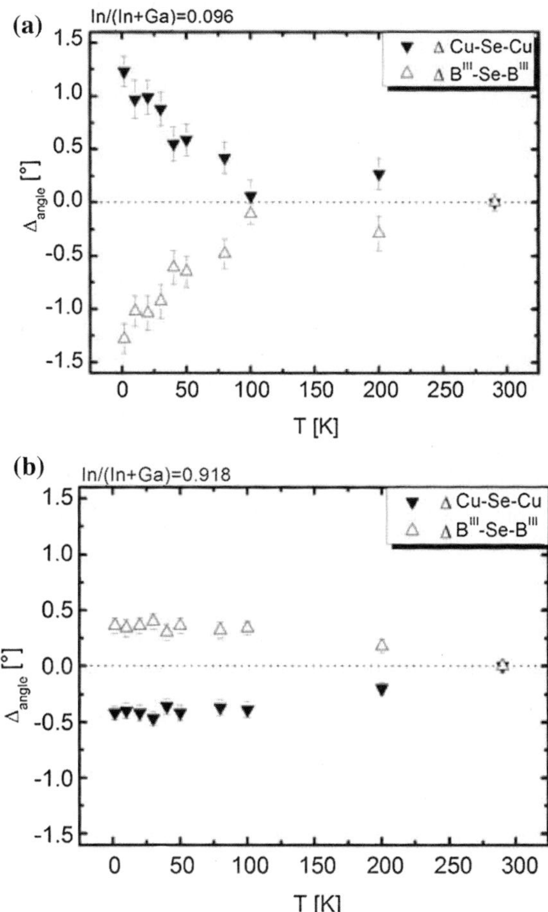

5.5 Novel Materials for In-Reduced Thin Film Solar-Cell Absorbers

Chalcopyrite-based thin-film solar cells with the semiconductor $Cu(In,Ga)(S,Se_2)$ (CIGSe) as absorber material show the highest efficiencies among thin-film photovoltaics in the laboratory as well as in module production (see Part 1). This indium based thin-film technology has a huge potential for low-cost photovoltaic production, but the scarcity of indium (indium's abundance in the continental crust is estimated to be approximately 0.05 part per million). Indium can be jointly refined from trace concentrations in leading ores as zinc, copper, and lead or the material comes from recycled scrap. Increasing prices of indium could easily limit the production growth. There is also an increased demand for indium for use in other technologies, for example items such as: flat panel displays, solders, thermal-interface materials, batteries, compound semiconductors and light-emitting diodes. On the other hand, the worldwide reserve of viable indium is approximately 16 kton.[1] Nevertheless, the price of indium has varied dramatically in recent years from US\$94/kg in 2002 to over US\$1,000/kg in 2006. The price in 2012/2013 was about US\$580/kg [35]. This reflects the scarcity of supply and the dependence on the small number of production facilities worldwide.

In order to secure the long term development of compound semiconductor based thin-film solar cells, the search for the replacement of indium is advisable. Semiconductors suitable as absorber materials in thin-film solar cells should fulfil a number of requirements, the most important amongst them being a high absorption-coefficient (in the range of 10^5 cm^{-1}) and a band-gap energy in the optimal range of about 1.4 eV. Here multinary compounds, like solid solutions between non-isotype binary II-VI and ternary I-III-VI$_2$ compounds, are potential candidates. Moreover, there are a variety of chalcogenide minerals available as a source of novel, indium-free absorber materials. Replacing indium in CuIn(S,Se)$_2$ by the abundant elements zinc and tin yields the quaternary compound Cu$_2$ZnSn(S,Se)$_4$, which is also a direct band-gap p-type semiconductor with an absorption coefficient higher than 10^4 cm^{-1} [36].

5.5.1 Cation Distribution in 2(ZnX)-CuB^{III}C$^{VI}_2$ Mixed Crystals

The formation of solid solutions between the non-isotype compounds ZnC and CuB^{III}C$_2$ (B-Al, Ga, In; C-S,Se) enables the band-gap energy to be changed from the large value of the binary wide-gap semiconductor ZnC to the band gap of the chalcopyrite-type end member CuInC$_2$ [37]. The possible application of 2(ZnS)-CuInS$_2$ mixed crystals as absorbers in thin-film solar cells was introduced by Bente et al. [38]. However, these solid-solution series are characterized by a relatively-large miscibility gap, where a tetragonal and a cubic phase coexists [37, 39–41], thus the applicable range is limited.

For tetragonal Zn$_{2x}$(CuB)$_{1-x}$C$_2$ mixed crystals, the question of the distribution of the three cations Zn^{2+}, Cu$^+$ and B^{3+} on the two cation positions of the chalcopyrite-type structure arises. Like in CIGSe, the cation distribution influences the opto-electronic properties of the solid-solution compounds. The Cu–Zn differentiation problem existing in X-ray diffraction due to a nearly-equal atomic form-factor f for Cu$^+$ and Zn^{2+}, can be solved by neutron diffraction because of the different neutron scattering cross-sections for Cu and Zn, which are based on the different neutron scattering lengths (b_{Zn} = 5.68 fm, b_{Cu} = 7.718 fm [18]).

The most detailed analysis was performed for chalcopyrite-type $Zn_{2x}(CuIn)_{1-x}S_2$ mixed crystals [42]. It should be noted that the study was performed using powder samples with stoichiometric composition (the chemical composition was determined by wavelength-dispersive X-ray spectroscopy using an electron microprobe system). For the interpretation of the average neutron-scattering lengths of the cation sites $4a$ and $4b$ ($\bar{b}_{4a}^{exp}$ and $\bar{b}_{4b}^{exp}$), derived from the cation SOFs determined by Rietveld analysis of neutron powder diffraction data, the principle of the conservation of tetrahedral bonds (CTB) for ternary ABC_2 chalcopyrites [43] was applied. In the chalcopyrite-type structure a displacement of the anion from its ideal position (¼, i.e. the middle of the cation tetrahedron) by $u - ¼$ (u is the anion x-coordinate) can be observed. Hence different bond lengths $R_{AC} \neq R_{BC}$ result, which in turn cause different-sized anion tetrahedra AC_4 and BC_4, resulting in a tetragonal deformation $\eta = c/2a$ parallel to the crystallographic c-axis. The bond lengths are [43]

$$R_{AX} = \left[u^2 + \left(1 + \eta^2\right)/16\right]^{1/2} a$$
$$R_{BX} = \left[(u - 1/2)^2 + (1 + \eta)^2 / 16\right]^{1/2} a \tag{5.6}$$

The parameters η and u are considered as the degrees of freedom of the chalcopyrite-type structure [Jaffe, Zunger 84]. The Abrahams-Bernstein relation [44]

$$u = \frac{1}{2} - \sqrt{\frac{c^2}{32a^2} - \frac{1}{16}} \tag{5.7}$$

correlates the tetragonal distortion u with the lattice parameters a and c. However, there is a limitation: only one of the anion tetrahedra is assumed to be deformed, whereas the other is taken as regular.

According to the CTB, the degrees of freedom (η and u) would attain values that simultaneously minimize the difference between the bond lengths R_{AC} and R_{BC} and the sums of the elemental radii as

$$R_{AC}(a,\eta,u) - r_A - r_c = 0 \quad \text{and} \quad R_{BC}(a,\eta,u) - r_B - r_C = 0 \tag{5.8}$$

By applying Eqs. (5.6) and (5.7) the solutions for Eq. (5.8) can be written as

$$\alpha^2 = \frac{12\alpha^2}{2\beta + \alpha - \left[(2\beta + \alpha)^2 - 18\alpha^2\right]^{1/2}} \tag{5.9}$$

$$\eta^2 = \frac{8(\beta - \alpha)}{3\alpha^2} \tag{5.10}$$

Here α is the bond-mismatch parameter and β the mean-square-bond

$$\alpha = R_{AX}^2 - R_{BX}^2 = (r_A + r_X)^2 - (r_B + r_X)^2$$
$$\beta = R_{AX}^2 + R_{BX}^2 = (r_A + r_X)^2 - (r_B + r_X)^2$$

(5.11)

The CTB model can be extended to quaternary chalcopyrite-type compounds, assuming the covalent radii in the equations above as the average radius of the cations on the two cation positions, according to

$$\bar{r}_A = Zn_A r_{Zn^{2+}} + Cu_A r_{Cu^+} + In_A r_{In^{3+}} \text{ and}$$
$$\bar{r}_B = Zn_B r_{Zn^{2+}} + Cu_B r_{Cu^+} + In_B r_{In^{3+}}$$

(5.12)

Here Zn_A, Cu_A, and In_A are the mole fractions of the cations on the Wyckoff position $4a$ (A) and $4b$ (B) according to the cation-distribution model. These fractions correspond to the total amount of Zn, Cu, and In in $(2ZnS)_x(CuInS_2)_{1-x}$ (i.e. $Zn_A + Zn_B = 2x$). Thus the average cation-radii are influenced by the cation distribution.

For the calculation of the average neutron-scattering lengths of the cation sites $4a$ and $4b$ ($\bar{b}_{4a}^{calc}$ and $\bar{b}_{4b}^{calc}$), a certain cation distribution has to be assumed. A first comparison with the experimentally-determined average neutron-scattering lengths show, that Zn is not statistically distributed on the sites $4a$ and $4b$. Thus three different aspects have to be taken into account for modelling the cation distribution:

(i)   Zn is non-statistically distributed
(ii) If Zn prefers the $4a$ position a Cu_{In} anti-site is enforced and if Zn prefers the $4b$ position an In_{Cu} anti site is enforced (enforced anti-sites)
(iii) Independent of the Zn distribution, Cu_{In} and In_{Cu} anti-sites may exist (spontaneous anti-sites)

The evaluation criteria for the cation distribution in tetragonal $(2ZnS)_x(CuInS_2)_{1-x}$ mixed crystal were formulated as:

(1) $u(exp) = u(calc)$ ($u(exp)$ is determined by Rietveld analysis of the powder diffraction data, $u(calc)$ applying the CTB rule)
(2) $\bar{b}_{4a}^{exp} = \bar{b}_{4a}^{calc}$ and $\bar{b}_{4b}^{exp} = \bar{b}_{4b}^{calc}$

Applying both criteria, and taking into account the aspects (i)–(iii), the cation distribution was evaluated in two steps. First, the possible cation distributions fulfilling criterion (1) were derived. As can be seen from Fig. 5.11, a variety of different cation distributions are possible.

In the second step criterion (2) is also taken into account. The graphical solution is shown in Fig. 5.12. It becomes clear, that Zn occupies the $4b$ site preferentially, enforcing ~ 1.8–4.5 % InCu. Moreover, there is a small fraction of spontaneous Cu-In anti-sites (i.e. Cu_{In} and In_{Cu}).

Taking into account the experimental error of the average neutron-scattering lengths $\bar{b}_{4a}^{exp}$ and $\bar{b}_{4b}^{exp}$ it can be deduced, that 27.5 % of the Zn occupies the $4a$ site,

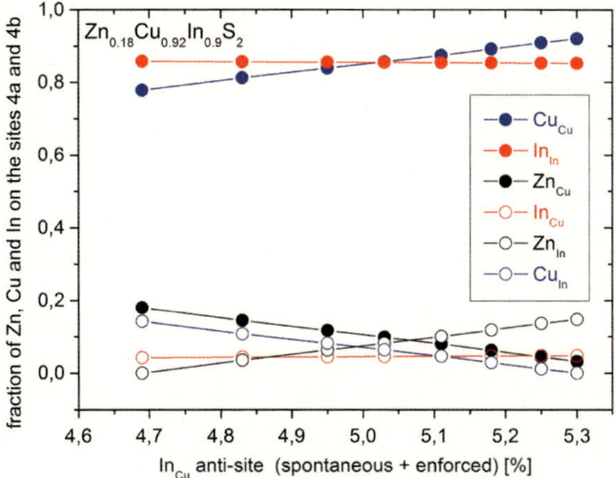

Fig. 5.11 Possible cation distributions fulfilling the criterion u(exp) = u(calc) for the sample $Zn_{0.18}Cu_{0.92}In_{0.90}S_2$. *Closed symbols* refer to the 4a position, *open symbols* to the 4b position

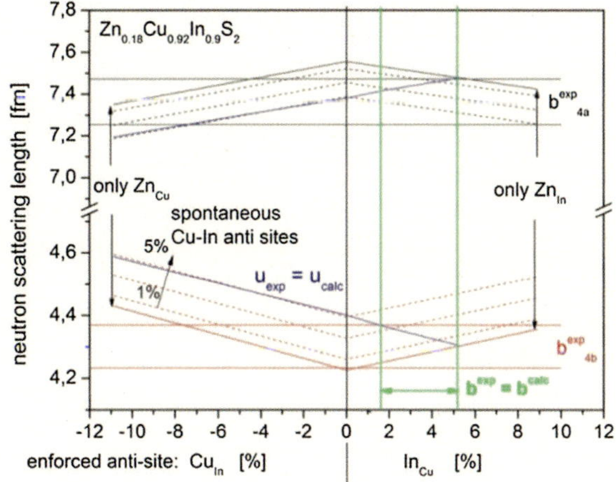

Fig. 5.12 Average neutron-scattering lengths calculated for possible cation distributions taking into account a non-statistical Zn distribution as well as enforced and/or spontaneous Cu–In anti-sites. Both limits (all Zn would be Zn_{Cu} or Zn_{In}) narrow the possible cation distributions. The *dotted lines* give the calculated average neutron-scattering length for the sites 4a and 4b for different cation distributions (with the experimental error), taking into account both enforced and spontaneous Cu-In anti-sites. The *blue lines* mark the distributions which fulfil criterion (1). Both criteria only meet within the region marked by the *green vertical lines*

whereas the rest of the Zn occupies the 4b site. This leads to 3.6 % enforced In_{Cu} anti-site. These results are in a good agreement with the cation distribution evaluated from X-ray powder diffraction data (($(Zn + Cu)_{Cu}$ = 0.91(3); In_{Cu} = 0.09(3); $(Zn + Cu)_{In}$ = 0.14 and In_{In} = 0.86(3)), but here Cu and Zn could not be

Table 5.3 Cation distribution in $Zn_{2x}(CuIn)_{1-x}S_2$ mixed crystals (the values are given as mole-fractions)

	Zn_{Cu}	Cu_{Cu}	In_{Cu}	Zn_{In}	Cu_{In}	In_{In}
$Zn_{0.11}Cu_{0.95}In_{0.94}S_2$	0.0173	0.9500	0.031	0.0927	0.0	0.9090
$Zn_{0.18}Cu_{0.92}In_{0.9}S_2$	0.0495	0.9035	0.047	0.1305	0.0165	0.853

distinguished. The cation distribution in $Zn_{2x}(CuIn)_{1-x}S_2$ mixed crystal is summarized in Table 5.3.

The CTB rule was only applied for the sulfide mixed-crystals due to the relatively-well known cation radii in sulfides [23]. For the selenide and telluride $Zn_{2x}(CuIn)_{1-x}C_2$ mixed crystals the differences between $\bar{b}_{4a}^{exp}$ and $\bar{b}_{4a}^{calc}$ as well as $\bar{b}_{4b}^{exp}$ and $\bar{b}_{4b}^{calc}$, assuming a statistical Zn distribution in the calculation of the average neutron-scattering lengths, were considered [42, 45]. Here, a non-statistical distribution of Zn on the both cation sites was also found. With increasing ZnX-content in $Cu_{0.5}In_{0.5}X$ there is a propensity for a more statistical distribution of the cations, indicating a tendency for disorder in the cation substructure.

It can be assumed that the non-statistical distribution of Zn and the associated Cu-In anti-sites are related to the limited solubility of ZnX in $Cu_{0.5}B_{0.5}X$. This fact can be discussed within the framework of formation energies of intrinsic point-defects in copper chalcopyrites. The Cu-In anti-site occupancy, resulting in Cu_{In} and In_{Cu}, are the defects with the lowest formation energies (in $CuInSe_2$: 1.3 eV for Cu_{In} and 1.4 eV for In_{Cu} [46]). Thus, these defects can be formed relatively easily. Nevertheless, the formation energies of Zn_{Cu} or Zn_{In} are not known. If one of the occupancies were energetically unfavourable, the solubility would be affected.

Using the approach of the average neutron-scattering length and the CTB rule it was possible to determine the cation distribution for various stoichiometric chalcopyrite-type $Zn_{2x}(CuB)_{1-x}C_2$ compounds. It was clearly shown that Zn tends to occupy the $4b$ site preferentially, resulting in the formation of In_{Cu} and Cu_{In} defects, resulting in a partially disordered chalcopyrite-type crystal structure.

5.5.2 Point Defects in $Cu_2ZnSn(S,Se)_4$ Kesterite-Type Semiconductors

Substitution of indium by zinc and tin in $CuIn(S,Se)_2$ leads to the quaternary compound semiconductor $Cu_2ZnSn(S,Se)_4$ (CZTSSe). The record efficiency of thin-film solar cells using a CZTSSe absorber layer is above 10 % [47].

Both compounds, CZTS and CZTSe, belong to the family of tetrahedrally-coordinated adamantine semiconductors [48]. Here each anion is tetrahedrally coordinated by four cations (two copper, one zinc and one tin), whereas each cation is coordinated by four anions (sulfur or selenium). Thus, the structure is characterized by a well-defined framework of tetrahedral bond arrangments, which is advantageous for the properties of the material.

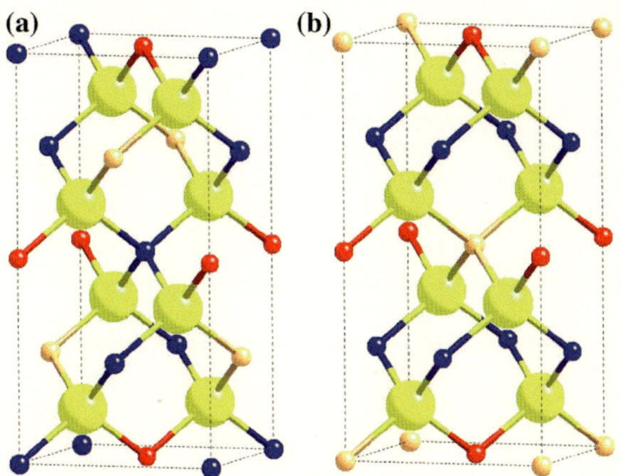

Fig. 5.13 Kesterite-type (**a**) and stannite-type (**b**) crystal structure (*blue* Cu, *orange* Zn, *red* Sn, *yellow* S, Se)

For the quaternary $A_2^{I}B^{II}C^{IV}X_4^{VI}$ chalcogenides with A-Cu; B-Zn; C-Si, Ge, Sn and X-S, Se, different crystal structures are discussed in literature: the kesterite-type structure (space group $I\bar{4}$), the stannite-type structure (space group $I\bar{4}2m$), as well as the wurtz-stannite (space group $Pmn2_1$) and the wurtz-kesterite type structure (space group Pc). The same tetrahedral metal-coordination (2Cu, one II- and one IV-element surrounding each S-atom) is possible in all four space groups. A clear decision can only be made by a detailed structure analysis for each compound.

At room temperature CZTS adopts the space group $I\bar{4}$. The structure can be described as a *cubic close-packed* array of sulfur atoms, with metal atoms occupying one half of the tetrahedral voids. CZTSe was reported to crystallize in the space group $I\bar{4}2m$ in a topologically-identical structure, but different cation distribution of A^{I} and B^{II} among the positions (0, 0, 0), (0, ½, ¼), and (0, ¼, ¾) [49]. Both structures are represented in Fig. 5.13.

Neutron powder diffraction and the method of the average neutron-scattering length were used to clarify possible differences of the cation distribution in CZTS and CZTSe, especially with respect to the electronically-similar elements copper and zinc [50, 51].

According to the general formula for the calculation of the average neutron-scattering length (see also Sect. 5.3.2)

$$\bar{b}_j = A_j \cdot b_A + B_j \cdot b_B + V_j \tag{5.13}$$

where A and B are two different cations, V represent possible vacancies and j stands for the Wyckoff position, the following equations were derived for the calculation of the experimental average neutron-scattering lengths $\bar{b}_j^{exp}$

Table 5.4 Cation distribution models used in the Rietveld analysis

CZTS				CZTSe		
Model 1	Model 2	Model 3	Model 4	Model 1	Model 2	Model 3
$I\bar{4}$	$I\bar{4}$	$I\bar{4}$	$I\bar{4}2m$	$I\bar{4}2m$	$I\bar{4}2m$	$I\bar{4}$
$2a$–Cu	$2a$–Cu	$2a$–Zn	$2a$–Zn	$2a$–Zn	$2a$–Cu	$2a$–Cu
$2c$–Zn	$2c$–Zn	$2c$–Cu	$4d$-Cu	$4d$-Cu	$4d$-Zn	$2c$–Zn
$2d$-Cu	$2d$-Cu	$2d$-Cu				$2d$-Cu

CZTS models 1 and 4 are the kesterite-type and stannite-type structure, Models 2 and 3 are the kesterite-type structure but with modified cation-distribution, respectively. CZTSe models 1 and 3 are the stannite-type and kesterite-type structure, respectively. Model 2 corresponds to the stannite structure, but with modified cation-distribution. Tin was always assumed to occupy the $2b$ position

Kesterite type structure (j = 2a, 2c, 2d):

$$\bar{b}_{2a}^{exp} = occ_{2a} \cdot b_{Cu}$$
$$\bar{b}_{2c}^{exp} = occ_{2c} \cdot b_{Zn}$$
$$\bar{b}_{2d}^{exp} = occ_{2d} \cdot b_{Cu}$$

Stannite type structure (j = 2a, 4d):

$$\bar{b}_{2a}^{exp} = occ_{2a} \cdot b_{Zn}$$
$$\bar{b}_{4d}^{exp} = occ_{4d} \cdot b_{Cu}$$

$$(5.14)$$

The cation site-occupancy values occ_j resulted from the Rietveld analysis of the neutron diffraction data. Because the study was performed using stoichiometric powder samples (the chemical composition of the samples was determined by wavelength-dispersive X-ray spectroscopy), vacancies were not taken into account.

The Rietveld analysis of the neutron diffraction data was performed by applying different cation-distribution models within both structure types as the starting model for the crystal structure (see Table 5.4). The cation site occupancies were taken as free parameters in the refinement. It was found that the refined cation site-occupancy values differ from their nominal values (Table 5.4), with the exception of the tin. Thus, one can conclude that the site is not only occupied by the initially-supposed model cation, but also by a mixture of different elements. For example, a decrease of the average neutron-scattering length of the $2d$ site in the kesterite-type structure can be attributed to a partial Zn occupation due to $b_{Cu} > b_{Zn}$.

The refined cation site occupancies were used to calculate the experimental average neutron-scattering lengths $\bar{b}_j^{exp}$ (see Fig. 5.14). The first result obtained was a confirmation of the occupancy of the $2b$ site by tin in both compounds. Concerning the sites $2a$, $2c$ and $2d$ for the space group $I\bar{4}$ as well as $2a$ and $4d$ for the space group $I\bar{4}2m$, a similar picture for both compounds appears. Irrespective of the structure model used in the Rietveld analysis, the average neutron-scattering length of the $2a$ position indicates that this position is occupied by copper only. The average neutron-scattering lengths of the sites $2c$ and $2d$ indicate a mixed occupancy of these both sites by copper and zinc. Approximately 50 % of the zinc site $2c$ is occupied by copper and vice versa concerning the copper site $2d$, forming

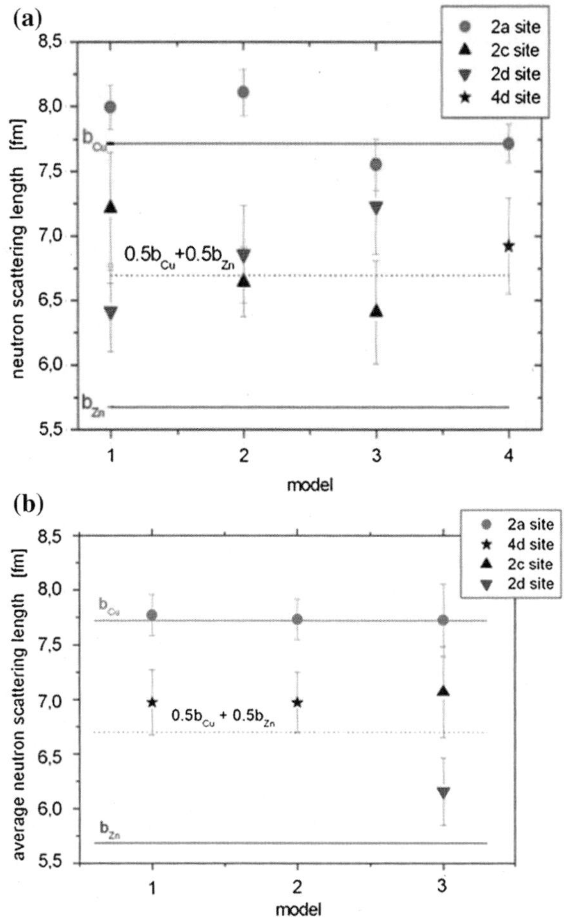

Fig. 5.14 Experimental average neutron-scattering length of the cation sites (the $2b$ site is excluded) evaluated from the corresponding site occupancy values determined by Rietveld analysis for CZTS (**a**) and CZTSe (**b**)

Cu_{Zn} and Zn_{Cu} anti-sites. Nevertheless, this disorder is limited to the positions $2c$ and $2d$ and hence to the lattice planes at $z = 1/4$ and $3/4$. The $2a$ site is always occupied by copper only, and there are no indications of zinc occupying that site.

These facts bring us to the conclusion that both compounds, CZTS and CZTSe, adopt the kesterite-type structure, but exhibit Cu-Zn disorder in the (001) lattice planes at $z = 1/4$ and $1/3$. This disorder creates Cu_{Zn} and Zn_{Cu} anti-site defects. These results are supported by *ab initio* calculations carried out on these systems [52, 53]. The authors report that the Cu_{Zn} anti-site acceptor can be predicted as the most probable defect and thus can be easily formed. Moreover, the kesterite-type structure was indicated as the ground-state structure for both CZTS and CZTSe, though the stannite-type structure has only a slightly-lower binding energy (CZTS: 2.9 meV/atom [52], 1.3 meV/atom [53], CZTSe: 3.8 meV/atom [52], 3.3 meV/atom [53]).

Using the approach of the average neutron-scattering length, and applying different structural models with different cation distributions in the structural description in the Rietveld refinement procedure, the crystal structure of CZTS and CZTSe was determined and intrinsic point-defects were identified experimentally. In contradiction with earlier X-ray diffraction studies, the kesterite-type structure was also proved for CZTSe.

References

1. T. Unold, C.A. Kaufmann, in *Chalcopyrite Thin-Film Materials and Solar Cells*, vol. 1, 1st edn. Comprehensive Renewable Energy, vol 1. Photovoltaic Solar Energy (Elsevier, Oxford, 2012)
2. R. Scheer, H.W. Schock, *Chalcogenide Photovoltaics: Physics, Technologies, and Thin Film Devices* (Wiley-VCH, Weinheim, 2011)
3. S.R. Kodigala, Chapter 8—$Cu(In_{1-x}Ga_x)Se_2$ and $CuIn(Se_{1-x}S_x)_2$ thin film solar cells, in *Thin Films and Nanostructures*, vol. 35, ed. by K. Subba Ramaiah (Academic Press, New York, 2010), pp. 505–679
4. S. Siebentritt, M. Igalson, C. Persson, S. Lany, Prog. Photovoltaics **18**, 390 (2010)
5. S.B. Zhang, S.H. Wei, A. Zunger, H. Katayama-Yoshida, Phys. Rev. B **57**, 9642 (1998)
6. S.H. Wei, S.B. Zhang, A. Zunger, J. Appl. Phys. **85**, 7214 (1999)
7. R. Caballero, C.A. Kaufmann, V. Efimova, T. Rissom, V. Hoffmann, H.W. Schock, Prog. Photovoltaics **21**, 30 (2012)
8. M. Bodegard, O. Lundberg, J. Malmstrom, L. Stolt, High voltage Cu(In,Ga)Se-2 devices with Ga-profiling fabricated using co-evaporation, in *Conference Record of the Twenty-Eighth IEEE Photovoltaic Specialists Conference—2000* (2000), pp. 450–453
9. D. Rudmann, D. Bremaud, A.F. da Cunha, G. Bilger, A. Strohm, M. Kaelin, H. Zogg, A. Tiwari, Thin Solid Films **480**, 55 (2005)
10. S. Niki, M. Contreras, I. Repins, M. Powalla, K. Kushiya, S. Ishizuka, K. Matsubara, Prog. Photovoltaics **18**, 453 (2010)
11. P. Jackson, D. Hariskos, E. Lotter, S. Paetel, R. Wuerz, R. Menner, W. Wischmann, M. Powalla, Prog. Photovoltaics **19**, 894 (2011)
12. A.M. Gabor, J.R. Tuttle, D.S. Albin, M.A. Contreras, R. Noufi, A.M. Hermann, Appl. Phys. Lett. **65**, 198 (1994)
13. D. Schmid, M. Ruckh, F. Grunwald, H.W. Schock, J. Appl. Phys. **73**, 2902 (1993)
14. H. Monig, C.H. Fischer, R. Caballero, C.A. Kaufmann, N. Allsop, M. Gorgoi, R. Klenk, H.W. Schock, S. Lehmann, M.C. Lux-Steiner, I. Lauermann, Acta Mater. **57**, 3645 (2009)
15. C. Stephan, S. Schorr, M. Tovar, H.W. Schock, Appl. Phys. Lett. **98**, 091901 (2011)
16. S.B. Zhang, S.H. Wei, A. Zunger, Phys. Rev. Lett. **78**, 4059 (1997)
17. L. Gütay, D. Regesch, J.K. Larsen, Y. Aida, V. Depredurand, A. Redinger, S. Caneva, S. Schorr, C. Stephan, J. Vidal, S. Botti, S. Siebentritt, Phys. Rev. B **86**, 045216 (2012)
18. A.J. Dianoux, G. Lander, *Neutron Data Booklet* (Institut Laue-Langevin, Grenoble, 2001)
19. H.M. Rietveld, J. Appl. Crystallogr. **2**, 65 (1969)
20. H.M. Rietveld, Acta Crystallogr. **22**, 151 (1967)
21. L.B. McCusker, R.B. Von Dreele, D.E. Cox, D. Louer, P. Scardi, J. Appl. Crystallogr. **32**, 36 (1999)
22. A. Furrer, J. Mesot, T. Strässle, in *Neutron Scattering in Condensed Matter Physics*, vol. 4. Series on Neutron Techniques and Applications (World Scientific, London, 2009)
23. R.D. Shannon, Acta Crystallogr. A **32**, 751 (1976)

24. C. Stephan, T. Scherb, C.A. Kaufmann, S. Schorr, H.W. Schock, Appl. Phys. Lett. **101**, 101907 (2012)
25. T.H.K. Barron, J.G. Collins, G.K. White, Adv. Phys. **29**, 609 (1980)
26. M. Blackman, Philos. Mag. **3**, 831 (1958)
27. A. Bienenstock, G. Burley, J. Phys. Chem. Solids **24**, 1271 (1963)
28. H. Neumann, P. Deus, R.D. Tomlinson, G. Kuhn, B. Hintze, Phys. Status Solidi A **84**, 87 (1984)
29. T.F. Smith, G.K. White, J. Phys. C. Solid State **8**, 2031 (1975)
30. H.G. Bruhl, H. Neumann, G. Kuhn, Solid State Commun. **34**, 225 (1980)
31. P. Deus, H. Neumann, G. Kuhn, B. Hinze, Phys. Status Solidi A **80**, 205 (1983)
32. H. Neumann, Cryst. Res. Technol. **18**, 659 (1983)
33. S. Schorr, D. Sheptyakov, J. Phys. Condens. Matter **20**, 104245 (2008)
34. J.C. Phillips (ed.), *Bonds and Bands in Semiconductors* (Academic, New York, 1973)
35. www.geodexminerals.com. Accessed 4 March 2013
36. K. Ito, T. Nakazawa, Jpn. J. Appl. Phys. **27**, 2094 (1988)
37. S. Schorr, G. Wagner, J. Alloy. Compd. **396**, 202 (2005)
38. K. Bente, T. Döring, Chemie der Erde **51** (1991)
39. S. Schorr, Thin Solid Films **515**, 5985 (2007)
40. G. Wagner, S. Lehmann, S. Schorr, D. Spemann, T. Doering, J. Solid State Chem. **178**, 3631 (2005)
41. L. Roussak, G. Wagner, S. Schorr, K. Bente, J. Solid State Chem. **178**, 3476 (2005)
42. S. Schorr, M. Tovar, N. Stusser, K. Bente, Phys. B **350**, E411 (2004)
43. J.E. Jaffe, A. Zunger, Phys. Rev. B **29**, 1882 (1984)
44. S.C. Abrahams, J. Bernstei, J. Chem. Phys. **55**, 796 (1971)
45. S. Schorr, M. Tovar, D. Sheptyakov, L. Keller, G. Geandier, J. Phys. Chem. Solids **66**, 1961 (2005)
46. J. Klais, Untersuchung intrinsischer Fehlstellen in nicht-stöchiometrischen CuInSe$_2$ Kristallen durch Positronenlebensdauer- und elektrische Messungen. Technische Universität Bergakademie Freiberg (1999)
47. D.A.R. Barkhouse, O. Gunawan, T. Gokmen, T.K. Todorov, D.B. Mitzi, Prog. Photovoltaics **20**, 6 (2012)
48. B.R. Pamplin, J. Phys. Chem. Solids **25**, 675 (1964)
49. I.D. Olekseyuk, L.D. Gulay, I.V. Dydchak, L.V. Piskach, O.V. Parasyuk, O.V. Marchuk, J. Alloy. Compd. **340**, 141 (2002)
50. S. Schorr, H.J. Hoebler, M. Tovar, Eur. J. Mineral. **19**, 65 (2007)
51. S. Schorr, Sol. Energ. Mat. Sol. C **95**, 1482 (2011)
52. S.Y. Chen, X.G. Gong, A. Walsh, S.H. Wei, Appl. Phys. Lett. **96**, 021902 (2010)
53. C. Persson, J. Appl. Phys. **107**, 053710 (2010)

Chapter 6
Organic Solar Cells

Mohamed Zbiri, Lucas A. Haverkate, Gordon J. Kearley,
Mark R. Johnson and Fokko M. Mulder

Abstract Organic-based photoconverters are subject to a considerable interest due
to their promising functionalities and their potential use as alternatives to the more
expensive inorganic analogues. We introduce the basic operational mechanisms,
limitations and some ideas towards improving the efficiency of organic solar cells
by focusing on probing the morphological/structural, dynamical, and electronic
aspects of a model organic material consisting of charge-transfer discotic liquid-
crystal system hexakis(n-hexyloxy)triphenylene/2,4,7 trinitro-9-fluorenone (HAT6/
TNF). For the electronic ground-state investigations, neutron-scattering techniques
play a key role in gaining deeper insight into structure and dynamics. These
measurements are complemented by Raman and nuclear magnetic resonance
probes, as well as resonant Raman and UV-vis spectroscopies that are used to
explore the low-lying excited states, at the vibronic level. Synergistically, numerical
simulations, either classical via empirical force fields, or first-principles via density
functional theory, are used for the analysis, interpretation and predictions.

6.1 Introduction

Research on organic devices is focused mainly on three concept organic photovoltaic
(OPV) designs in the solid-state phase by using either: (i) vacuum-deposited molecules by evaporation of two or more n-and p-dopants [1], (ii) solution-processed
polymeric macromolecules [2], and, (iii) the Grätzel concept [3] to build dye-sensi-

M. Zbiri (✉) · M.R. Johnson
Institut Max von Laue-Paul Langevin, 71 avenue des Martyrs,
CS 20156, 38042 Grenoble, France
e-mail: zbiri@ill.fr

L.A. Haverkate · F.M. Mulder
Faculty of Applied Sciences, Reactor Institute Delft, Delft University of Technology,
Mekelweg 15, 2629 JB Delft, The Netherlands

G.J. Kearley
Australian Nuclear Science and Technology Organisation,
Lucas Heights, NSW, Australia

© Springer International Publishing Switzerland 2015
G.J. Kearley and V.K. Peterson (eds.), *Neutron Applications in Materials
for Energy*, Neutron Scattering Applications and Techniques,
DOI 10.1007/978-3-319-06656-1_6

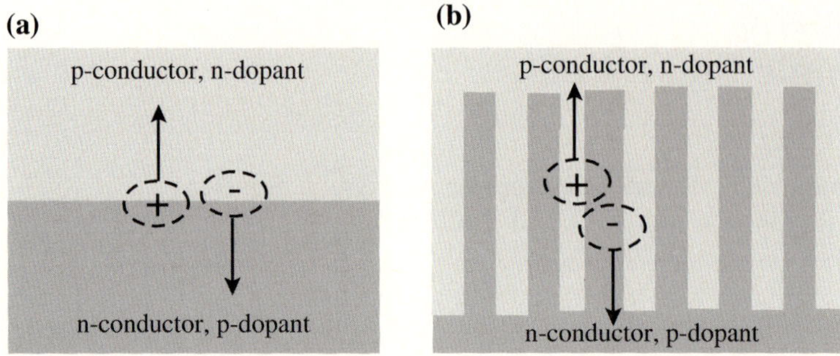

Fig. 6.1 Schematic representation of a p-n heterojunction (**a**) and a bulk heterojunction (BHJ) of interpenetrating networks of p-and n-dopants (**b**)

tized devices based on mesoporous nanocrystalline thin-films of TiO_2 covered by a dye layer. In terms of architectures, solar cells based on cases (i) and (ii) are similar and can be set either into a heterojunction or a bulk heterojunction (BHJ) device (Fig. 6.1).

The former represent the "standard" way of building a p-n junction and the donor-acceptor (DA) interface at which the exciton dissociation into charge carriers should take place. The latter architecture is based on a more elaborated technique of enhancing the p-n junction of setup (i) and making it as an interpenetrating network of the n-and p-dopant materials. It should be noted that in the case of polymer-based OPV both heterojunction and BHJ setups are made by a solution process because of the impossibility of evaporation due to the large macromolecular weight. The Grätzel-type solar cell is based on a different device architecture comprising mesoporous dye-covered TiO_2 films connected through a pore-penetrating liquid electrolyte or an organic n-dopant material, to a transparent counter electrode. Each of these OPV concepts presents different advantages and limitations in terms of efficiency and production costs. The high-vacuum deposition of small molecules ensures good processability and reproducibility of the device in contrast to cases (ii) and (iii). However, it is an expensive and complex technology which limits its use. In this respect, polymer-based OPVs are better because they can be fabricated under ambient conditions. Moreover, polymers in this case can be made soluble and spray-deposited on plastic substrate, which considerably increases their flexibility with obvious benefits for commercialization [4]. The dye-sensitized (Grätzel) solar cell performs the best [5] for power-conversion efficiency (PCE), but because a heat treatment is needed for TiO_2 a flexible realization of this architecture is limited.

6.2 Performance, Efficiency, and Limitations of OPVs

The performance of an OPV is characterized by the current-voltage dependence (I–V curve) depicted in Fig. 6.2.

The directly-measurable parameters that allow a quantitative analysis of the performance of a PV cell are:

(i)   The open-circuit voltage V_{oc} corresponding to zero-current in the cell, which is dependent on the offset of the energy levels (highest occupied molecular orbital of the donor: $HOMO^D$ and lowest unoccupied molecular orbital of the acceptor: $LUMO^A$) of the DA interface of the cell (Fig. 6.3). The short-circuit current I_{sc} which is related to the photo-generated charges. The fill factor (FF) which is related to the ratio of the red and grey areas in Fig. 6.2, which reflects the quality of a photovoltaic (PV) cell, or more precisely, it is determined by the mobilities of charge carriers [4, 9].

(ii) The incident-photon-to-current efficiency (IPCE) is defined as the ratio of the number of incident photons and the number of photo-induced charge carriers which can be generated. Unlike the internal quantum efficiency, IPCE accounts for losses by reflection, scattering, and recombination.

(iii) The power conversion efficiency (PCE), mentioned above, depends on the FF, V_{oc}, I_{sc}, and the incident light intensity. PCE is maximal for larger values of the first three parameters. A standard AM1.5 spectrum is used for PV characterizations.

(iv) The PCE can be decomposed in terms of different "sub-efficiencies" related to each step towards the generation of the photocurrent. The overall performance of an OPV depends on the sequence of steps mentioned above and illustrated in Fig. 6.3.

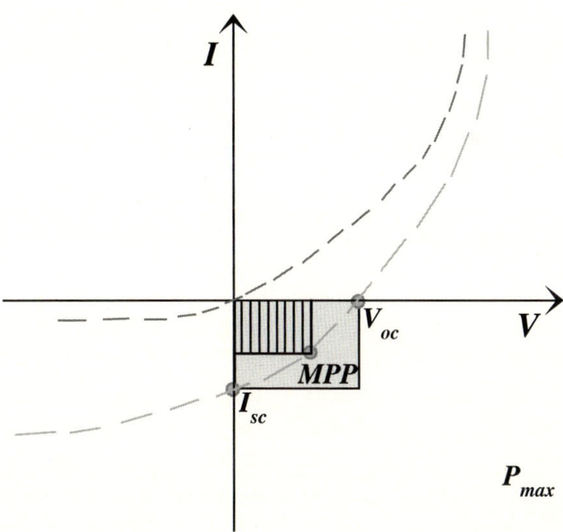

Fig. 6.2 The I–V curves characteristics of PV cells with (*short dash*) and without (*long dash*) solar illumination. A diode-like behaviour is observed in the absence of illumination. The short-circuit current I_{sc} (V = 0) and the open-circuit voltage V_{oc} (I = 0) are shown. MPP and P_{max} denote the maximum power point and maximum power output of the PV cell, respectively

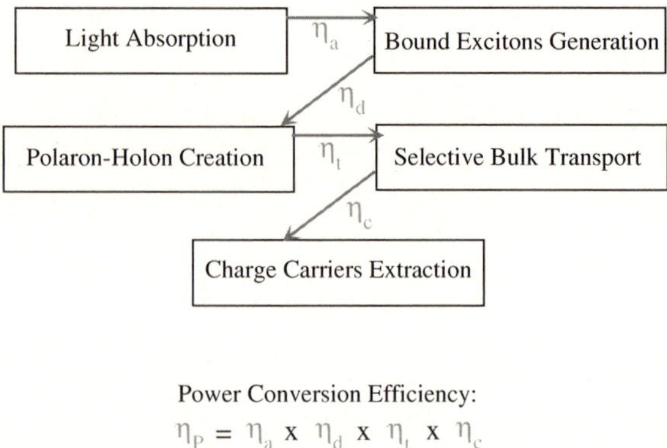

Power Conversion Efficiency:

$$\eta_P = \eta_a \times \eta_d \times \eta_t \times \eta_c$$

Fig. 6.3 Schematic representation of the main fundamental steps describing the photo-current conversion by an organic solar cell along with the related efficiencies in terms of which the power conversion efficiency (PCE) parameter, η_p, is shown

The incident light, with energy falling inside the HOMO-LUMO gap, is absorbed, leading to the creation of a strongly-bound Frenkel exciton by promoting an electron from its ground state to a higher energy level. The exciton is of a strongly "bound" nature because the electron-hole pair is subject to a strong mutual attractive interaction [6]. In OPVs additional steps are then required for electron-hole extraction, which is in contrast to inorganic PVs where the free carriers are released immediately. Therefore the transport processes are much more complex in OPVs due to the polaronic nature of the carriers. In the absence of an applied electric field, the exciton diffuses inside the organic semiconductor and becomes spatially dissociated into bound, positive and negative carriers (a polaron-holon pair). The effective dissociation occurs at the DA interface followed by transport of the polarons and holon to the relevant electrodes where they are extracted generating the final photocurrent (Fig. 6.4). During its diffusion the exciton may recombine before it reaches the DA interface, with a consequent loss of the absorbed energy by dissipation. Indeed, the diffusion length of excitons to the DA interface is much shorter than the optical absorption length. This is of prime importance in understanding the different loss mechanisms that should be avoided to improve the dissociation and collection of charges before recombination.

Figure 6.5 shows different loss mechanisms, either by exciton recombination or by charge-carrier trapping, which may occur during the different steps towards photocurrent production explained above. Loss by recombination can even occur after a successful exciton dissociation recombination, and trapping during the charge transport or collection at the DA interface and electrodes, is a further loss. After a recombination, the loss leading to dissipation of the absorbed solar-energy can be either radiative or non-radiative, the former being detected via fluorescence and/or phosphorescence, whilst the later via a phonon creation. When compared with inorganic devices, the organic analogous have two important limitations: first, their narrow absorption-window, and second, the tendency for thermal motion to perturb

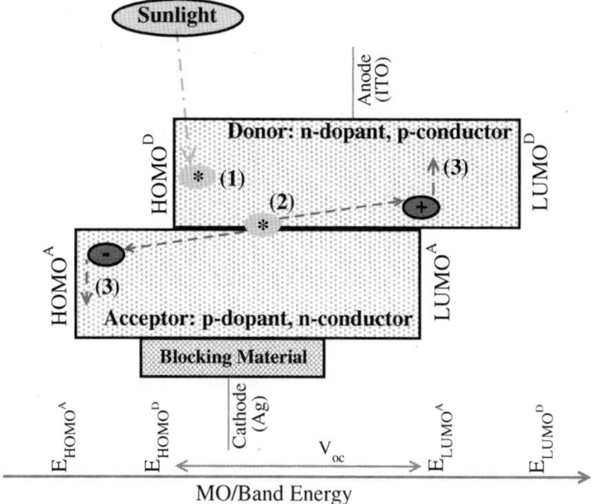

Fig. 6.4 Schematic illustration of a layered setup, operational mechanism, and relevant energy level diagram of an organic heterojunction solar cell with a DA interface. Relevant energy levels are the HOMO and the LUMO-HOMO^D and LUMO^D for the donor, and HOMO^A and LUMO^A for the acceptor. The V_{oc}, which is the maximum possible voltage, is directly proportional to the energy difference between HOMO^D and LUMO^A. The sequence of numbers *(1)*, *(2)*, and *(3)* represent the different steps upon sunlight absorption leading to a photo-excited electron, and towards charge-carriers generation, dissociation, transport and finally collection of the polaron (−) and the holon (+) at the cathode made from silver (Ag) and at the anode made from indium tin oxide (ITO), respectively. Step (2) marks the dissociation at the DA interface. The role of the additional layer of the blocking material is to shield the photoactive area of the OPV device from the metal electrodes

Fig. 6.5 Schematic representation of the different steps describing the photo-current generation in organic solar cell and the loss mechanisms which might occur at each step

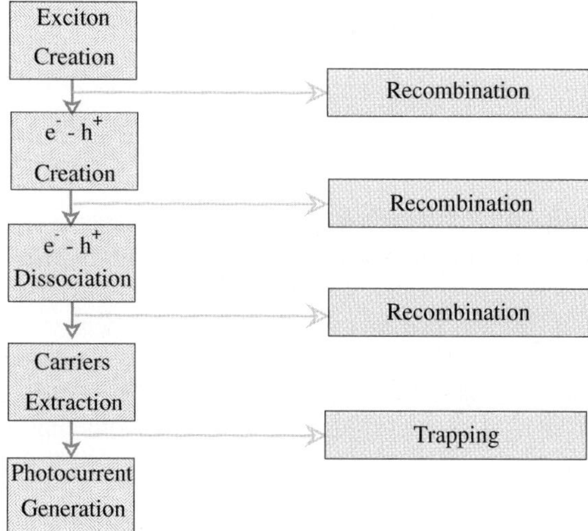

the rather soft charge-transport system. Therefore, efficient energy-conversion for practical applications requires significant improvements of charge transport and transfer processes. Energy loss during the photogeneration process should be minimized by maximizing the number of excitons that dissociate into free charge-carriers, rather than simply recombining without contributing to the photocurrent. A rich research field has been stimulated aimed at extending the spectral sensitivity of OPVs to cover a broader wavelength region and therefore the band-gap tuning.

In this context, it has been proposed that stacking different OPVs would help to achieve this goal. The problem of inefficient exciton-dissociation could be solved thanks to the introduction of tandem[1] device architectures and BHJ [7–11]. The interpenetrating networks of n-and p-semiconductors (Fig. 6.1) improve the charge separation and transport to their respective electrodes for collection.

6.3 Discotic Liquid Crystals

The polymer-based concept for building OPVs presents a good balance between technical complexities and production costs as compared to the high-vacuum deposited small molecule and Grätzel-device concepts; presented earlier. However, the conjugated or resonating double-bonds of conducting polymers, which are polar in nature, are very sensitive to the shorter wavelength of the solar spectrum and these bonds can be broken if the energy of the solar light lies within a certain ultraviolet (UV) range. Therefore, new types of organic systems are required to overcome this limitation, and make efficient use of a wide part of the solar spectrum. Discotic liquid crystals (DLCs) are expected to fill this role as an alternative to the conducting polymers. The name "discotic" describes materials formed from disk-like molecules (discogens) which possess a central planar aromatic core to which peripheral aliphatic chains are attached. Thermal fluctuations of the chains, or tails, are sufficient to suppress inhomogeneously-distributed structural traps, giving rise to a "liquid-like" dynamic disorder of the tails. The cores can be formed by triphenylene, coronene or phthalocyanine molecules (Fig. 6.6).

When Chandrasekhar, observed the thermotropic DLCs for the first time in 1977 [12] he explained that the term "liquid crystal" signifies a state of aggregation that is intermediate between the crystalline solid and the amorphous liquid [13]. Materials that form liquid-crystalline phases are called mesogens and different phases can be distinguished within the liquid-crystalline state. These are nematic, smectic, cholesteric, and discotic phases. Figure 6.7 depicts schematically the nematic and columnar phases. The nematic phase has a one-dimensional orientational ordering and is subject to positional disorder. The columnar hexagonal phase however, exhibits both orientational and positional ordering, where the molecular core-on-

---

[1] A difference is marked in the literature between "stacked" and "tandem" PVs. The former refers to the case of layers made from the same material, whereas in the latter there are two different materials.

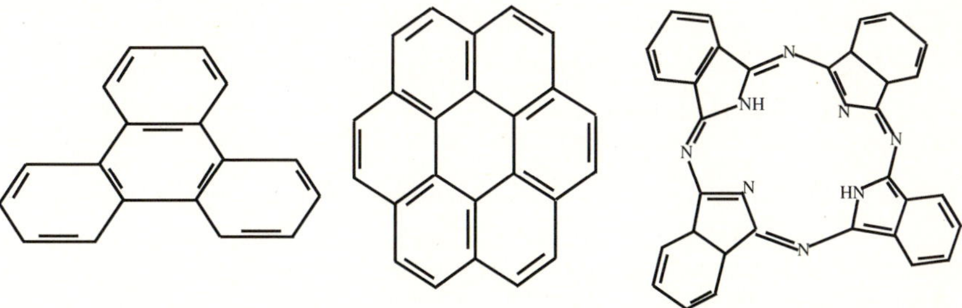

Fig. 6.6 Molecular schematics of triphenylene (*left*), coronene (*middle*) and phthalocyanine (*right*)

Fig. 6.7 Schematic representation of different liquid crystalline phases with increasingly-enhanced orientational and positional orderings: nematic discotic (*left*), columnar hexagonal (*middle*), and helical columnar (*right*)

core stacking forms a hexagonal array of columns. The conductivity in this phase is due to the stacking between neighbouring disc-shaped molecules as a consequence of the π-π overlap originating from the delocalized π-orbitals above and below each aromatic core and interactions between the aliphatic chains. These chain-chain interactions ensure self-assembly of the discotic phase rather than a herringbone arrangement that characterizes systems with short chains. The π-π overlap, which defines the co-facial distance/separation between two discs in a stack, provides a one-dimensional pathway for charge migration (conductivity) along the column direction (stacking axis) and is normally thought to be the only part of the charge-transport system. The distance between the columns on the other hand is controlled by the length of the aliphatic chains forming the molecular tails. The highly anisotropic character of the electrical conductivity having its axial component greater than the in-plane makes these molecular wires attractive, offering features such as charge transfer for molecular-conducting devices including photovoltaic applications. Further, the columnar hexagonal positional ordering can be enhanced towards a helical type ordering (Fig. 6.7) which improves further the conductivity.

6.3.1 Case Study: Structure and Dynamics of a Discotic Liquid Crystal HAT6 and Its Charge-Transfer Complex with TNF Acceptor

From a "molecular" point of view, each electronic state of the organic molecule has a well-defined potential energy hyper-surface (PES), with a minimum corresponding to the molecular geometry. There is therefore also a set of characteristic molecular vibrations and the electronic relaxation process is mediated by the molecular vibrations of the corresponding electronic state. One aspect of the energy losses concerns the hot-carrier relaxation of the molecule through excited electronic-states, and the excited state via which the electronic charge is finally transferred to the neighbouring molecule. The electric potential related to an exciton is greater for the higher excited states. In analogy with relaxation to the ground state, relaxation between electronic states is driven by the molecular vibrations in the corresponding electronic states. The carrier relaxation-dynamics are of fundamental importance to understanding the processes underlying carrier transport in both organic and inorganic devices. The advantage of an organic system as a model for research is the rich molecular vibrational-spectrum that can be studied computationally and experimentally. Determining the PES of ground and excited electronic-states would provide insights into the variation in molecular geometry and vibrations in these states, with a view to ascertaining whether such changes are significant and measureable and further, whether the structure and vibrations can be chemically tailored to improve the performance of OPVs.

Organic systems, like hexa-peri-hexabenzocoronene with a perylene dye have been successfully used in photodiodes with efficiencies as high as 34 %. However, these systems are too complex for high-level, molecular-modelling studies of vibrations in ground and excited electronic-states. These vibrations play a key role in properties such as: light-harvesting, exciton transport, primary charge-separation, and electron/hole conduction. In this context, the columnar discotic liquid-crystals based on triphenylene derivatives like hexakis(alkyloxy)triphenylenes (HATn), are attractive representative model systems for organic photovoltaic devices. Works on ground-state (valence-band) electron-phonon coupling of HAT6 (HATn with n = 6) have shown that structure and dynamics play a central role in the charge-transfer process of this material, and it was noted that the dynamics of the aromatic cores and the alkyl tails have an important effect on the electronic properties [14–18]. Perhaps surprisingly, it was found that the electronic properties are affected not only by the dynamics of the aromatic cores, but also by the dynamics of the alkyl tails. HAT6 is a convenient model system (Fig. 6.8) because it has the optimum side-chain length giving the broadest mesophase range. Between five and seven carbons in the alkyl tail are required for the columnar phase to form. Further increase or decrease of the tail length reduces the temperature range of the hexagonal columnar mesophase formation. From a practical point of view, it is generally thought that the

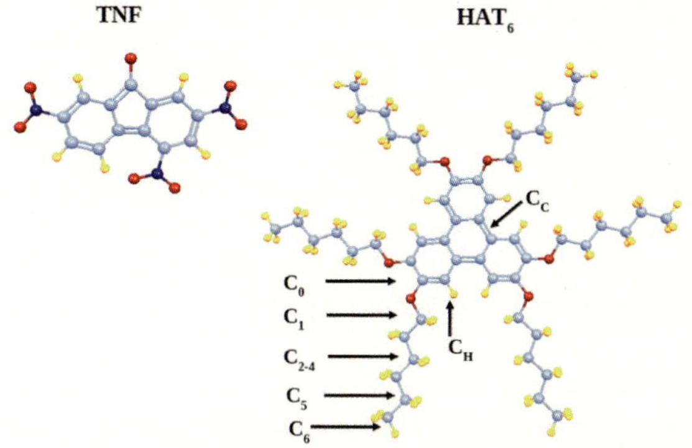

Fig. 6.8 Illustration of HAT6 (D_{3h} symmetry) and TNF, including the labeling of the HAT6 carbons used for the NMR analysis

BHJ setup leads to optimal performance of DLC solar cells. An interesting design route to achieve this device architecture is to dilute the columnar liquid crystalline phase with a non-discogenic electron acceptor, such as 2,4,7-trinitro-9-fluorenone (TNF) (Fig. 6.8).

A HAT6/TNF diamond-like carbon (D-LC) charge-transfer (CT) system is then realized. The following sections present selected results that highlight the use of neutron powder diffraction (NPD) and quasielastic neutron scattering (QENS) measurements, as well as simulations based on classical molecular-dynamics (MD) and ab initio density functional theory (DFT), to probe structure and dynamics of the HAT6/TNF D-LC-CT system. We also show that this type of study, in which both the nuclear and electronic aspects of structure and dynamics are mixed, extensive work from other techniques must be considered. Consequently, whilst the main thrust is neutron scattering, results from nuclear magnetic resonance, Raman, resonant-Raman, UV-visible, and infrared (IR) measurements are also presented and discussed. We emphasise here that neutron-scattering methods can bring interesting structural and dynamical information for HAT6 systems, largely because of their abundant H-atoms. In contrast to X-rays, neutron scattering from H atoms is strong and by deuterating the HAT6 sample (HAT6D) the related sites and their dynamics are highlighted differently.

6.3.1.1 Dependence of the Charge Transfer on the Structural Fluctuations and Conformations in HAT6

The fundamental steps are:

(1) Determining the PES of the stacked HAT6 molecules and extracting the corresponding equilibrium parameters in terms of co-facial separation (D) and twist angle (θ) between the monomers,

(2) Calculating charge-transfer integrals (CTIs) by modelling a defined structural-disorder.

(3) Evaluating how these quantities vary with the imposed conformational fluctuations.

Three degrees of freedom can be modelled in practice at the ab initio level[2] to mimic structural fluctuations: the co-facial separation D, the twist angle θ, and the lateral slide or offset L. The three-dimensional PES highlights (Fig. 6.9) an energy minimum at $\theta \sim 30°$ and $D \sim 3.5$ Å, L being constrained to zero. This agrees well with previous work dedicated to similar systems with smaller aliphatic tails. Following the estimation of the equilibrium parameters of stacked HAT6 molecules, the next step is to vary these parameters in order to mimic structural disorder, and to estimate how the HAT6-HAT6 interaction is affected. This is achieved by evaluating CTIs[3] which indicates the importance conformational fluctuations in the columnar phase. This is shown in Fig. 6.9 (middle left and right, and bottom). The dependence of the charge transfer J on the twist angle θ between two stacked HAT6 molecules is evaluated at a fixed distance $D = 3.5$ Å, determined from PES calculations.[4] Due to the D_{3h} point-group symmetry, the angular dependence of J is periodic.

When $\theta \neq n\frac{\pi}{3}$, there is a reduction in point-group symmetry from D_{3h} to C_3 and if $\theta = n\frac{\pi}{3}$ then the symmetry is lowered from D_{3h} to C_{3v}. Increasing the twist angle from 0 to 60° results in a decrease of the interaction of individual HAT6 molecules. The dependence of the charge transfer J on the co-facial separation D between two stacked HAT6 molecules, with a fixed twist angle, $\theta = 30°$, is found to decreases exponentially. J increases rapidly with increase of the co-facial separation D until reaching the long-range interaction limit at higher D, which is the monomeric zero

---

[2] Calculations based on standard semi-local, hybrid or meta functionals lead to a repulsive PES, i.e. no minimum is found. The combination of hybrid and meta contributions in PBE1KCIS is presently adequate for treating weak interactions. For reliability, detailed benchmark calculations were made for several non-bonded systems and are reported in [20–22] (and references therein). It is worthwhile to note that DFT-based methods can nowadays be improved in an efficient way like wave function-based approaches [23].

[3] The quantitative fragment-orbital approach and a symmetry-adapted linear combination of the HOMOs of the two HAT6 molecules are used. The corresponding computational procedure [24, 25] can be summarized as follows: Molecular orbitals (MOs) are firstly calculated for each single HAT6 forming the dimer in the specific orientation. Subsequently, MOs of the two stacked molecules are then expressed as a linear combination of the MOs of the monomer HAT6 (fragment), φi, leading to the overlap matrix S, the eigenvector matrix C, and the eigenvalue vector E. The relation $hKS = SCEC^{-1}$ provides CTIs, $< \varphi i|hKS|\varphi j >$. This procedure allows exact and direct calculations of J as the diagonal elements of the Kohn-Sham Hamiltonian hKS. A second approach is adopted to estimate J, which is called the dimer approach and is based on a zero spatial overlap assumption and can be used in some limited cases where the overlap between the interacting individual units forming a stacked system is negligible. It consists of the use of the half energy splitting between the HOMO and HOMO-1 to get, qualitatively, the effective CTI.

[4] Results of the fragment approach are compared to those obtained using the dimer approach. There is a significant difference between the methods due to the non-zero overlap neglected in the latter.

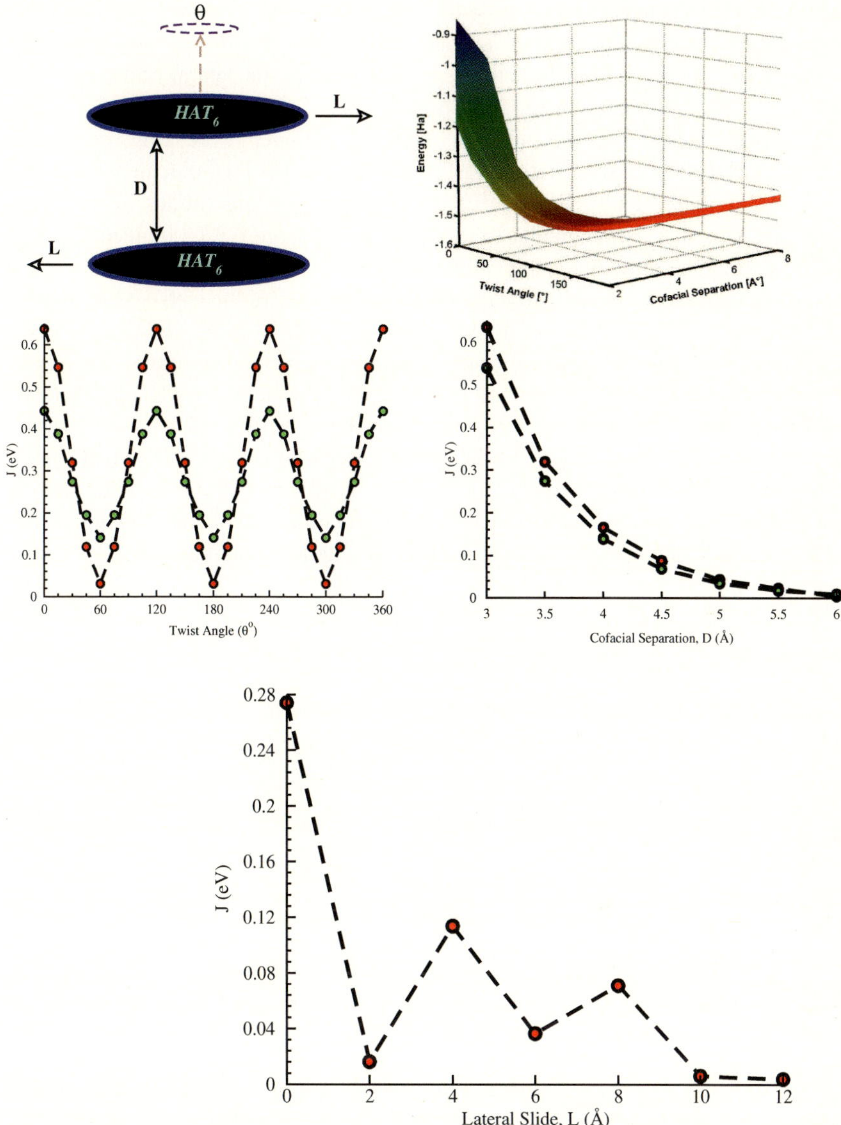

Fig. 6.9 *Top left* The modelled structural fluctuations between stacked HAT6 molecules in the form of twist angle θ, co-facial separation *D*, and lateral slide *L*. *Top right* Three-dimensional potential-energy surface of the stacked HAT6 molecules along the twist angle θ and the co-facial separation *D*, with a zero offset (*L* = 0 Å). *Middle* (*left* and *right*) and *bottom* Calculated CTIs (J) and their dependence on θ, *D*, and *L*, respectively

overlap limit. Therefore the charge-transport description, in terms of only the neighbouring electronic coupling, is adequate.

The third degree of freedom to be investigated is the lateral slide, or offset, L between the columnar stacked HAT6 molecules. Such a fluctuation can noticeably perturb charge-transfer processes through symmetry breaking, and hence spatial overlap. The offset L is achieved by sliding one HAT6 molecule with respect to the other along a C_2 axis. The complex nodal structure (nodes of the wavefunction), which is perpendicular to the plane of the large HAT6 molecule, is well reflected through consecutive and damped oscillations of local maxima and minima, which correspond to a constructive/destructive character of the overlap between the individual HAT6 molecules. A higher offset results in a zero overlap, and hence zero charge-transfer. In the presence of dynamic and/or static structural fluctuations the mobility, and therefore the conductivity, should scale approximately quadratically with the charge-transfer integral as has been reported for similar organic materials [19–21]. Thus, the above data can be used to quantitatively obtain charge-transport properties in triphenylene derivatives. It follows that this level of analysis can be extended to more realistic large-scale structural fluctuations, obtained from NPD measurements and MD simulations in the liquid crystalline phase.

6.3.1.2 Towards a More Realistic Morphological Study of HAT6

Recent work on hexabenzocoronene (HBC) derivatives and semi-triangular discotic molecules showed that classical MD simulations are of key importance in determining the influence of structure and dynamics on the conductivity of DLCs [22–25]. MD simulations are capable of predicting the conductivity of the mesophases and reveal how local conformation, disorder, and dynamics affect efficient charge-transfer along the one-dimensional column pathway. However, to realize the possibility for rational design of compounds with optimal structure-mobility relationships, it is necessary to verify that the MD simulations describe the real liquid-crystalline phase correctly. DLC structures from MD simulations were successfully compared to experiments on lattice constants, density, phase-transition temperatures, order parameter, and mutual orientation (twist angle) in specific cases. Furthermore, a more thorough and efficient analysis such as Rietveld refinement of crystalline structures can be performed. HAT6 exhibits a rather limited charge-carrier mobility of about 10^{-4} cm^2 V^{-1} s^{-1} as measured with pulse-radiolysis time-resolved microwave conductivity [26]. The question is to relate this to the much better mobilities in larger molecules such as HBC.

By comparing classical MD simulations with NPD measurements, and probing dynamics using QENS [27] along with considering the above first-principles DFT calculations (Sect. 6.3.1.1), we can go beyond the identification of the local molecular conformation, structural defects, and thermal motions in HAT6 by investigating and discussing their effect on the charge transport.

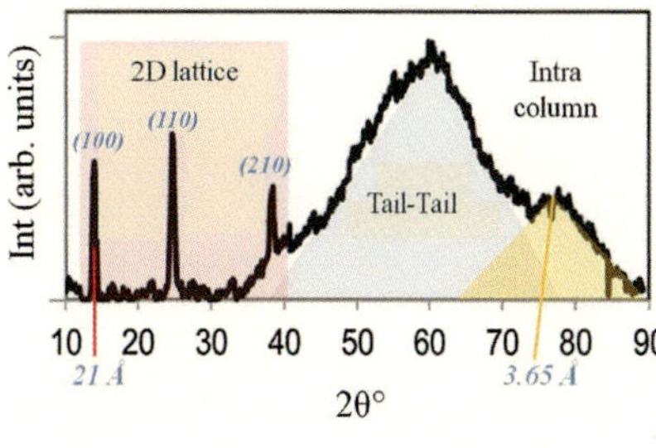

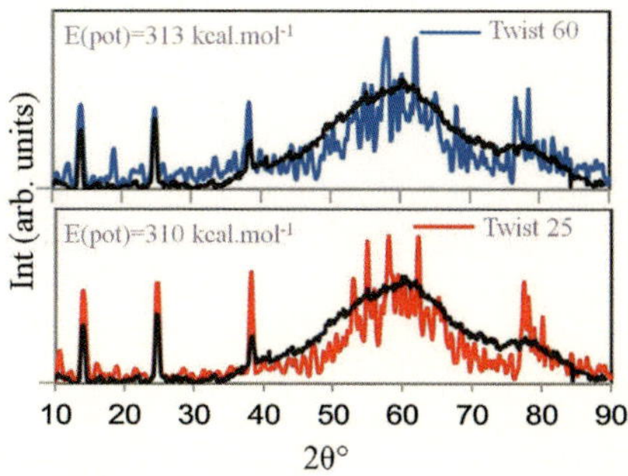

Fig. 6.10 *Left* ND pattern of HAT6D at 345 K, with the two-dimensional lattice of columns and distributions in tail-tail and intra-column distances, as indicated. *Right* Simulated diffraction patterns of MD models TWIST60 and TWIST25 compared with experiment (*black line*). In the model TWIST60, the nearest neighbour molecules in a column are rotated by 60° around their principal axis, while molecules in one two-dimensional layer have a similar orientation. In the model TWIST25 a unit of three columns is built by taking a twist angle of 25° for the first column, then taking the same mutual rotation angle, but in the opposite direction for the second column, constraining the twist angle of the third column to 5° in order to avoid superposition of the tails

In the columnar phase, the neutron diffraction (ND) pattern of a discotic liquid crystal can be conveniently subdivided into three regions, as illustrated in the case of the neutron-diffraction pattern of deuterated HAT6 (Fig. 6.10, left).[5] The red area indicates a region with three sharp peaks originating from the two-dimensional hexagonal lattice, the (100) peak corresponding to an average column-column distance. The blue region originates from the broad distribution in tail-tail distances. The intra-columnar distances correspond to the broad yellow shoulder around 3.65 Å. The tail-tail region and shoulder disappear almost completely for the fully-protonated sample. Thus, the shoulder represents intra-columnar distances between whole molecules rather than only the core-core separation.

The crystal structure predicted by the MD simulations[6] reproduces the three regions in the diffraction pattern (Fig. 6.11b). The dynamical behaviour of the liquid-crystalline phase is included in the comparison of the MD simulations with the experimental observations. The dynamic behaviour was included by calculating the anisotropic displacements (Fig. 6.11b). The experimentally-observed

---

[5] Neutron powder diffraction on HAT6 and HAT6D was performed using the D16 diffractometer at the Institut Laue Langevin in France, using wavelength of 4.54 Å to get a good compromise between d-spacing range and angular resolution.

[6] The MD force field employed is COMPASS which is a second-generation force field that generally achieves higher accuracy by including cross terms in the energy expression to account for such factors as bond, angle, or torsion distortions caused by nearby atoms. A periodic hexagonal super-cell consisting of 72 HAT6D molecules (10,368 interacting atoms) was used. The 72 molecules were arranged in 12 columns, each column consisting of 6 molecules (Fig. 6.11c).

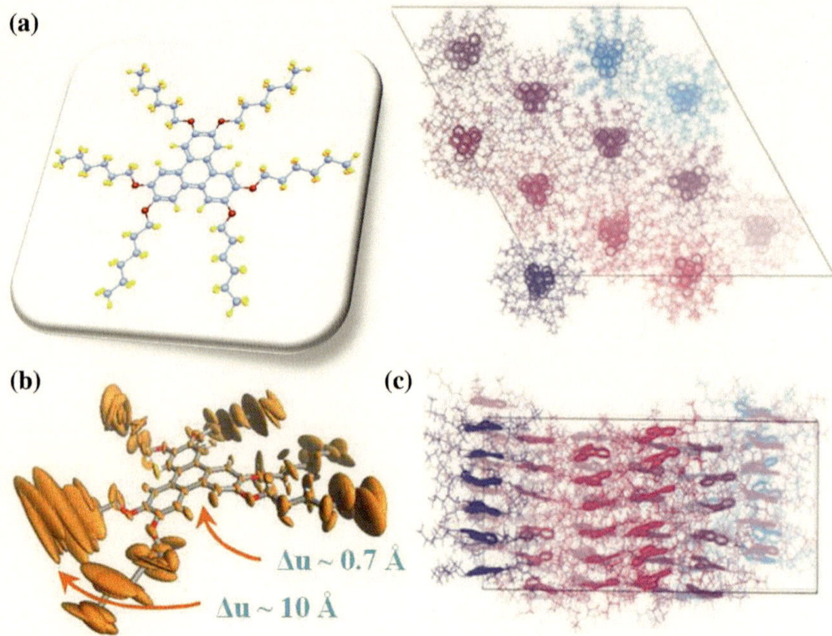

Fig. 6.11 **a** HAT6 in its D_{3h} configuration, **b** anisotropic displacements for a HAT6 molecule ranging from 0.7 (aromatic core) to 10 Å (tail end), and **c** Snapshot of a model MD simulation

diffraction pattern is intermediate between the simulated patterns of the models. The order in intra-columnar distances and the intensity of the (210) peak is overestimated in both cases. Despite these differences, both the diffraction pattern of TWIST60 and TWIST25 show remarkable agreement with the experimental pattern, while the relaxed structures stem from entirely different starting configurations (details of the models TWIST60 and TWIST25 are given in the caption of Fig. 6.10). Both models converge to comparable minima in configuration space, close to the actual liquid-crystalline structure, considering the agreement with the experimental diffraction pattern. The competition between the mutual van der Waals interactions of the cores with the steric repulsion of the tails causes a high disorder in the core-core distances, rather than a uniform shift of the core-core distance distribution to higher distances. The tails tend to orient toward the open spaces formed by these core-core defects, since the whole-molecule peak lies at a higher distance than the core-core peak. The inferred values of the twist-angle distribution are found to be around 36°, which is close to the to the DFT-estimated minimum-energy twist angle of 30° [28].

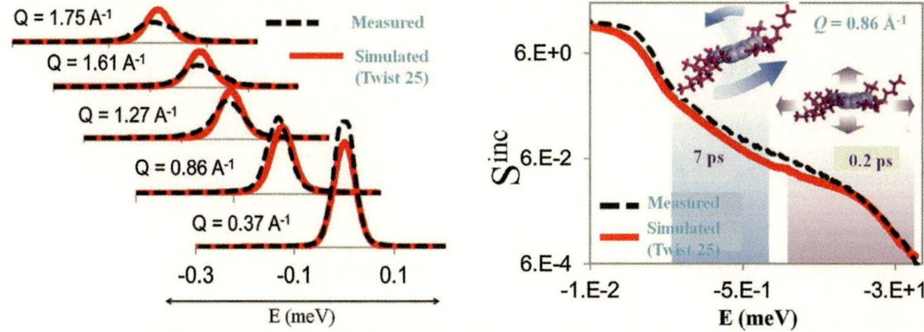

Fig. 6.12 *Left* Comparison of the incoherent scattering function from the MD simulation with the measured function at 348 K. *Right* Incoherent scattering function at 345 K and Q = 0.86 Å^{-1} on a log scale for both experiment and MD simulation. The *shaded areas* indicate the regions of dominant rotational (7 ps) and translational (0.2 ps) motion

To characterize the dynamical processes on the picosecond timescale, the MD simulations are compared with QENS[7] experiments [27]. In the analyses of the experiments, it was consistently found that in addition to the elastic peak, at least two Lorentzian functions are required to fit the data. The peak widths corresponded to time scales of about 0.2 and 7 ps timescale. The incoherent-scattering function can be extracted from the MD simulation and compared with the measured function (Fig. 6.12, left). The agreement is satisfactory, and the essential observation is that the MD models compare well with the measured temporal and spatial dynamics (Fig. 6.12, right). By examining the simulation trajectories more closely the underlying thermal motions can be characterized. Further, the typical amplitudes of molecular motion on the 0.2 and 7 ps can be estimated.

Figure 6.13 illustrates the characteristic translational and tilt motions for a single molecule on the 0.2 ps timescale. The twist-angle deviations are negligible on this timescale with amplitudes smaller than 0.1°. Apparently, the tilt movement of the molecular core is too fast to be followed by the aliphatic tails. On the 7 ps timescale, the rotational motions are much more prominent: the aliphatic tails start to follow the rotations of the core. A step further towards the understanding of the hot carriers relaxation would be to follow closely the dynamics of both the core and tails at the vibrational level and investigate how the molecular vibrations in the ground and excited electronic-states behave.

---

[7] The dynamics on the picosecond timescale were determined by directly comparing the QENS spectra of the two MD models TWIST25 and TWIST60 with the experiments [27]. QENS has the advantage that neutrons follow both the temporal and spatial characteristics of atomic motion via a well-characterized interaction with the atomic nuclei. Consequently, it is fairly straightforward to calculate the expected spectral profiles by using the atomic trajectories from the MD simulations. If these are in acceptable agreement with the observed spectra, the time scales of motion can be assigned to the underlying mechanisms. Further details about the experiment and the theoretical basis can be found in reference [27].

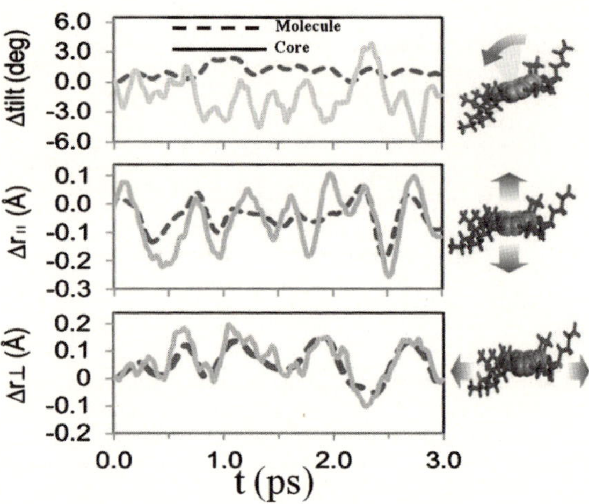

Fig. 6.13 Sub-picosecond translational and tilting motion for a typical molecule extracted from the MD trajectory, for both the molecular centre of mass (CM) including the tails (*broken lines*) and the movement of the aromatic core CM alone (*solid lines*)

6.3.1.3 A View of the Vibrational Dynamics of HAT6

In this section, an introductory study of molecular vibrations and, indirectly, relaxation dynamics, of single molecule HAT6 is made by extending previous ground-state computational work to excited states, while retaining the molecular tails [15, 16, 28, 29]. The model calculations are validated by IR and UV absorption measurements of HAT6 in solution at room temperature [30]. The goals are three-fold:

(i)   To determine the electronic excitation with the largest oscillator strength in HAT6;
(ii) Modelling the molecular structure of the selected excited state;
(iii) Understanding effects of electron-phonon coupling by comparing the vibrations of the ground state (GS) and the selected excited state (ES1).

However, a thorough investigation of the vibronic and electronic aspects of the DLC-CT HAT6-TNF is presented in Sect. 6.3.1.5.

It has been found [30] that the HOMO-to-LUMO transition dominates the targeted ES1 with the largest oscillator strength (~ 1). The calculated excitation-energy using the time-dependent DFT matches perfectly the measured one at 3.76 eV. The most important feature is that ES1 can be approximated as simply a HOMO-to-LUMO excitation, the HOMO-to-LUMO contribution in ES1 being $\sim 80\,\%$. As far as changes in the PES in going from GS to ES1 are concerned, these can be explored directly by comparing the structural parameters of the two electronic-states. Results of the structural analyses can be summarized as follows: the structural distortion of ES1, compared to GS, corresponds to geometrical changes mainly in the aromatic core of HAT6. The oxygen atoms are of central importance,

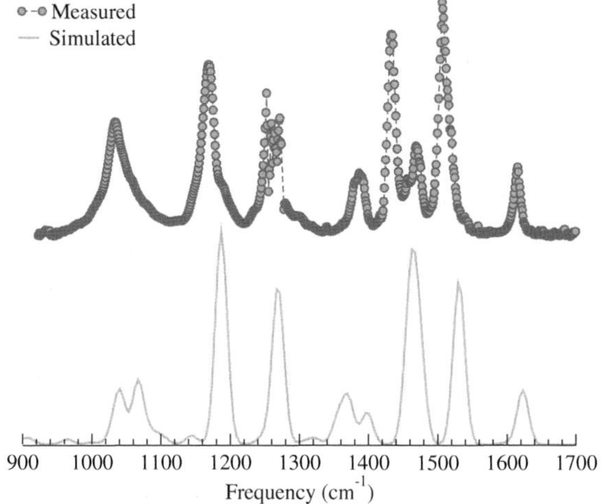

Fig. 6.14 Measured and calculated IR spectra of HAT6. The accessible low-frequency range shown (900 cm^{-1}) corresponds to the experiment. Below this value the signal-to-noise ratio becomes very small

since they connect the aromatic core to the alkyl tails, and due to their electronegativity, they could play an electronic role in the charge-transport process. Bond angles involving the oxygen atoms show changes that are comparable in amplitude to those within the aromatic core. The alkyl tails are, as expected, less sensitive to the ES1 structural distortion, but nevertheless, a change is found in the chain structure [30]. These changes reflect the different minima of the PES of GS and ES1, but gradients of the PES around local minima also change, and these are probed by the molecular vibrations. The HAT6 molecule with 144 atoms has 426 normal modes. The D_{3h} symmetry of the molecule results in six irreducible representations (irreps) with the following distribution of modes; 42 (A'1), 41 (A'2), 166 (E'), 30(A"1), 29 (A"2), 118 (E"). The modes corresponding to E' and E" are doubly degenerate. Modes with A"2 and E' symmetry are IR active. Figure 6.14 compares the calculated IR spectrum with that measured. The agreement is very good allowing the excited state of the molecule and its vibrations to be studied with more confidence.

Figure 6.15 shows the calculated IR spectra for GS and ES1 over the whole spectral range (195 active modes) and, in more detail, in the restricted spectral range from 550 to 1,700 cm^{-1} (118 active modes). It is in this range that the most striking differences are observed between GS and ES1 and, indeed, differences occur throughout this part of the spectrum. However, we have selected bands with the strongest IR intensities (and high dipole strengths), these being denoted: I, II, III, and IV. Modes I–III are composed of single peaks (pairs of degenerate modes), whereas mode IV is composed of 5 peaks/frequencies. With the exception of II, the intensities of the modes are greater in ES1 than in GS. The frequency shifts in going from GS to ES1 are about 10 cm^{-1} for modes II–IV but for mode I the shift is 29 cm^{-1}. The related atomic displacements for GS and ES1 modes in the four

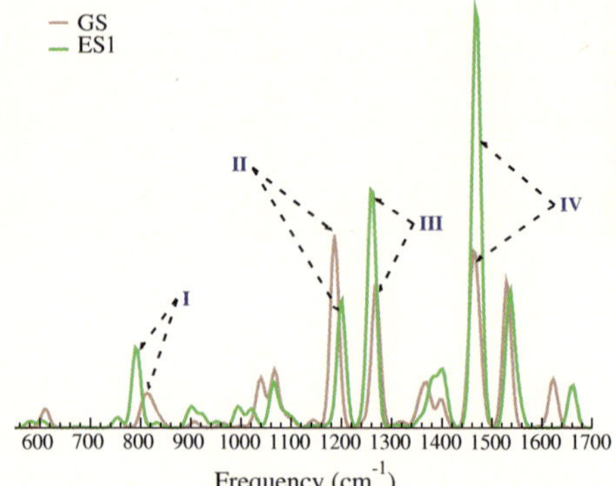

Mode	Freq [cm^{-1}]	Int [km mole^{-1}]	Dip Str [10^{-40} esu^2 cm^2]
GSI	821	149	724
GSII	1187	607	2042
GSIII	1269	456	1432
GSIV	1461	525	1820
ES$_1$I	792	258	1300
ES$_1$II	1200	408	1355
ES$_1$III	1262	764	2415
ES$_1$IV	1468	1010	3250

Fig. 6.15 *Left* Simulated IR vibrational spectra of HAT6 in the GS and ES1, the peak width (full-width at half-maximum) having been constrained to 20 cm^{-1}. The main vibrational modes are indicated as I, II, III and IV. *Right* Calculated frequencies (Freq), IR intensities (Int) and dipole strengths (Dip Str) for the selected ground state and ES1 vibrations

frequency bands can be analysed. For mode I, which has the biggest frequency shift, the GS mode is an out-of-plane mode, whereas the ES1 mode is an in-plane mode. Accordingly, the displacements of the alkoxy tails are significantly different, being polarised in the planes perpendicular to the tail directions in GS and being polarised along the tails in ES1. For modes II–IV, the frequency shifts are smaller and the similarity of modes between GS and ES1 is stronger, with mode III being almost identical in the two electronic states. In these three bands, all modes show in-plane polarisation of the molecular cores. Mode II involves deformations of the aromatic cores and C–O stretches, with corresponding responses in the alkoxy tails, and the displacement patterns change between GS and ES1. For mode IV, the main difference between GS and ES1 involves the pattern of C–C stretches in the aromatic core, but there is a notable difference in the chain displacements. It is evident from this investigation of GS and ES1 molecular vibrations that core and tail vibrations are coupled.

In the intermediate frequency range that has been considered, deformations of the aromatic core are accompanied by complex displacements of the alkoxy tails, and these are not simple rigid-body motions. Accordingly, the QENS study reported in Sect. 6.3.1.2, in which diffusive dynamics on the ps timescale were probed, proved in a consistent way, that motion of the alkyl tails is driven by the core dynamics. The alkyl chain (beyond the O atom) is also found to play a significant role in the molecular distortion and change in molecular vibrations upon electronic excitation. The vibrational spectrum in the excited electronic state, which has the strongest oscillator strength, is considerably different from that in the ground state, despite the size and overall flexibility of the system. This result, supported by the good agreement with measured IR and UV spectra, is encouraging for gaining insight into the technologically important relaxation processes within the conduction band that lead to serious efficiency losses in solar-cell applications. After all, most energy present in the incoming light is dissipated and converted to heat due to fast relaxation from higher excitations towards the lower band edge. In this context Sect. 6.3.1.5 focuses also on some resonant Raman measurements which are related to the excited-state molecular structure via the vibrations that constitute the atomic shifts between the ground and excited electronic-state structures.

6.3.1.4 Morphology of the Discotic Charge-Transfer System HAT6-TNF

Although pure DLCs generally show a poor absorption in the visible spectral domain, mixtures of the electron-donating discoids with non-discogenic electron acceptors could exhibit absorption bands in the visible due to the formation of a CT complex. In many cases CT complexation even causes a considerable increase in the stability of the columnar mesophase. For good performance of a photovoltaic device the donor and acceptor molecules must form separate columns, i.e., enable charge separation and subsequent charge transport along the columnar wires. The position of the electron acceptors within the columnar mesophases is still contro-versial. Acceptor molecules such as TNF have been reported to be sandwiched between discotic molecules within the same column [31–34] but also inter-columnar, i.e., between the columns within the aliphatic tails of the discotic mol-ecules [29, 35]. Only the inter-columnar juxtaposition could provide a morphology with separate continuous columns for electron and hole transport. Another issue is that the characterization of (photo-induced) electron transfer and relaxation pro-cesses in self-assembled aggregates such as DLCs and DLC-CTs is in its infancy. The addition of electron acceptors such as TNF has been shown to increase the conductivity of DLCs. On the other hand, it has been proposed that recombination processes limit the hole photocurrent in DLC-CT compounds. Charge carriers in CT compounds are supposed to be trapped and readily annihilated through rapid, phonon-assisted relaxation and recombination processes.

In this section the morphology issue is elucidated by considering the prototypical discotic CT compound HAT6-TNF (Fig. 6.8), where HAT6 is used as electron-donating discoid and TNF as electron acceptor. HAT6-TNF forms a CT compound exhibiting a stable columnar phase from below room-temperature and up to 237 °C. The high symmetry and moderate molecular size of discogens such as HAT6 make these systems attractive for exploring the effects of increasing molecular complexity by comparing their photo-physical properties with those of the fundamental building blocks: benzene, and large poly-aromatic hydrocarbons. For discotic liquid crystal-line CT-complexes it is generally accepted that intermolecular charge-transfer occurs in the excited state, but not in the ground state. Mixtures of the electron-donating discoids with non-discogenic electron-acceptors exhibit absorption bands in the visible region due to excited-state charge-transfer. Support for these indications can be found from a combination of NMR and Raman spectroscopy measurements. Furthermore, the electronic transitions involved in the CT-band of HAT6-TNF can be characterized by combining UV-visible absorption and resonant Raman

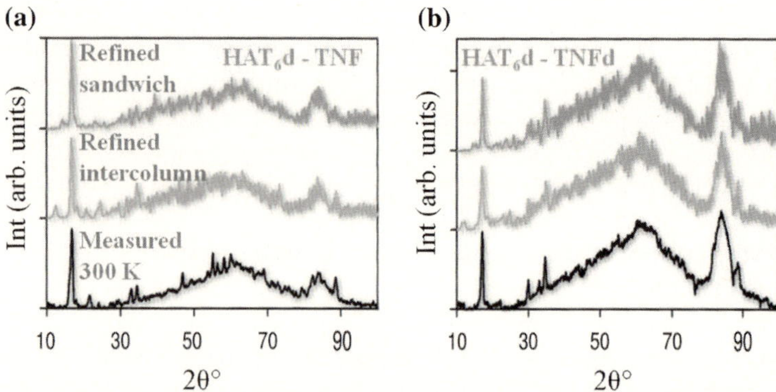

Fig. 6.16 Comparison between the neutron diffraction patterns of the refined sandwich (*top*) and inter-columnar (*centre*) models and the measurements at 300 K (*bottom*) for the CT complex HAT6TNF with deuterated HAT6 (HAT6D-TNF) (**a**), and both deuterated HAT6 and TNF (HAT6DTNFD) (**b**)

spectroscopies. Additionally, the UV-visible and Raman measurements are accompanied by DFT calculations, to identify the vibrational modes that assist charge-carrier relaxation in the hot-band of HAT6 and in the CT-band of HAT6-TNF.

The ND pattern of HAT6-TNF (Fig. 6.16) is characteristic for a columnar mesophase, with sharp reflections in the small 2θ region ((100), (010), etc.) originating from the two-dimensional columnar lattice, a broad liquid-like band from the distribution in tail-tail distances, and a broad (001) peak from the intra-columnar distances [36]. Two observations are important. First, there is no superstructure peak visible with a double co-facial distance as would result in an intra-columnar juxtaposition of TNF where TNF and HAT6 alternate, doubling the cell dimensions in the z-direction. This essentially makes such configurations unlikely. Second, the large decrease in lattice parameters is not accompanied by a similar increase in density that would result when simply shrinking the HAT6 cell to the new dimensions.[8]

These observations indicate that the columnar morphology has drastically changed in the charge-transfer compound. For a hexagonal columnar-structure with the TNF sandwiched between the HAT6 molecules, the column-column distance should be about 17 % smaller compared to pure HAT6. This only appears possible if the HAT6 and TNF are also alternately packed in the hexagonal plane. Other distributions result in energetically unfavourable inter-digitation of the aliphatic tails, such as the often suggested alternating intra-columnar packing with the discotic molecules positioned in the same hexagonal plane. The intra-columnar juxtaposition of TNF, on the other hand, is only consistent with the experiments if the HAT6 columns are tilted on an oblique lattice. Such an arrangement, with discotic molecules slid laterally, has already been observed for highly polar $HAT2\text{-}NO_2$

---

[8] A skeletal density of 1.08 ± 0.01 g cm^{-3} is observed for HAT6D-TNFD, corresponding to an increase in density of about 7 % with respect to HAT6.

molecules. However, the tilted HAT6 columns leave such small spaces within the tail region that the TNF molecules should mainly adopt a vertical orientation.

The sandwich (with HAT6 and TNF alternating in the horizontal plane) and intra-columnar (with tilted HAT6) models were further analysed with Rietveld refinement. Remarkably, the refinement does not favour either of the juxtapositions of TNF over the other. Both models reproduce the characteristic features of the experimental diffraction patterns with comparable agreement. In addition, the effect of deuteration on the diffraction patterns also fits well with the measurements in both cases, particularly considering that no extra refinement step has been made; i.e., only the deuterium atoms of TNF were replaced with protons. Even the orientation dependences of the diffraction patterns on macroscopically aligned samples show little difference between the two models. As expected, the two-dimensional lattice peaks have maximum intensity when the diffraction beam is perpendicular to the column director, and the intra-columnar peak is at a maximum for the parallel orientation.

Clearly, it is difficult to determine the juxtaposition of TNF using only diffraction. Nevertheless, the sandwich and inter-columnar models only fit with the observed density and diffraction patterns under the conditions illustrated in Fig. 6.17. As anticipated, the HAT6 columns in the inter-columnar model are tilted, with a co-facial slide of about 3.5 Å between two neighbouring molecules in a column. We defined d *column–column* as the average separation between the column directors of two neighbouring columns, which is significantly larger for the inter-column juxtaposition of TNF. The minimal distance dHAT6*core*–TNF between the core of HAT6 and TNF predominantly determines the charge-transfer

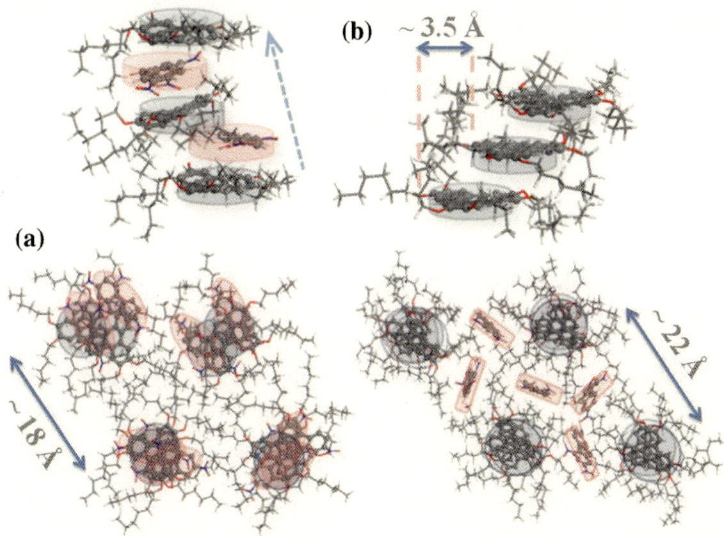

Fig. 6.17 Illustration of the sandwich (**a**) and the intra-columnar (**b**) model structures after Rietveld refinement. *Elliptical pink shape* TNF; *grey disk* HAT6

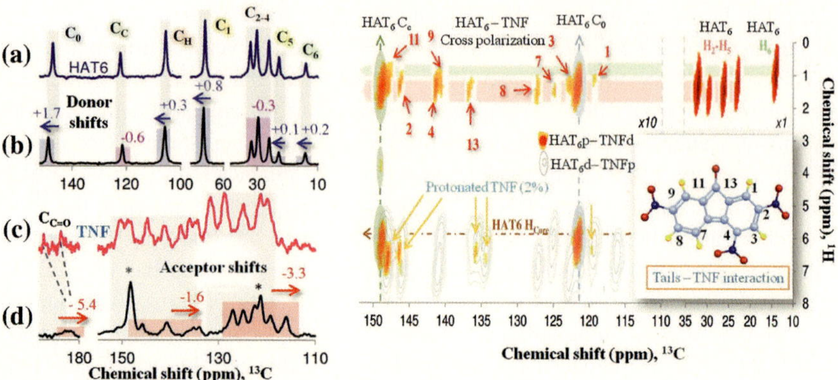

Fig. 6.18 *Left* Solid state CP-MAS ^{13}C NMR spectra for HAT6 (**a**), HAT6-TNFD (**b**), TNF (**c**), and HAT6-TNF (**d**). The temperature and CP mixing time are 358 K, 10 ms for (**a, b**), 300 K, 2 ms for (**c**), and 300 K, 5 ms for (**d**). The HAT6 carbon assignment in (**a**) follows the labelling of Fig. 6.8. The *blue* (*red*) *arrows* indicate the downfield (up-field) shifts of HAT6 (TNF) peaks in the composite. *Right* ^{1}H^{13}C two-dimensional hetero-nuclear correlation spectra for HAT6-TNFD (*coloured* contours) and HAT6D-TNF (*grey* contours) at 290 K with a CP mixing time of 10 ms. The red numbers indicate cross-polarization between protons on the tail of HAT6 (H2H5 and H6) and specific TNF carbons shown in the *inset*. The *circles* in the *inset* surrounding the numbered carbons illustrate the strength of these interactions and the possible HAT6 hydrogens involved (*pink* for H2H5, *green* for H6). The HAT6 C_c and C_o carbon and the H*core* hydrogen chemical shifts are indicated by the *green*, *blue*, and *brown dashed lines*, respectively. The signals marked with *yellow arrows* are due to imperfect deuteration of TNF

behaviour of the complex. For the inter-columnar arrangement dHAT6*core*–TNF is difficult to estimate, since there is considerable freedom left in the refinement of the TNF position. We refined several inter-column models with different initial vertical positions of TNF from which we estimated that dHAT6*core*–TNF should be within the range of 4–10 Å. Typically, the closest distance between TNF and HAT6 for the intra-columnar juxtaposition involved a CH C atom of HAT6 and TNF NO_2 group.

Figure 6.18 (left) shows the solid state ^{13}C cross-polarization (CP) magic-angle spinning nuclear magnetic resonance spectra of liquid-crystalline HAT6, TNF, and the CT complex. On the basis of these spectra of the un-complexed samples, all the peaks in the CT-complex spectrum are assigned to specific HAT6 or TNF carbons. In the CT compounds, however, the chemical shifts of the HAT6 and TNF carbons are changed significantly. All the TNF carbon signals are shifted up-field, reflecting a stronger local magnetic-field for the TNF in the mixture compared with the pure compound. In contrast to TNF, the HAT6 lines show both downfield and up-field shifts.

Chemical-shift changes in charge-transfer complexes have been attributed to partial electron-transfer from donor to acceptor molecules in the electronic ground-state. According to Mulliken's theory, partial transfer of electron density occurs from the HOMO of the donor to the LUMO of the acceptor in the electronic GS. This is observed for the HAT6 donor. The anticipated general shift-sign (lower electron density, lower field) that would result from partial electron-transfer to the acceptor TNF. For HAT6, the largest down-field shifts are observed for ^{13}C nuclei in the outer part of the aromatic core, which is consistent with the spatial distribution of the HOMO [37].

A clear conclusion on the juxtaposition of TNF can be drawn from the two-dimensional ^{1}H^{13}C hetero-correlation nuclear magnetic resonance measurements. The signals indicated with red arrows in Fig. 6.18 (right) result from coherence transfer between the HAT6 tail-proton spins H1H6 and specific TNF carbons labelled in the inset. The TNF should be, at least for the major part, within the tail region of the HAT6 molecules. On the other hand, no interaction between TNF and the core of HAT6 was observed. The proton H_{core} attached to the HAT6 core only shows cross-polarization with HAT6 carbons. For the differently deuterated sample, HAT6D-TNF, very little coherence transfer between TNF protons and any of the HAT6 carbons was observed. The absence of a significant reverse CP from TNF to the HAT6 tails is likely due to the different number of protons involved (12 for HAT6 tail protons H_2H_5, 1 for TNF protons). The combination of a rapid rigid-like CP build-up and the narrowing of the ^{1}H line widths are indicative of selective averaging or quenching of weak longer range homo-nuclear ^{1}H^{1}H dipolar interactions by anisotropic motion of the HAT6 and TNF molecules in the liquid crystalline phase. Larger molecular displacements of HAT6, related to dynamic defects in the liquid-crystalline phase, occur on time scales up to milliseconds. These motions can quench the long-range dipolar interactions and contribute to the narrowing of the ^{1}H lines in the data sets without decoupling, while allowing at the same time for the high CP rates for the ^{13}C in the aromatic core by strong short-range hetero-nuclear dipolar interactions.

The nuclear magnetic resonance analyses seem to indicate that charge transfer from the HAT6 core to TNF already takes place in the ground state of the complex. Excited-state or ground-state charge-transfer from the HAT6 core to TNF requires that the electron acceptor should not be too far from the aromatic core. For the inter-columnar CT-complex structure the diffraction analyses resulted in HAT6*core*-TNF distances of 4–10 Å, mostly involving the optically active NO_2 groups of TNF. To ensure a sufficient orbital overlap for CT electron delocalization, most of the TNF molecules should be in the lower part of this range. The consistent analysis reveals a CT-complex morphology with dynamically disordered TNF molecules that are vertically oriented between the HAT6 columns, i.e., within the aliphatic-tail region. What does this mean for PV applications?

A promising observation is that there is a hole-conducting column present that is well separated from the electron acceptors. The columnar morphology has changed drastically in the composites, with a time-averaged tilted orientation and smaller average distances between the neighbouring HAT6 molecules within the column than in pure HAT6. In the CT complex the hole transport through the column will thus still be possible, while the CT process enables efficient charge separation. The liquid-crystalline structure and its facile alignment over macroscopic distances is an important asset for device realization.

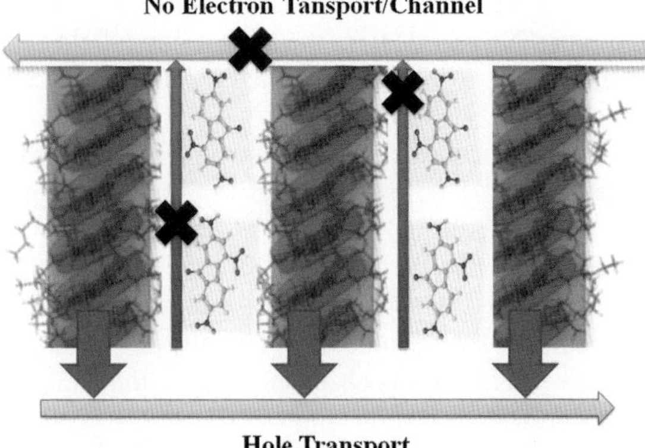

Fig. 6.19 Illustration of a HAT6-TNF BHJ PV device with self-assembled columns at the molecular nanoscale

The future PV application of CT complexes such as HAT6-TNF thus relies on improving the electron transport channel (Fig. 6.19). For instance, by searching for better acceptors that self-assemble into a separate channel, designing molecularly-connected donor and acceptor groups. The persistence of the hole conducting HAT6 column in the CT complex is promising for future application in organic PV systems. The major challenge towards such an application is to achieve a morphology that enables a BHJ PV device architecture.

6.3.1.5 Electronic and Vibronic Properties of the Discotic Charge-Transfer System HAT6-TNF

As discussed in Sect. 6.2, one of the loss mechanisms to which OPVs might be subject can be via the strong electron-phonon coupling inherent to the molecular nature of the device, which limits the efficiency of charge separation. Upon photo-excitation, strongly bound exciton-states are formed that first need to dissociate before charge-transport to the electrodes can occur. Dissociation takes place at the DA interface, where intermediate CT states are formed with the hole on the donor, and electron at the acceptor molecule. It is established that charge separation is mediated by the higher lying vibronic-states of the CT manifold [38, 39]. As introduced in Sect. 6.3.1.3, fundamental knowledge about the electronic and vibrational properties of the excited-state levels and relaxation pathways is a key topic in further improving the working of OPVs.

For self-assembled aggregates such as DLCs and DLC-CT complexes, the characterization of photo-induced electron-transfer and relaxation processes is in its infancy. The addition of electron acceptors has been shown to increase the conductivity of DLCs. On the other hand, it has been proposed that recombination processes limit the hole photocurrent in such DLC-CT compounds [40].

An important step towards the characterization of the influence of molecular vibrations on the charge-carrier relaxation in self-assembled DLCs can be made by studying the prototypical discotic electron-donating discoid HAT6 and its 1:1

mixture with electron acceptor TNF, combining synergistically Raman, resonant Raman, UV-visible and nuclear magnetic resonance techniques, as illustrated in reference [41]. From this one can draw the following conclusions:

(i)   The lowest electronic-transition in the CT complex is due to charge-transfer from the HAT6 core to TNF, with a strong involvement of the nitro groups.

(ii)   There are strong indications for a weak ground-state electron-transfer in the HAT6-TNF complex.

(iii)   Both nuclear magnetic resonance chemical-shift changes and Raman frequency-shifts are seen and are consistent with a weak electron-transfer from the HAT6 core to TNF leading to a delocalized redistribution of the charge on TNF.

(iv)   The resonant Raman spectra of both HAT6 and the CT-complex show a strong enhancement of the modes related to the benzenes forming the triphenylene core. This is characterized by a doubly-degenerate electronic-state which is vibronically coupled (pseudo-Jahn-Teller) to a nondegenerate state. It should be noted that this Jahn-Teller mode provides a significant contribution to the reorganization energy of HATn, being a limiting factor for hole transport along the columnar stacks [42].

(v)   The hot-carrier relaxation processes in the CT-band in the visible-light region are relatively slow compared to the fast relaxation within the original UV absorption band of pure HAT6, which will be relevant concerning the efficient separation of charge in organic PV-devices.

6.4   Conclusions

Inorganic photovoltaics possess a major advantage over organic solar cells by providing the highest power-conversion efficiencies. But in terms of the fabrication flexibility, production costs and market accessibility, organic photovoltaics are also potential candidates for the future generation of solar cells. This requires understanding the dynamic and electronic properties so that optimization routes can be devised that will increase their efficiency and boost their performance. This chapter illustrated how neutron scattering, in combination with other techniques, can be used to extract this fundamental aspect of organic photovoltaics working principles. In this chapter different structural and dynamical aspects of an organic molecular-model for photovoltaic applications are presented, including an evaluation of the known limitations. The chapter provides an example of how the logical sequential advance of experiment, theory and understanding lead towards improvements. Different experimental techniques and theoretical methods are used, in a quite complementary way— if not synergetic. Neutron-scattering techniques are shown to play an important role in this field by probing efficiently structure and dynamics in the presented hydrogenated materials. Numerical simulations are almost mandatory in this field to aid the interpretation and analysis of the different measurements including the other experimental techniques (IR, UV, Raman, and nuclear magnetic resonance methods).

References

1. I. Bruder, M. Karlsson, F. Eickemeyer, J. Hwang, P. Erk, A. Hagfeldt, J. Weis, N. Pschirer, Sol. Energy Mater. Sol. Cells **93**, 1896 (2009)
2. J. Peet, Nat. Mater. **6**, 497 (2007)
3. B. ORegan, M. Grätzel, Nature **353**, 6346 (1991)
4. I. Bruder, PhD Thesis "Organic solar cells: correlation between molecular structure, morphology and device performance", Max-Planck-Institut für Festk¨orperforschung (2010)
5. M. Grätzel, Prog. Photovoltaics **8**, 171 (2000)
6. E.A. Silinsh, V. Capek, *Organic Molecular Crystals-Interaction, Localization and Transport Phenomena* (American Institue of Physics, New York, 1994)
7. D. Cheyns, H. Gommans, M. Odijk, J. Poortmans, P. Heremans, Sol. Ener. Mater. Sol. Cells **91**, 399 (2007)
8. M. Hiramoto, M. Suezaki, M. Yokoyama, Chem. Lett. **19**, 327 (1990)
9. A. Yakimov, S.R. Forrest, Appl. Phys. Lett. **80**, 1667 (2002)
10. J. Dreschel, B. Mannig, F. Kozlowski, M. Pfeiffer, K. Leo, H. Hoppe, Appl. Phys. Lett. **86**, 244102 (2005)
11. G. Yu, J. Gao, J.C. Hummelen, F. Wudl, A.J. Heeger, Science **270**, 1789 (1995)
12. S. Chandrasekhar, B.K. Sadashiva, K.A. Suresh, Pramana **9**, 471 (1977)
13. S. Chandrasekhar, *Liquid Crystal* (Cambridge University Press, Cambridge, 1977)
14. A.M. Van de Craats, M.P. De Haas, J.M. Warman, Synth. Met. **86**, 2125 (1997)
15. O. Kruglova, F.M. Mulder, L.D.A. Siebbeles, G.J. Kearley, Chem. Phys. **330**, 333 (2006)
16. O. Kruglova, F.M. Mulder, A. Kotlewski, S.J. Picken, S. Parker, M.R. Johnson, G.J. Kearley, Chem. Phys. **330**, 360 (2006)
17. G.J. Kearley, F.M. Mulder, S.J. Picken, P.H. Kouwer, J. Stride, Chem. Phys. **292**, 185 (2003)
18. P. Etchegoin, Phys. Rev. E **56**, 538 (1997)
19. K. Senthilkumar, F.C. Grozema, F.M. Bickelhaupt, L.D.A. Siebbeles, J. Chem. Phys. **119**, 9809 (2003)
20. B.R. Wegewijs, L.D.A. Siebbeles, N. Boden, R.J. Bushby, B. Movaghar, O.R. Lozman, Q. Liu, A. Pecchia, L.A. Mason, Phys. Rev. B **65**, 245112 (2002)
21. M.A. Palenberg, R.J. Silbey, M. Malagoli, J.L. Bredas, J. Chem. Phys. **112**, 1541 (2000)
22. D. Andrienko, J. Kirkpatrick, V. Marcon, J. Nelson, K. Kremer, Phys. Status Solidi B **245**, 830 (2008)
23. J. Kirkpatrick, V. Marcon, J. Nelson, K. Kremer, D. Andrienko, Phys. Rev. Lett. **98**, 227402 (2007)
24. V. Marcon, T. Vehoff, J. Kirkpatrick, C. Jeong, D.Y. Yoon, K. Kremer, D.J. Andrienko, Chem. Phys. **129**, 094505 (2008)
25. A. Troisi, D.L. Cheung, D. Andrienko, Phys. Rev. Lett. **102**, 116602 (2009)
26. A.M. Van de Craats, M.P. De Haas, J.M. Warman, Synth. Met. **86**, 2125 (1997)
27. F.M. Mulder, J. Stride, S.J. Picken, P.H.J. Kouwer, M.P. de Haas, L.D.A. Siebbeles, G.J. Kearley, J. Am. Chem. Soc. **125**, 3860 (2003)
28. L.A. Haverkate, M. Zbiri, M.R. Johnson, B. Deme, F.M. Mulder, G.J. Kearley, J. Phys. Chem. B **115**, 13809 (2011)
29. O. Kruglova, E. Mencles, Z. Yildirim, M. Wubbenhorst, F.M. Mulder, J.A. Stride, S.J. Picken, G.J. Kearley, ChemPhysChem **8**, 1338 (2007)
30. M. Zbiri, M.R. Johnson, L.A. Haverkate, F.M. Mulder, G.J. Kearley, Aust. J. Chem. **63**, 388 (2010)
31. D. Markovitsi, H. Bengs, H.J. Ringsdorf, Chem. Soc. Faraday Trans. **88**, 1275 (1992)
32. K. Praefcke, J.D.J. Holbrey, Incl Phenom. Mol. Recognit. Chem. **24**, 19 (1996)
33. Y. Kamikawa, T. Kato, Org. Lett. **8**, 2463 (2006)
34. V. Percec, M.R. Imam, M. Peterca, D.A. Wilson, R. Graf, H.W. Spiess, V.S. Balagurusamy, P.A. Heiney, J. Am. Chem. Soc. **131**, 7662 (2009)
35. P.H.J. Kouwer, W.F. Jager, W.J. Mijs, S.J. Picken, Macromolecules **35**, 4322 (2002)

36. L.A. Haverkate, M. Zbiri, M.R. Johnson, B. Deme, H.J.M. de Groot, F. Lefeber, A. Kotlewski, S.J. Picken, F.M. Mulder, G.J. Kearley, J. Phys. Chem. B **116**, 13098 (2012)
37. M. Zbiri, M.R. Johnson, G.J. Kearley, F.M. Mulder, Theor. Chem. Acc. **125**, 445 (2010)
38. J. Lee, K. Vandewal, S.R. Yost, M.E. Bahlke, L. Goris, M.A. Baldo, J.V. Manca, T. Van Voorhis, J. Am. Chem. Soc. **132**, 11878 (2010)
39. R.D. Pensack, J.B. Asbury, J. Phys. Chem. Lett. **1**, 2255 (2010)
40. K.J. Donovan, K. Scott, M. Somerton, J. Preece, M. Manickam, Chem. Phys. **322**, 471 (2006)
41. L.A. Haverkate, M. Zbiri, M.R. Johnson, E.A. Carter, A. Kotlewski, S. Picken, F.M. Mulder, G.J. Kearley, J. Chem. Phys. **140**, 14903 (2014)
42. V. Lemaur, D.A. Da Silva Filho, V. Coropceanu, M. Lehmann, Y. Geerts, J. Piris, M.G. Debije, A.M. Van de Craats, K. Senthilkumar, L.D.A. Siebbeles, J.M. Warman, J.L. Bredas, J. Cornil, J. Am. Chem. Soc. **126**, 3271 (2004)

Part II
Energy Storage

Energy storage is a key enabling technology to the larger area of sustainable energy. For transport applications, the storage needs to mirror the energy density of petrol and diesel, and because sustainable energy-sources such as solar and wind are variable in nature, the buffering capacity of storage is also of importance. The main energy-storage technologies are batteries, capacitors, pumped storage, hydrogen, flywheels, and compressed (or liquefied) gasses. Each has its technological and logistical challenges, with Chaps. 7 and 8 examining the two major areas of lithium-ion batteries and hydrogen storage, respectively.

The central theme in both battery and hydrogen-storage materials is increasing energy density gravimetrically and volumetrically, in a way that enables the energy to be stored and retrieved in a time and energy-efficient way. It is difficult for either battery or hydrogen storage materials to compete with the energy density of petrol and diesel. Neutron techniques of analysis here are used to understand the interactions that hold the energy-carrier in place, but can then release it when it is required for use. In these chapters the role of in situ studies is particularly important and this is an area in which the penetration of neutrons excels. It is sufficiently important that neutron-scattering centres provide custom made gas-handling and potentiostat/galvanostat equipment that can be used by several different groups.

Chapter 7 shows a large volume of study has been undertaken to determine the crystal structure of electrode materials to determine the position and environment of lithium in complex and multiphase electrode materials, even as the battery is charged and discharged. However, disordered and amorphous materials often have the desired materials properties, and it is illustrated that neutron scattering is also relevant to these by using the total scattering technique and spectroscopy. Although lithium is better suited to study by neutron scattering than many other techniques, it is nevertheless challenging, particularly when compared to hydrogen (Chap. 8). First, the absorption cross-section of one of the naturally-occurring lithium isotopes is high, and whilst this can be overcome, and even exploited by isotopic selectivity, this is expensive. Secondly, not only would we like fast diffusion for better batteries, but it would make the study of diffusion by neutron scattering easier. Current neutron instrumentation can access lithium diffusion when used at high resolution, but this limits the neutron counting-rate, making the experiments more difficult, and in some cases impractical. Fortunately, in some examples it is reasonably straightforward to derive the dynamics of the lithium by studying the response (or driving dynamics) of the environment, in this case it is the environment that has a strong neutron signal.

Hydrogen storage (Chap. 8) has probably been the most important area of sustainable-energy research for neutron scattering due to the special interaction between neutron and hydrogen and its isotope deuterium. In common with battery materials, there has been a large volume of work using diffraction to determine the position and environment to be derived of the H-atoms (or molecules), much of which existed prior to hydrogen-storage applications. The interplay between atomistic modelling and neutron scattering pervades sustainable-energy research, but is particularly important in hydrogen storage due to the ease with which the dynamics of the model can be compared with neutron-spectroscopy results. This enables very detailed information about how hydrogen interacts with all the components of its environment, and whilst this is also possible for the battery materials above, the experimental difficulties associated with lithium make it much more challenging. Hydrogen is a light element and quantum effects in its dynamics are pronounced. Because of the way in which the neutron spin interacts with the nuclear spin of hydrogen, neutrons provide a very sensitive measure of these quantum dynamics, which is in turn very sensitive to its environment. These dynamics are not only of interest for tuning the interaction of hydrogen with its storage material, but are also of fundamental interest in their own right and much of the theory for understanding the measured neutron-scattering spectra had already been established because of this. There is a considerable body of analogous work for methane, which whilst of considerable technological importance, is not covered in this book. The analogy is strongest for storage in metal-organic framework and clathrate materials, and in most cases the extension of the work presented here for hydrogen to the methane case is straightforward.

Chapter 7
Lithium-Ion Batteries

Neeraj Sharma and Marnix Wagemaker

Abstract The effort of material scientists in the discovery, understanding, and development of Li-ion batteries largely depends on the techniques available to observe the relevant processes on the appropriate time and length scales. This chapter aims at demonstrating the role and use of different neutron-scattering techniques in the progress of Li-ion battery electrode and electrolyte properties and function. The large range in time and length scales offered by neutron-scattering techniques is highlighted. This illustrates the type of information that can be obtained, including key parameters such as crystal structure, Li-ion positions, impact of nano-particle size and defects, ionic mobility, as well as the Li-ion distribution in electrodes and at electrode-electrolyte interfaces. Special attention is directed to the development of in situ neutron-scattering techniques providing insight on the function of battery materials under realistic conditions, a promising direction for future battery research.

7.1 Li-Ion Batteries for Energy Storage

Electrochemical energy storage is attractive, having very high storage efficiencies typically exceeding 90 %, as well as relatively-high energy densities. Li-ion battery technology provides the highest energy densities of commercialized battery-technologies and has found widespread use in portable electronic applications. Application of Li-ion batteries in electrical vehicles and as static storage media is emerging, however, improved performance and reduced cost, combined with safety

N. Sharma (✉)
School of Chemistry, The University of New South Wales, Sydney, NSW 2052, Australia
e-mail: neeraj.sharma@unsw.edu.au

M. Wagemaker
Faculty of Applied Sciences, Radiation, Science and Technology Department, Delft
University of Technology, Mekelweg 15, 2629 JB Delft, The Netherlands
e-mail: m.wagemaker@tudelft.nl

© Springer International Publishing Switzerland 2015

139

G.J. Kearley and V.K. Peterson (eds.), *Neutron Applications in Materials for Energy*, Neutron Scattering Applications and Techniques,
DOI 10.1007/978-3-319-06656-1_7

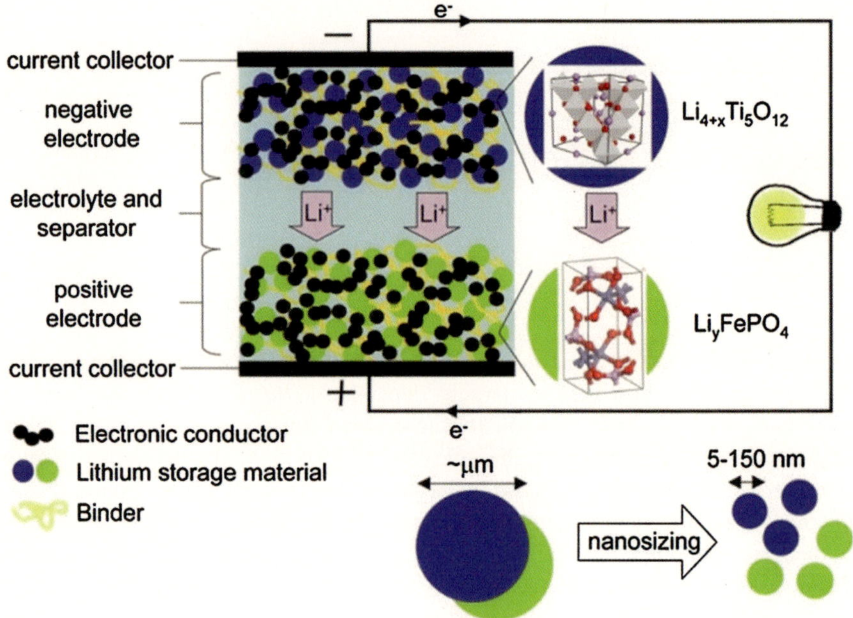

Fig. 7.1 Schematic illustration of the operating principle of a Li-ion battery. Reprinted with permission from (M. Wagemaker, F.M. Mulder, Acc. Chem. Res. **46**, 1206 (2013)) [37]. Copyright (2013) American Chemical Society

enhancements, are required. This has initiated a worldwide research effort for Li-ion electrode and electrolyte materials that combine desirable properties such as high energy and power density, low cost, high abundance of component elements, and electrochemical stability [1–4]. In the current generation of Li-ion batteries, insertion materials that reversibly host Li in the crystal structure form the most important class of electrodes. Although the future of Li-ion batteries looks bright, it should be noted that the availability of a number of relevant transition metals and possibly Li itself is a topic of interest [2].

In a Li-ion battery two insertion-capable electrodes[1] with a difference in Li chemical potential (change in free energy upon Li addition) are in contact through an electrolyte (an ionic conductor and electronic insulator) and a separating membrane, see Fig. 7.1. The Li will flow from the insertion material in which Li has a high chemical-potential towards the electrode in which Li has a low chemical-potential. Only Li-ions can flow through the electrolyte and the charge compensation requires electrons to follow via the external circuit which can be used to power an application. By applying a higher electrical potential than the spontaneous equilibrium open circuit polarization the process can be reversed. High energy-density requires a large

---

[1] Research exists in conversion-alloying electrodes of Li containing compounds, but this chapter is focused on the insertion materials.

specific-capacity of ions in both electrodes and a large difference in chemical potential. High power (and fast insertion/extraction) requires both electrons and Li-ions to be highly mobile throughout the electrode materials and electrolyte.

7.1.1 Demands and Challenges

The success of Li-ion batteries is based on their high volumetric and gravimetric energy-density available for storage. This has enabled the realization of small portable devices like mobile telephones and laptops. However, important demands also include material and production costs, safety, cycle life (number of charge/ discharge cycles), and high (dis)charge rates. Addressing all these demands makes it very challenging to find better electrode and electrolyte materials to improve Li-ion batteries [2, 4]. To gain more insight into the challenges in battery research it is useful to express the performance-related demands in more specific material properties of the electrodes and electrolytes. With respect to the general demands of cost and safety we merely state that batteries require abundant and cheap materials that are intrinsically stable during battery operation. For more extensive literature we refer the reader to some excellent reviews [2, 4–6].

The gravimetric and volumetric energy density of a battery is CV_{OCP}, where C is the specific or volumetric capacity ($mAhg^{-1}$ or $mAgcm^{-3}$), respectively, and V_{OCP} is the open circuit, or equilibrium potential of the battery. Large gravimetric energy-density is required for automotive applications, and large volumetric energy densities are essential in mobile electronic equipment. It is important to realize that the specific Li capacity of the electrode materials is not necessarily the only decisive factor, and that the actual capacity of an electrode also depends significantly on the electrode morphology [7–9]. Large gravimetric energy-density requires dense electrodes and hence low porosities. In this context nano-sizing of electrode materials, aimed at higher storage capacities and higher rates, typically carries the disadvantage of low tap densities (powder packing) leading to less dense electrodes and compromising both volumetric and gravimetric energy-density.

The power density is the product of the specific current and the voltage that the battery can deliver. The battery voltage is defined by the difference in potential between the electrodes and the current via the internal resistance of the battery. High power densities, allowing for fast (dis)charge, require low internal resistance of the battery. The various charge-transport phenomena inside a battery contribute to this internal resistance, including the electronic conductivity through the electrodes, the ionic conductivity through the electrode and electrolyte, and finally the charge transfer through the interface between the electrode and electrolyte. Generally, the Li ion and electronic conductivity through the electrodes are assumed to be rate limiting in Li-ion batteries. Note that because the electrode is porous the ionic conductivity of the electrodes includes both the transport of ions through the solid-state electrode as well as through the electrolyte dispersed in the electrode pores. The solid-state transport of ions through electrodes depends on the specific electrode host material and is typically orders of magnitude slower than in liquid electrolytes. The overall Li-ion conductivity within the electrodes also depends strongly on the electrode morphology, such as characterized by the porosity [10] and the interconnectivity of the pores [11, 12].

High energy and power density are conflicting demands, as evidenced by the impact of electrode morphology and electrode thickness on these. High porosities, such as in nanostructured electrodes, generally lead to fast ionic transport throughout the electrodes responsible for high (dis)charge rates, and hence high power densities. However, the downside of large porosities is the larger volume required to store the same amount of energy, resulting in smaller volumetric energy-density. Because in most cases either electronic or ionic conductivity through the electrodes is rate limiting [7–9], thin electrodes can be charged faster than thick electrodes. Hence, the power density of a battery can be improved by the use of thinner electrodes. However, building a battery from thin electrodes leads to a smaller amount of active electrode material per gram of battery because of the relatively larger amount of current collector, electrolyte, and packing materials.

The cycle life of batteries is determined by a combination of the chemical and physical processes that occur during (dis)charging in the electrolyte and electrodes of a battery. Generally, electrodes undergo structural changes upon Li insertion and extraction. Large volumetric changes upon Li insertion and extraction lead to mechanical failure of the electrode. The result is generally that part of the active material is electronically disconnected from the electrode and therefore inactive. Consequently, small structural changes, such as occurring in $Li_4Ti_5O_{12}$ spinel (where the change in the unit-cell volume, $\Delta V_{unit\text{-}cell}$, ≈ 0.1 %) [13], and moderate structural changes, such as occurring in $LiFePO_4$ ($\Delta V_{unit\text{-}cell} \approx 6.5$ %) [14], contribute to batteries with long cycle-life. The other extreme is the alloying reaction of Li with silicon up to the $Li_{4.4}Si$ composition resulting in a volume expansion >300 %, which leads to mechanical failure of the electrodes containing relatively large silicon particles within only a few charge/discharge cycles [15]. Nevertheless, the very high capacity associated with this type of alloying reaction has motivated chemists and material scientists to develop smart strategies to maintain the mechanical coherence of these electrodes [15–17].

The second factor that is important for the cycle life, and also to safety, is the thermodynamic stability of the electrolyte with respect to the positive and the negative electrodes (Fig. 7.2). The electrodes have electrochemical potentials μ_A (anode) and μ_C (cathode) equal to their Fermi energies ε_F. If μ_A is above the lowest unoccupied molecular orbital (LUMO) of the electrolyte, the anode will reduce the electrolyte. This reduction decomposes the electrolyte unless a passivating layer, generally referred to as the solid electrolyte interface (SEI) layer, is formed. The SEI layer often forms in the first few cycles and can act to electronically insulate the anode from the electrolyte, preventing further reduction. The SEI growth tends to stop after a few cycles, resulting in stable battery performance. Similarly, if μ_C is below the highest occupied molecular orbital (HOMO) of the electrolyte, the cathode will oxidize the electrolyte unless a passivating SEI layer is formed, electronically insulating and preventing further oxidation. Therefore, not only should the

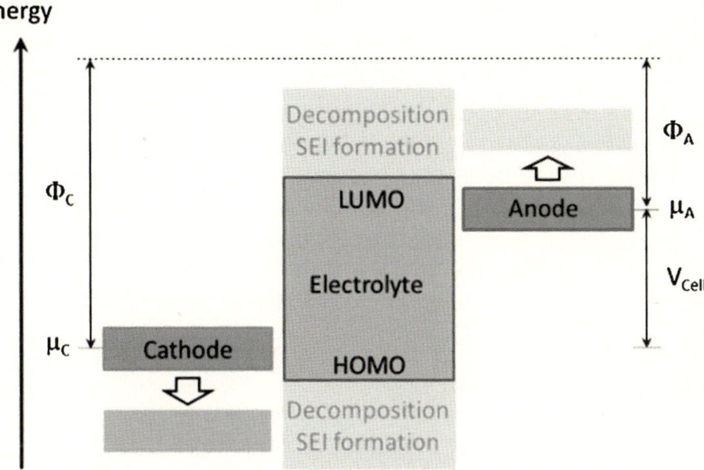

Fig. 7.2 Schematic energy diagram of an electrolyte as well as the cathode and anode work functions, Φ_C and Φ_A, respectively (equal to the electrode electrochemical potentials, the difference of which is the open-cell voltage of the battery, V_{Cell}). The difference between the LUMO and the HOMO is the stability window of the electrolyte. If the electrode electrochemical potentials fall outside this stability window the electrolyte will decompose, which SEI layer formation can passivate, leading to the kinetic stability of the electrolyte and making the light areas (*lower left* and *upper right*) accessible

electrolyte provide a wide stability window in terms of voltage, but the gap between its LUMO and HOMO must be larger than the difference between the chemical potentials of the electrodes. That is, the electrolyte voltage window should be positioned such that LUMO $> \mu_A$ and HOMO $< \mu_C$. The large open-cell voltage (V_{Cell}) of Li-ion batteries (e.g. 3.6 V) that is responsible for their high energy and power densities requires electrolyte stabilities that exceed that of water ($E_{LUMO}-E_{HOMO} \approx 1.3$ V) or water-containing electrolytes, leading to the application of non-aqueous electrolytes. Such stability demands do not apply if an electronically-insulating layer is formed upon electrolyte reduction or oxidation which can passivate further reactions. An example of passivation of SEI layers is the graphite or carbon-based anodes that work at about 0.5 V below the stability of typical carbonate electrolyte solutions, and hence 0.5 V above the LUMO of the electrolyte. The first few cycles with these anodes and carbonate-based electrolyte solutions results in the reduction of the electrolyte which leads to the formation of a stable SEI layer that prevents further reduction of the electrolyte. Importantly, the SEI layer does not grow significantly in subsequent cycles. Therefore, strategic combinations of electrodes, electrolytes, and SEI layers can result in better-performing batteries.

Improving Li-ion batteries with respect to energy and power density, manufacturing costs, safety, and cycle life is clearly a formidable challenge. A common starting point is to lower costs and use environmentally-benign materials for both the electrodes and the electrolyte, noting that the cathode can account for as much as 40 % of the cost of the battery. The electrolyte should have high Li-ion conductivity over a practical ambient-temperature range and a large stability (potential) window allowing use of large differences in electrode potentials. The electrodes should work within the stability window of the electrolyte and allow fast (dis)charge of a large reversible capacity. This not only depends on the electrode material and composition but also on the electronic and ionic 'wiring' of the electrodes determined by the electrode morphology. The development of new electrode and electrolyte materials and the improvement of existing materials requires a fundamental understanding of the Li insertion and transport mechanisms that involve both structural and kinetic phenomena. Probing these is a tremendous challenge given the complexity of the system and the difficulty of probing a light element such as Li, particularly under in situ conditions.

7.1.2 Development Using Neutron Scattering

The work of material scientists in the discovery, understanding, and development of Li-ion batteries largely depends on the techniques available to observe the relevant processes on the appropriate time and length scales. This chapter aims at demonstrating the role and use of different neutron-scattering techniques in gaining insight into Li-ion battery electrode and electrolyte function. This is not an exhaustive review of the neutron-scattering work on Li-ion battery materials or on the materials themselves, but an attempt to demonstrate the range of possibilities of neutron scattering in Li-ion battery materials research.

The role of neutron scattering in battery research is mainly based on the sensitivity of neutrons to Li compared to X-rays and electrons. The coherent neutron-scattering cross section of Li often allows the determination of the Li positions, atomic displacement parameters (ADPs), and occupancies using diffraction where X-rays do not, making it possible to elucidate Li-ion insertion/extraction mechanisms. The relatively-large coherent neutron-scattering cross section of Li provides sufficient contrast that Li distributions can be studied using imaging and, in thin-films, reflectometry. The incoherent neutron-scattering cross section of Li allows examination of Li mobility. Measuring Li diffusion directly is difficult as Li has only a moderate incoherent neutron-scattering cross section, so each material has to be considered individually to determine if the neutron-scattering signal originates from Li. Often measuring host material dynamics, e.g. anion and hydrogen group dynamics or lattice vibrations using quasi-elastic neutron scattering (QENS) and inelastic neutron scattering (INS), can provide information on Li dynamics. The diffusion pathway can be additionally corroborated or even independently obtained using the abovementioned large coherent cross-section of Li by considering the anisotropic contribution in the displacement parameter because of the deviation from harmonicity due to thermal motion at elevated temperatures. The relatively large neutron absorption cross-section of Li enables neutron depth profiling to determine Li distributions in materials. In these neutron-based techniques used to study Li-ion battery materials, the effective cross-sections of samples under study can be tuned through the isotopic composition. Naturally occurring Li is composed of 7.5 % ^{6}Li and 92.5 % ^{7}Li. The larger coherent neutron-scattering cross section and lower incoherent neutron-scattering and absorption cross-section of ^{7}Li make it possible to improve data quality by tuning compositionally (isotopically) samples according to the neutron investigation technique being used.

7.2 Electrodes

Since the commercialization of the Li-ion battery by SONY Corporation in 1991, research has focused on identifying better electrode and electrolyte materials. SONY combined $LiCoO_2$ as a positive electrode material with a carbonaceous material as a negative electrode, and $LiPF_6$ in a carbonate solution as the electrolyte. Initially, research focused on replacing the relatively expensive $LiCoO_2$ with other transition-metal oxides where the most important structural groups include spinel and layered transition-metal oxides. Layered Li transition-metal oxides, $LiMO_2$ with M = Mn, Co, and Ni, represent one of the most successful classes of positive-electrode materials. The layered topology offers easily-accessible two-dimensional ion diffusion pathways. In particular, the $LiCo_{1/3}Ni_{1/3}Mn_{1/3}O_2$ composition [18] results in high capacity, safety, and lower material costs than $LiCoO_2$.

An important alternative to $LiCoO_2$ is the spinel $LiMn_2O_4$ with the inexpensive and environmentally-benign Mn, which functions with the $Mn^{3+/4+}$ redox couple. $LiMn_2O_4$ operates at 4.1 V versus Li/Li^+ offering excellent safety and high power due to its three-dimensional lattice, in principle allowing three-dimensional Li-ion diffusion. Substitution of Mn by M = Co, Cr, Cu, Fe, and Ni, has led to the discovery of the high-voltage spinels $LiM_{0.5}Mn_{1.5}O_4$ and $LiMMnO_4$ with potentials between 4.5 and 5 V versus Li/Li^+ [19], exemplified by spinel $LiNi_{0.5}Mn_{1.5}O_4$ operating at 4.7 V versus Li/Li^+ [20, 21]. However, Mn-based spinels have been plagued by capacity fade, generally considered to be the result of Mn dissolution into the electrolyte and Jahn-Teller distortion of Mn^{3+}. Mn dissolution has been largely inhibited by substitution of dopants in the spinel structure [22].

The introduction of $LiFePO_4$ [14] has initiated research on polyanion-based positive electrodes with the structural formula $LiM(XO)_4$ (M = Fe, Mn, Co, and X = S, P, Si). The strong covalent X-O bonds result in the large polarization of oxygen ions towards the X cation, leading to larger potentials compared to oxides. Phosphates, in particular $LiFePO_4$, have been extensively explored because of their favourable electrode properties, reasonably high potential and capacity, stability against overcharge or discharge, and their composition of abundant, cheap, and non-toxic elements. To date, almost a thousand papers have been devoted to the understanding and improvement of conductivity in the semiconductor $LiFePO_4$. These have encompassed studies of surface-conductive phases [23, 24], modification of crystallite size [25], and elegant fundamental mechanistic and modelling studies [26–28]. Compared to the phosphates, the silicates ($LiMSiO_4$) exhibit lower electrode-potentials and electronic conductivity [29]. Other promising groups of positive-electrode materials include fluorophosphates with the A_2MPO_4F (A = Alkali metal) stoichiometry [29].

For many years graphite has been the main negative-electrode material allowing Li-ion intercalation between graphene sheets at ~ 0.3 V versus Li/Li$^+$. The low potential results in a high battery-voltage, partly responsible for the high energy and power density of Li-ion batteries. The main disadvantage of the low voltage of a negative electrode is the restricted stability of the commercial carbonate-based electrolyte solutions at low potentials versus Li/Li$^+$. Depending on the electrolyte-electrode combination a kinetically stable SEI layer can be formed. For graphite this was achieved by the addition of ethylene carbonate [30] which lead to the success of the LiCoO$_2$–C battery. Efforts to develop low-voltage negative electrodes with capacities that exceed that of C have concentrated on reaction types other than insertion reactions. Metals such as Si and Sn form alloys with Li in so-called conversion (or alloying) reactions [31–34]. Although the (gravimetric) capacities of these reactions can reach almost 10 times that of graphite, the volumetric energy-density is not significantly improved. Moreover, the large volume changes inherently related to this type of reaction make it a challenge to reach good cycleability. The oxide insertion hosts offer much better stability, highlighted by the Li$_4$Ti$_5$O$_{12}$ spinel [35] that has almost no volume change upon Li insertion. Other titanium oxides that attract considerable attention as negative electrodes are anatase, brookite, and rutile TiO$_2$, particularly in the nanostructured form [36, 37] and in combination with carbonaceous materials providing good with electronic-conductivity [16, 36]. The disadvantage of titanium oxides is their relatively high negative-electrode potentials, reducing the overall working voltage of the battery and thereby reducing both energy and power density. For example, titanium oxides operate at ~ 1.6 V versus Li/Li$^+$, over four times higher in voltage than graphite. In this context, an interesting oxide exceeding the graphite capacity at similarly low potentials is the layered transition-metal oxide Li$_{1+x}$V$_{1-x}$O$_2$ [38].

One of the key strategies that improved electrode performance in general is the nano-sizing of the electrode crystal particles. The nano-sized electrode particles reduce both solid state (Li-ion) diffusion and electron conduction. The latter is achieved by mixing and coating with conducting phases (i.e. with carbonaceous materials). However, numerous recent observations indicate that nano-sizing electrode particles also has a large impact on electrode materials properties [39, 40] creating both opportunities and challenges for enhanced Li-ion storage. Changes in properties that are observed upon nano-sizing include smearing of the voltage profile [41–43], changing solubility limits and phase behaviour [41, 44–46], unexpected kinetics [47], and larger capacities [45, 48–51]. The downside of the large surface-area of nanostructured materials is the relative instability of nano-materials, which can promote electrode dissolution and the increased reactivity towards electrolytes at commonly used voltages, e.g. below 1 V versus Li/Li$^+$, which may adversely affect performance. Another potential disadvantage is the lower packing density of electrodes, leading to lower volumetric energy-densities.

Among the materials that benefit from the possibilities of nano-sizing are the relatively stable transition-metal oxides and phosphates operating well within the stability window of the electrolyte.

7.2.1 Crystal Structure

In Li-ion battery electrode research, neutron powder diffraction (NPD) is one of the first techniques used for structural characterization of synthesized materials, often in combination with X-ray diffraction. NPD also plays an essential role in understanding the insertion mechanisms that may induce phase transitions and/or solid-solution behaviour, both of which may depend strongly on temperature, particle size, doping, and the chemical or electrochemical conditions.

The coherent neutron-scattering cross-sections of Li is relatively greater than that obtained for many transition-elements and greater than the analogous coherent scattering cross-sections of Li for X-ray or electron scattering, making it possible to determine Li-ion positions and occupancies. Thanks to these and other advantages of NPD, the technique has played a key role in the development of the large diversity of Li-ion battery electrode materials that exist to date.

Knowledge of the Li-induced phase transitions in C is primarily based on X-ray diffraction [52]. Although the crystal stages I and II in C are proven, long standing debate exists concerning LiC_{18} and no evidence of the higher-order lithiated stages exists based on X-ray diffraction studies. Neutron diffraction proved the existence of the LiC_{18} stage [53] and showed deviations from the common structural picture of lithiated C, including a charge-discharge dependent structural evolution and the appearance of higher-ordered stages [54].

Neutron diffraction has been used to determine the lithiated structures of titanium oxides including Li_xTiO_2 anatase [55–57], rutile [58], brookite [59], and $Li_{1+x}Ti_2O_4$ and $Li_{4+x}Ti_5O_{12}$ spinel [55]. $Li_4Ti_5O_{12}$ spinel is a state-of-the-art Li-ion battery electrode material [35, 60–63] operating at 1.56 V versus Li and is suitable as an anode using high-voltage cathode materials. $Li_4Ti_5O_{12}$ can be lithiated up to the composition $Li_7Ti_5O_{12}$ and possibly higher, with end members having a spinel structure adopting the cubic space group *Fd-3m*. This material is attractive as a Li-ion insertion electrode because the 'zero strain' property results in excellent cycle life: upon lithiation from the initial $Li_4Ti_5O_{12}$ to the 'fully-lithiated' $Li_7Ti_5O_{12}$ there is almost no change in lattice parameters (0.2 %) [35, 60–65]. In the defect spinel $Li_4Ti_5O_{12}$ all the energetically-favourable tetrahedral 8*a* sites are occupied by Li. Additionally, 1/6 of the 16*d* sites are also randomly occupied by Li while the remaining 5/6 of the 16*d* sites are occupied by Ti atoms, and this can be represented as $[Li_3]^{8a}[Li_1Ti_5]^{16d}[O_{12}]^{32e}$. Lithiation leads to occupation of all the octahedral 16*c* sites and emptying of the tetrahedral 8*a* sites to reach the lithiated $Li_7Ti_5O_{12}$ composition, which can be represented as $[Li_6]^{16c}[Li_1Ti_5]^{16d}[O_{12}]^{32e}$.

In lithiated anatase, neutron diffraction data obtained to relatively high momentum transfers were used to resolve the split Li-ion positions within the material's distorted oxygen octahedra [56]. In composite anatase $TiO_2/Li_4Ti_5O_{12}$, neutron diffraction showed that the TiO_2 phase was lithiated before $Li_4Ti_5O_{12}$, as expected from the lower potential of $Li_4Ti_5O_{12}$ versus Li/Li^+, enabling tuned Li insertion/extraction based on the choice of voltage range [66]. The increase in curvature of the voltage profile and larger capacities for nano-sized materials appears to be a general observation for the various TiO_2 polymorphs such as the tetragonal anatase [42, 45, 48, 58, 67] and rutile [48], orthorhombic brookite [68] and monoclinic $TiO_2(B)$ [50, 69]. The sensitivity of neutron diffraction for Li-ions has been decisive in revealing the altered thermodynamics of nano-sized titanium oxides. In anatase the Li solubility increases systematically when particle sizes are reduced leading to a phase-size diagram [45]. In addition, a second phase transition from the known Li-titanate phase towards tetragonal $LiTiO_2$ was discovered, which was later confirmed by neutron diffraction in anatase nano-tubes [70]. Higher Li-ion solubility was also observed in nano-structured rutile using neutron diffraction [71], suggesting similar size effects. A remarkable finding in spinel nano-sized $Li_4Ti_5O_{12}$ is that of increased capacity with decreasing particle size, exceeding the maximum composition observed for the micron-sized $Li_7Ti_5O_{12}$. Neutron diffraction proved the increase in capacity to be due to simultaneous Li occupation of both $8a$ and $16c$ sites, providing an atomic-scale explanation for the larger capacity of the nano-sized materials [43].

A very interesting negative electrode, exceeding the graphitic anode capacity, at similarly low potentials, is the layered transition-metal oxide $Li_{1+x}V_{1-x}O_2$ [72]. Interestingly, $LiVO_2$ does not allow Li insertion whereas $Li_{1+x}V_{1-x}O_2$ with $x > 0$ leads to a very low intercalation voltage close to 0.1 V with a capacity almost twice that of graphite. Neutron diffraction has shown that in $Li_{1.07}V_{0.93}O_2$ part of the octahedral V is replaced by Li, which allows additional Li-ions, responsible for the large capacity, to occupy the neighbouring tetrahedral sites that are energetically unfavourable in $LiVO_2$, and is supported by modelling studies [38].

Gummow et al. used neutron diffraction to show that the structure of the low-temperature cubic phase of $LiCoO_2$ is not ideally layered, and that 6 % of the Co reside in the octahedral $(8a)$ sites of the Li layers [73, 74]. The hexagonal structure of the high-temperature phase of $LiCoO_2$ was also determined from neutron diffraction, illustrating that Co and Li planes alternate in the ABCABC oxygen stacking. Aiming at higher capacity cathodes, $Li_{1+x}CoO_2$ has been prepared raising the question of where the additional Li resides [75]. Combined Rietveld refinement using both X-ray and neutron diffraction data excluded both Co in the Li site and the presence of tetrahedral Li and Co [76]. Based on this, it was deduced that excess Li replaces some Co and that the charge is compensated for by O vacancies [77]. In mixed-cation layered transition-metal oxides, such as the so called 'high capacity' $LiMn_{1/3}Co_{1/3}Ni_{1/3}O_2$, neutron diffraction continues to be an indispensable tool for determining the cation distributions which have been shown to depend on the synthesis conditions [78–80].

Neutron diffraction has played a pivotal role in understanding the complex insertion and phase transitions in spinel transition-metal oxides. Using NPD, Fong et al. [81] described the crystal structure of $Li_xMn_2O_4$ for $x = 1$ and 0.2 and Wills et al. [82] determined the crystal structure of $LiMn_2O_4$ at low temperatures as well as its magnetic properties. Neutron diffraction revealed the partial charge-ordering in spinel $LiMn_2O_4$ at 290 K that further hinders its use as a positive-electrode material in Li-ion batteries [83]. From neutron diffraction data, superstructure reflections were found (related to charge-ordering phenomena) at 230 K which in combination with electron diffraction patterns revealed a $3a \times 3a \times a$ super cell of the cubic room-temperature spinel representing the columnar ordering of electrons and holes [83]. Two of the five Mn sites correspond to well-defined Mn^{4+} and the other three sites are close to Mn^{3+} as derived from Mn–O bond length analysis. This charge ordering is accompanied by simultaneous orbital ordering due to the Jahn-Teller effect in the Mn^{3+} ions. Li-excess compounds $Li_{1+x}Mn_{2-x}O_4$ were found to provide better cycling performance than the stoichiometric $LiMn_2O_4$ as they minimize the extent of the Jahn-Teller distortion during cycling (i.e. increase the overall oxidation state of Mn during cycling). In addition, Li doping at octahedral $16c$ sites reduces the exothermicity of the Li insertion/extraction reactions by an amount similar to that associated with the dilution of the Mn^{3+} ion [84]. Neutron diffraction by Berg et al. [85] showed that Li occupies $16c$ sites in $Li_{1.14}Mn_{1.86}O_4$ which is also accompanied by charge-compensating vacancies at Mn $16d$ sites. Calculations also showed that the $16d$ sites should be favourable for Li at low Li contents while at higher contents, the $16c$ and mixed $16c$ and $16d$ site occupation is likely [86]. However, a recent study by Reddy et al. [87, 88] shows that at lower Li doping regimes, $x = 0.03$ and 0.06, the structural model containing Li at $16c$ sites still results in a better fit to the neutron diffraction data than models with $16d$ site Li occupation. In the work by Yonemura et al. [89] samples were synthesized in controlled atmospheres which led to the realization and quantification of O-deficient $LiMn_2O_4$ and Li-excess (O-deficient) $Li_{1+x}Mn_{2-x}O_{4-y}$ spinels. Using neutron diffraction data they determined the quantity of O, mixing of Li and Mn at the $8a$ and $16c$ sites, the interatomic bond distances, and the relationship between these crystallographic parameters.

In the high-voltage spinels, neutron diffraction enabled the transition-metal ordering in $LiNi_{0.5}Mn_{1.5}O_4$ that breaks the cubic Fd-$3m$ to cubic $P4_332$ symmetry to be determined and this was found to be dependent on the cooling rates used in the synthesis [90]. In X-ray diffraction data the small difference in atomic number between Mn and Ni makes it hard to quantify this ordering, whereas it is easily modelled using neutron diffraction [91, 92]. In the low-voltage plateau, using the large difference in the coherent neutron-scattering cross section of Mn and Ni, researchers determined that extensive migration of Ni and Mn was occurring in the spinel structure due to the loss of long range Ni-Mn ordering [92].

In the polyanion-based positive-electrode materials, neutron diffraction data contributes significantly in the characterization and understanding of the electrode properties. The higher potential of the Mn^{3+}/Mn^{2+} redox couple has initiated the synthesis of $LiMnPO_4$ and $LiMn_yFe_{1-y}PO_4$ materials. Neutron diffraction data helped reveal that the reduced activity of the Mn^{3+}/Mn^{2+} couple is related to the distortion of the MO_6 octahedra with $M = Mn^{3+}$, and this distortion was found to be much larger than the change in the unit cell [93], effectively prohibiting the Mn^{3+} to Mn^{2+} transition. Neutron diffraction was also used to characterize the cation distribution in related olivine structures with other transition metals and transition-metal mixtures such as $LiCo_yFe_{1-y}PO_4$ [94], Li_xCoPO_4 [95], and V-substituted $LiFePO_4$ [96, 97].

For tavorite, $LiFePO_4(OH)$, neutron diffraction showed both the Li and H to be located in two different tunnels running along the a and c-axes, the tunnels being formed by the framework of interconnected PO_4 tetrahedra [98]. Another promising class of tavorite-structured cathode materials are the fluorophosphates which exhibit good storage capacity and electrochemical and thermal stability. $LiFePO_4F$ exhibits a complex single-phase regime followed by a two-phase plateau at 2.75 V. Neutron diffraction in combination with X-ray diffraction was used to resolve the single phase end-member Li_2FePO_4F structure showing that Li-ions occupy multiple sites in the tavorite lattice [99]. Additionally, in the pyrophosphate-based positive electrode $Li_{2-x}MP_2O_7$ (M = Fe, Co), multiple Li sites were identified using neutron diffraction [100].

In general, the combination of X-ray and neutron diffraction has become the established approach to characterizing electrode materials in great detail. The following example concerning the extensively-studied olivine $LiFePO_4$ demonstrates the value of neutron diffraction in revealing the impact of dopants, defects, and particle size on $LiFePO_4$ structure and performance, thereby providing crucial understanding for the design of future electrode materials.

In the last decade $LiFePO_4$ has emerged as one of the most important positive electrodes for high-power applications owing to its non-toxicity and outstanding thermal and electrochemical stability [14]. The first-order phase transition, preserving its orthorhombic $Pnma$ symmetry, results in highly-reversible cycling at the 3.4–3.5 V versus Li/Li^+ voltage plateau with a theoretical capacity of 170 $mAhg^{-1}$. The olivine structure is built of $[PO_4]^{3-}$ tetrahedra with the divalent M ions occupying corner-shared octahedral "$M2$" sites, and the Li occupying the "$M1$" sites to form chains of edge-sharing octahedra. The magnetic structure of $LiFePO_4$ has been solved using neutron diffraction, with the appearance of extra reflections below the Néel temperature indicating antiferromagnetic behaviour at low temperatures for both end-members $FePO_4$ (Fe^{3+}) and $LiFePO_4$ (Fe^{2+}) [101].

In contrast to the well-documented two-phase nature of this system at room temperature, Delacourt et al. [102, 103] gave the first experimental evidence of a solid solution Li_xFePO_4 ($0 \leq x \leq 1$) at 450 °C, and in addition, the existence of two new metastable phases with compositions $Li_{0.75}FePO_4$ and $Li_{0.5}FePO_4$. These metastable phases pass through another metastable phase on cooling to room temperature where approximately 2 out of 3 Li-positions are occupied, again determined using neutron diffraction to be $Li_{\sim 0.67}FePO_4$ [103]. In $Li_{\sim 0.67}FePO_4$, the average Li–O bonds are longer than in $LiFePO_4$ due to the shortening of Fe–O bond lengths as shown in Fig. 7.3. It was suggested that this bond-length variation is the origin of the metastability of the intermediate phase, and thus of the two-phase mechanism between $LiFePO_4$ and $FePO_4$. Interestingly, this metastable phase appears to play a vital role in the high charge/discharge rate of the olivine material [104].

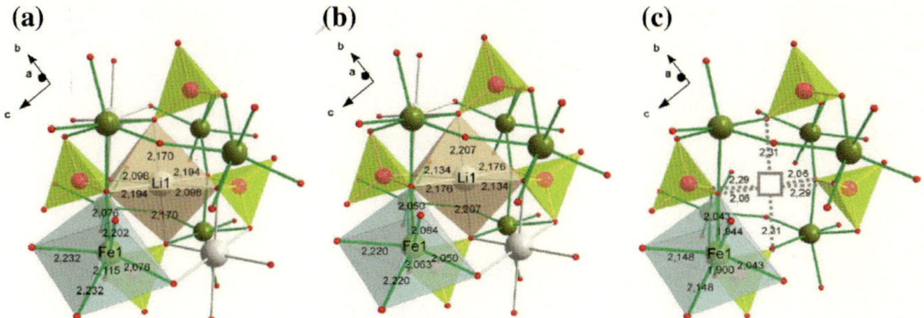

Fig. 7.3 View of the FeO_6 and LiO_6 octahedra for **a** $LiFePO_4$, **b** $Li_{\sim 0.67}FePO_4$, and **c** $FePO_4$, displaying the contraction of average Fe–O bond lengths from $LiFePO_4$ to $FePO_4$, together with the slight expansion of the $M1$ size (related to the average Li–O bond lengths in Li-containing phases). The models are based on combined structural refinements using neutron and X-ray diffraction data. Reprinted with permission from (C. Delacourt, J. Rodriguez-Carvajal, B. Schmitt, J.M. Tarascon, C. Masquelier, Solid State Sci. **7**, 1506 (2005)) [103]. Copyright (2005) Elsevier

The room-temperature miscibility gap in Li_xFePO_4 was determined by Yamada et al. [105] using NPD. These researchers found intermediate Li-poor $Li_{\alpha=0.05}FePO_4$ and Li-rich $Li_{1-\beta=0.89}$ phases, as shown in Fig. 7.4. This explains the compositional range over which the voltage is constant (plateau) and proves the presence of mixed-valence states of iron (Fe^{2+}/Fe^{3+}). These mixed-valence states provide ionic and electronic conductivity, an essential ingredient for the material's application as a Li-ion battery electrode.

An early report that led to intensive discussions suggested that the poor electronic conductivity of $LiFePO_4$ could be raised by 8 orders of magnitude by supervalent-cation doping, which was proposed to stabilize the minority Fe^{3+} hole carriers in the lattice [106]. It was only after detailed refinement of models against combined neutron and X-ray diffraction data that researchers were able to determine

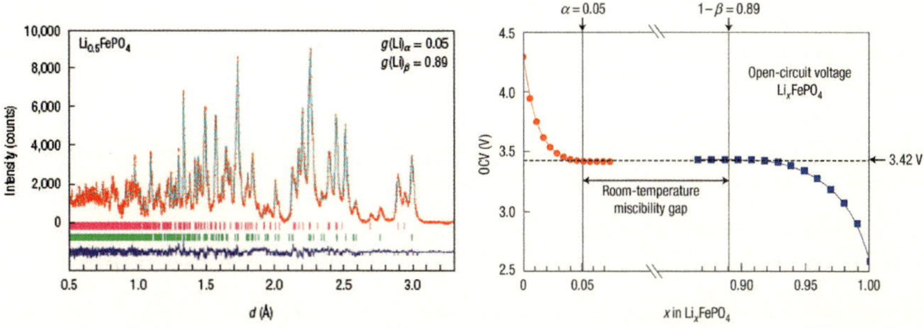

Fig. 7.4 *Left* Refinement using neutron diffraction data of $Li_{0.5}FePO_4$ resulting in solubility limits $\alpha = 0.05$ and $1-\beta = 0.89$ in the Li-poor triphylite and Li-rich heterosite phases, respectively. *Right* Open circuit voltage versus composition, where the vertical lines indicate the monophase/biphase boundary as determined from the Li site occupancies resulting from Rietveld refinement using neutron diffraction data. Reprinted by permission from Macmillan Publishers Ltd: (A. Yamada, H. Koizumi, S.I. Nishimura, N. Sonoyama, R. Kanno, M. Yonemura, T. Nakamura, Y. Kobayashi, Nat. Mater. **5**, 357 (2006)) [105]. Copyright (2006)

Fig. 7.5 a LiFePO$_4$ adopting *Pnma* symmetry with the split Li-ion (*medium grey*) position in the centre. **b** NPD data for LiFePO$_4$ and Li$_{0.96}$Zr$_{0.04}$FePO$_4$ (target composition) including the difference between the fits and data. The fit residuals are wR$_p$ = 1.7 % and R$_p$ = 1.9 %, as well as wR$_p$ = 1.7 % and R$_p$ = 1.8 %, respectively. **c** The same data as in (**b**) shown for a limited d-spacing range. Reprinted with permission from (M. Wagemaker, B.L. Ellis, D. Luetzenkirchen-Hecht, F.M. Mulder, L.F. Nazar, Chem. Mater. **20**, 6313 (2008)) [107]. Copyright (2008) American Chemical Society

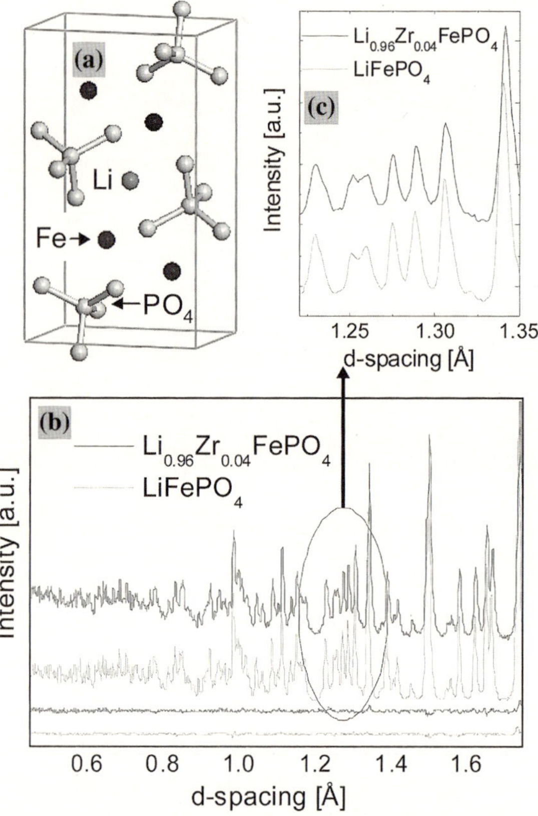

the positions and role of the dopants. In this case NPD provides contrast between Li and the dopants at the Li site (*M*1), and X-ray powder diffraction provides contrast between the Fe and many of the dopants at the Fe site (*M*2). Moreover, to determine three species (Li or Fe, dopants and vacancies) on crystallographic sites (*M*1 or *M*2) requires more than X-ray or neutron diffraction alone. The neutron diffraction pattern for one of the doped materials is shown in Fig. 7.5. Although the changes in the diffraction pattern upon doping are extremely small, the accuracy afforded by the data make it possible to conclusively locate the supervalent-cation dopants in LiFePO$_4$. Figure 7.6 shows that supervalent-cation doping of up to $\sim$3 % atomic substitution can be achieved in the LiFePO$_4$ lattice in bulk materials prepared by a solid-state route at 600 °C. The results show that the dopant resides primarily on the *M*1 (Li) site and that aliovalent-dopant charge is balanced by Li vacancies, with the total charge on the Fe site being +2.000 ($\pm$ 0.006), within the limit of experimental error [107]. It is thus expected that dopants may have little influence on the electronic conductivity of the material, which is confirmed by calculations [108]. Furthermore, the location of the immobile high-valent dopant within the Li channels is expected to hinder Li-ion diffusion assuming one-dimensional diffusion.

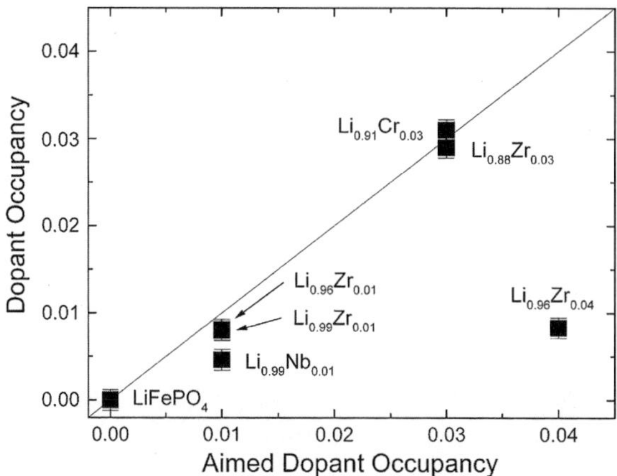

Fig. 7.6 Supervalent doping occupancies from refinements using combined X-ray and neutron diffraction data plotted versus the targeted dopant concentration. Reprinted with permission from (M. Wagemaker, B.L. Ellis, D. Luetzenkirchen-Hecht, F.M. Mulder, L.F. Nazar, Chem. Mater. **20**, 6313 (2008)) [107], Copyright (2008) American Chemical Society

The one-dimensional migration channels through the $LiFePO_4$ olivine structure means that the electrode performance can be severely influenced by defects. In the olivine structure the most favourable defect is predicted to be the Li–Fe anti-site pair, in which a Li-ion (at the $M1$ site) and a Fe ion (at the $M2$ site) are interchanged. These defects, with concentrations up to 8 %, were first observed in $LiFePO_4$ synthesized at low temperatures, leading to non-thermodynamically favoured materials [109]. Small concentrations of anti-site defects, as high as 1 %, were suggested to remain even up to solid-state synthesis temperatures as high as 600 °C [110]. In addition, combined neutron and X-ray diffraction has indicated that after fast hydrothermal synthesis crystalline-defective $Li_xFe_yPO_4$ coexists with amorphous Li/Fe-PO_4 structures. These techniques also showed that the Fe is included in the structure more rapidly from the amorphous precursor than Li, causing defects in the structure [111]. Anti-site defects are expected to play a decisive role in the Li-ion conductivity and Gibot et al. [112], using combined neutron and X-ray diffraction data, demonstrated that large concentrations (up to 20 %) of these anti-site defects in nanoparticles suppress the first-order phase transition normally observed in $LiFePO_4$ leading to a single-phase room temperature reaction upon (de)lithiation. More detailed insight into the correlation between particle size and Li-ion substoichiometry was obtained by the direct synthesis of substoichiometric $Li_{1-y}FePO_4$ nanoparticles [113]. Combined neutron and X-ray diffraction data of partially-delithiated substoichiometric olivines revealed segregated defect-free (where Li is extracted) and defect-ridden (where Li remains) regions, as shown in Fig. 7.7. This proved that both the anti-site defects obstruct Li^+ diffusion, explaining the detrimental electrochemistry and that the anti-site defects form clusters.

Further details of the anti-site clustering in $LiFePO_4$ were obtained using a combination of neutron diffraction with high-angle annular dark-field scanning

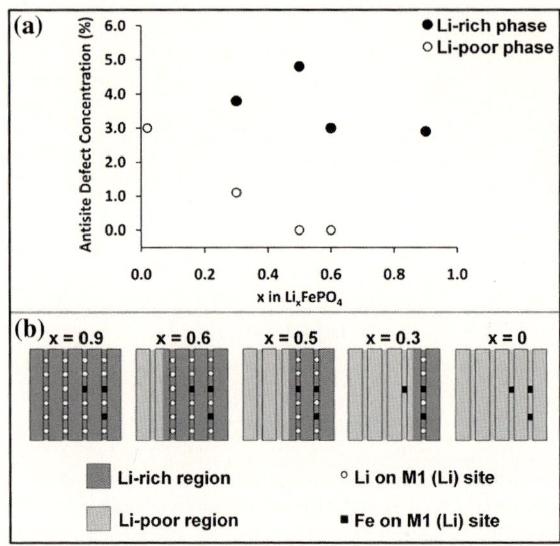

Fig. 7.7 Interpretation of the combined neutron and X-ray diffraction results for delithiation of Li$_{0.90}$FePO$_4$: Composition dependence and site disorder. **a** Evolution of the site-defect concentration in the Li-rich and Li-poor phases as a function of delithiation. **b** Overall schematic illustration of the phase segregation of the Li-rich and Li-poor regions of the crystallites with regions free of Fe anti-site defects delithiating before regions containing $M1$ site defects. Reproduced from (S.-P. Badi, M. Wagemaker, B.L. Ellis, D.P. Singh, W.J.H. Borghols, W.H. Kan, D.H. Ryan, F.M. Mulder, L.F. Nazar, J. Mater. Chem. **21**, 10085 (2011)) [113] with permission from The Royal Society of Chemistry

transmission electron microscopy and ab initio calculations, indicating that they form zig-zag type clusters, completely different from the structurally equivalent LiMnPO$_4$ where the anti-site defects appear to be randomly distributed [114].

Another topic that has been of great interest is the impact of the particle size on the intercalation properties. When insertion electrode materials are downsized to nanometer dimensions, voltage profiles change considerably reflecting a change in thermodynamics [37, 39]. First direct evidence of modified electrochemical-structural behaviour in nano-sized insertion electrodes was provided by neutron diffraction on TiO$_2$ anatase, which showed large changes in Li solubility in phases and a strongly-altered phase composition and morphology [45]. Also, the solubility limits during the insertion reaction in LiFePO$_4$ have been under active research, mainly using neutron diffraction as a direct probe [41, 44, 102, 115–120]. This research shows narrow solid-solution domains in micron size particles at room temperature [117] and a solid solution over the entire compositional range above 520 K [102, 121]. Yamada et al. [117] suggested that the extended solid-solution composition-ranges in small particles and a systematic decrease of the miscibility gap was due to strain based on Vegard's law [41]. Kobayashi et al. [44] isolated solid-solution phases, also supporting a size-dependent miscibility gap. Direct evidence of enhanced solubility in the end phases with decreasing primary crystallite-size was provided by a systematic

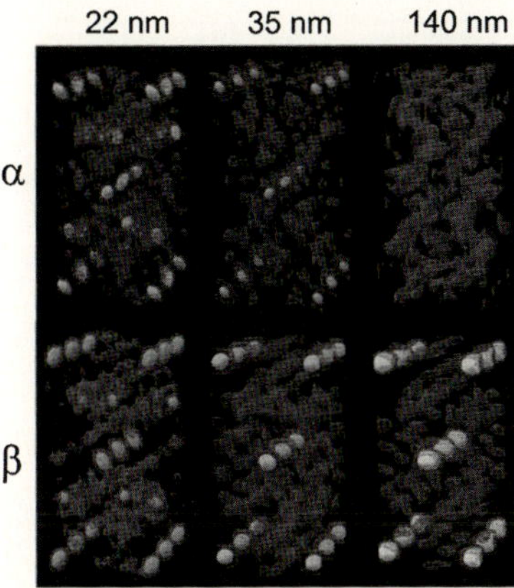

Fig. 7.8 The structural impact of nano-sizing illustrated by Fourier-density difference maps obtained from neutron diffraction. The maps are shown for both the Li-poor α-phase and the Li-rich β-phase in $Li_{0.5}FePO_4$ for the three different particle sizes indicated. The maps were obtained by the Fourier transform of the difference between the neutron diffraction data and the calculated diffraction pattern based on the structure with no Li present. Therefore, these density maps should show Li density. As expected for large particles, large Li density is observed in the Li-rich heterosite β-phase, and no density is observed in the Li-poor triphylite α-phase. Progressive particle-size reduction decreases observed Li density in the heterosite β and more evidently Li density increases in the triphylite α phase, indicating a reduction of the miscibility gap with decreasing particle size. Reprinted from (M. Wagemaker, D.P. Singh, W.J.H. Borghols, U. Lafont, L. Haverkate, V.K. Peterson, F.M. Mulder, J. Am. Chem. Soc. **133**, 10222 (2011)) [46]

neutron diffraction study of particle sizes between 22 and 130 nm [46]. The Fourier-density difference maps in Fig. 7.8 illustrate that the Li densities in the Li-poor and Li-rich phases increase and decrease respectively, with decreasing particle size. These observations could be reproduced by calculations based on a diffuse interface model [46, 122]. The diffuse interface introduces an energy penalty for a Li concentration-gradient creating a smoothly-varying Li concentration over an interface region with a width of ~ 10 nm, as shown in Fig. 7.9. The confinement of this interface layer in nano-sized particles moves the observed solubility away from the bulk values. Interestingly, neutron diffraction also proved that the solubility in both phases ($LiFePO_4$ and $FePO_4$) depends on the overall composition, especially in crystallites smaller than 35 nm. Furthermore, this observation could be explained quantitatively by the diffuse-interface model. By varying the overall composition the domain sizes of the coexisting phases change, in this case leading to confinement effects in the minority phase.

The ex situ neutron diffraction studies discussed above have contributed to our current state of understanding of electrode materials. This is in particular based on the sensitivity of neutrons for Li, the charge-carrying element in Li-ion battery electrodes. This is vital knowledge not only for the synthesis of new materials, but also for mechanistic understanding of the impact of supervalent doping, defects, composition, and particle size on the intercalation process as illustrated for olivine $LiFePO_4$.

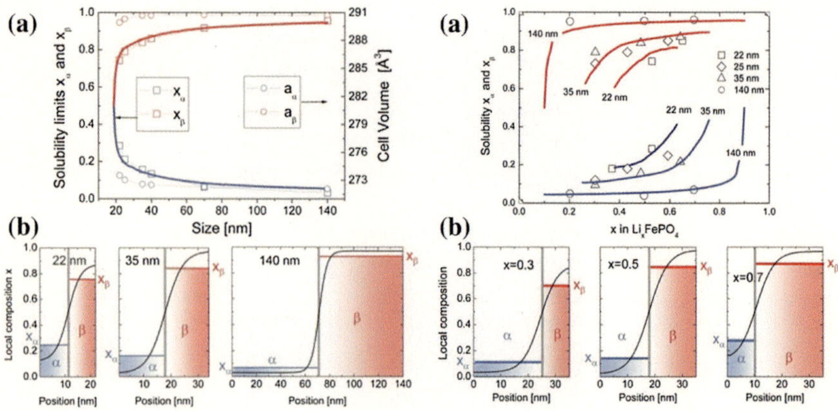

Fig. 7.9 Measured and calculated solubility limits as a function of particle size and overall composition. *Left* **a** *Symbols* Li occupancy for both the Li-poor triphylite α-phase $Li_{x_α}FePO_4$ and the Li-rich heterosite β-phase $Li_{x_β}FePO_4$ where $x_α$ and $x_β$ represent the average solubility limits as a function of particle size, having an overall composition $Li_{0.5}FePO_4$. $V_α$ and $V_β$ represent the corresponding unit-cell volumes. The size of the symbols is approximately the size of the error. *Lines* Calculated average compositions based on the diffuse interface model. **b** Calculated concentration profiles based on the diffuse interface model in the *a*-lattice direction for three different particle sizes at the overall composition $Li_{0.5}FePO_4$. *Right* Measured and calculated solubility limits as a function of overall composition. **a** *Symbols* Li occupancy derived from neutron diffraction data for both the Li-poor triphylite α-phase and the Li-rich heterosite β-phase representing the average solubility limits as a function of overall composition for different particle sizes. *Lines* Calculated average compositions based on the diffuse interface model. The size of the symbols is approximately the size of the error. **b** Calculated concentration profiles based on the diffuse interface model in the *a*-lattice direction for three different overall compositions all having the particle size 35 nm. Reprinted from (M. Wagemaker, D.P. Singh, W.J.H. Borghols, U. Lafont, L. Haverkate, V.K. Peterson, F.M. Mulder, J. Am. Chem. Soc. **133**, 10222 (2011)) [46]

7.2.2 Local Structure

To obtain the local structure in glassy, nano, disordered and amorphous materials, having unresolved, weak or broad signals, neutron total scattering, or/and inelastic neutron scattering (INS) or quasielastic neutron scattering (QENS) can be powerful tools. Total scattering allows extraction of the local structure in terms of interatomic distances, bond angles and coordination numbers. In this case scattering is detected over a wider Q-range and short-range interactions of a sample are probed and modelled. In addition, rotations and vibrations picked up by INS and QENS are very sensitive to local distortions and allow otherwise difficult to detect relevant species such as protons, OH, water and in rare cases Li to be studied.

The negative-anode carbon is a good example of where neutron total scattering, in conjunction with other neutron-based methods, has been able to quantify important, previously ill-defined, aspects of the material's function [20, 123, 124], as demonstrated with a range of low-crystallinity C negative electrodes. Additionally, C-based anodes can be analysed in the lithiated and delithiated states and over the course of phase transitions. Typical neutron total-scattering data for graphite is presented as a radial distribution function (or pair distribution function) as illustrated in Fig. 7.10, and peak positions are indicative of interatomic distances. Often, neutron total scattering is combined with INS data to provide supporting information concerning the short-range order in C. For further information on total scattering the reader is directed to a review on the structure and dynamics of ionic liquids [125], and total scattering is likely to become increasingly used as the range of nano-sized active electrode materials increase.

Total scattering and INS are particularly attractive for disordered C where conventional diffraction provides limited information and more generally for Li arrangements in C. Disordered C where a large amount of H is present can exhibit significant Li capacity (one 'excess' Li per H) and studies have investigated how Li is taken into these materials [123, 124, 126]. Studies have shown that these materials exhibit randomly-arranged graphene fragments of different sizes with edges terminated by a single H, similar to Si with H at the surface. The spectra also contain a boson peak, an indicator of disorder, and distinct similarities to polycyclic aromatic hydrocarbon (PAH) spectra exist, some of which feature two or three edge-terminating H. Additionally, comparison with PAH spectra allowed the determination of methyl groupings when higher H concentrations are used. The boson peak is at the same position in samples with different concentrations of H and changes in position and intensity with Li insertion. This shows that the Li interacts with the C environment, contrary to the idea of Li accumulation in voids. These findings agree with two models of Li insertion: One where Li resides on both sides of the graphene layers (the so-called 'house-of-cards' model) and the other where Li is bonded to the H-terminated C at the edge of the graphene layers (and reside in interstitial sites). INS data also illustrate that the Li–Li interlayer and intralayer

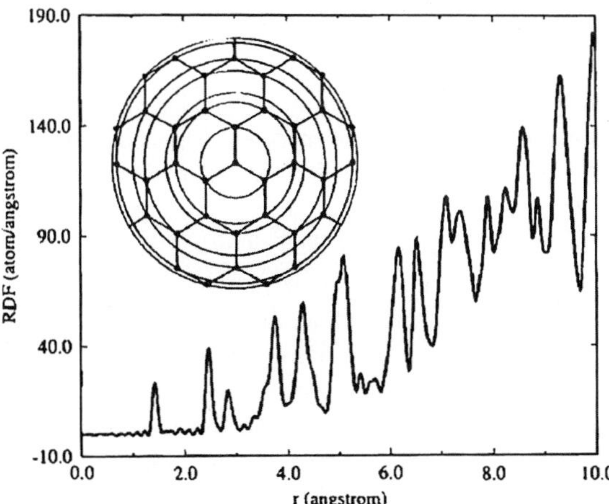

Fig. 7.10 Radial distribution functions (RDFs) of graphite and the in-plane honeycomb structure *inset*. Each concentric circle in the honeycomb structure produces a peak in the RDF. Reprinted with permission from (P. Zhou, P. Papanek, R. Lee, J.E. Fischer, W.A. Kamitakahara, J. Electrochem. Soc. **144**, 1744 (1997)) [124]. Copyright (1997), The Electrochemical Society

interactions are comparable in strength. Computer modelling showed that there is insignificant energy difference between interstitial Li and those that are bonded to the terminal H. Other models include the formation of covalent Li_2 molecules, but no evidence was found in support of these. The key aspect in these studies is that all models satisfy the observed capacity of LiC_6. Finally, QENS [124] was used to show that Li jumps between nearest or second-nearest neighbour interstitial sites.

Related work investigated the entropy of intercalation into C [127]. This study shows how the sign of entropy changes from low Li concentrations on initial charge ($x < 0.2$ in Li_xC_6) to higher concentrations ($x > 0.2$) indicating that multiple processes are occurring and that one of these is vibrational in origin. In graphite the entropy remains negative, but reduces in magnitude as lithiation progresses. Similar entropy information from INS data during lithiation of $LiCoO_2$ cathodes has also been reported [128].

Cathode materials pertinent to Li-ion batteries based on olivine $LiMPO_4$ have also been probed with INS, but for magnetic properties (low temperature) rather than Li-ion diffusion or lattice dynamic studies. Studies of $LiFePO_4$ [129], $LiNi_{1-x}$–Fe_xPO_4 [130], and $LiMnPO_4$ [131], show spin-wave dispersions and allow characterisation of magnetic-exchange interactions. Further INS work was motivated by the need to understand the electronic conductivity in $LiFePO_4$ and probed the thermodynamics and vibrational entropy of the phase transition in $Li_{0.6}FePO_4$ [121]. The oxidation state of Fe influences its neighbouring O atoms and the polyhedral distortions can characterize the motion of carrier hopping between Fe sites, which results in relaxations or displacements that can in turn be considered as the sum of longitudinal phonons. Similarly, occupation or vacancy of Li can result in distortions of atom positions and are expected to alter the frequency of phonons, in particular longitudinal optical phonons.

The phase evolution of $Li_{0.6}FePO_4$ as a function of temperature, via a two-phase transition to a disordered solid-solution transition at 200 °C [121], can shed light on the reaction mechanism during charge/discharge of this cathode. This is particularly pertinent as the two-phase or solid-solution mechanism of $LiFePO_4$ is a topical issue as discussed above. The difference in two-phase and solid-solution $LiFePO_4$ optical modes above 100 meV (higher energies) was found, with broadening evident for the solid-solution sample. The low-energy region features mostly acoustic lattice modes, translations and librations of PO_4 and translations of Fe. By comparison with infrared (IR) and Raman data, it was found that the PO_4 stretching vibrations are damped in the solid-solution sample. The difference in INS data of solid-solution and two-phase samples at higher energy mostly involve optical modes that can arise from motion of Li-ions, charge hopping between Fe-ions, and heterogeneities. The entropy was found to be larger in the solid-solution phase in conjunction with the subtle differences in the dynamics due to different optical modes. The similarity in two-phase and solid-solution phonon density of states (Fig. 7.11) agrees with the ease with which $LiFePO_4$ seems to undergo either transition, and the difficulty in pinning down the experimental evidence related to the reaction-mechanism evolution.

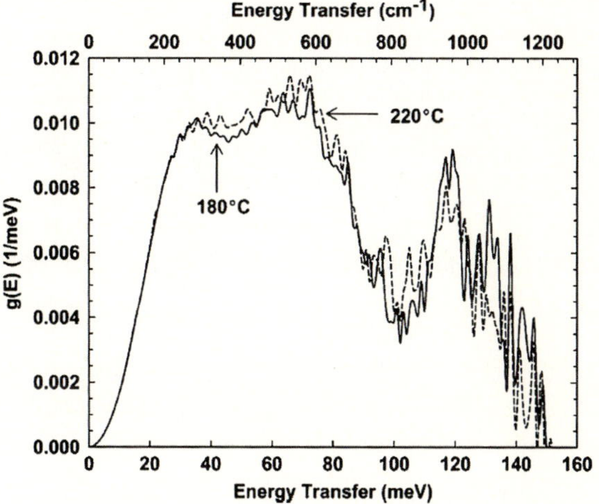

Fig. 7.11 The phonon density-of-states of $Li_{0.6}FePO_4$ at 180 °C (*solid*) and 200 °C (*dashed*). Reprinted with permission from (R. Stevens, J.L. Dodd, M.G. Kresch, R. Yazami, B. Fultz, B. Ellis, L.F. Nazar, J. Phys. Chem. B **110**, 22732 (2006)) [121]. Copyright (2006), American Chemical Society

Arguably the most studied materials using INS are manganese oxides and lithiated manganese oxides, predominantly due to the ease of using H as a probe for Li. These compounds are used for both primary and Li-ion batteries and ion-exchange methods have been used to show where Li may reside in these compounds. Although indirect, this information can provide further answers to some of the problems in this field of research. Attempts are also being made to use INS to provide comparisons between H and Li where H is used as a calibrated probe for Li [132].

One approach is to replace structural or surface water present on manganese oxides with protons, which can in turn be exchanged for Li to see how Li might displace water in these compounds. This was undertaken for spinel $Li_{1.33-x/3}Co_xMn_{1.67-2x/3}O_4$ [133] which shows, as is the case in many compounds of this family, that protons are inserted as hydroxyl groups giving a strong incoherent INS signal. The hydroxyl groups are located on the O atoms neighbouring the vacant $16d$ sites and aligned with the $8a$ sites in the spinel structure. Conversely, studies on undoped spinels have shown that the Li extraction from the $16d$ sites allows the insertion of protons. The main features of the INS spectra are strong $\gamma(OH)$ modes, a highly ordered proton site, a shoulder and smaller features between 300–700 cm^{-1} showing riding of protons on the oxide lattice and some librational water modes. The hydroxyl groups have characteristic signals around 908 cm^{-1} and their orientations are also determined using INS [134–137] of spinel-derivative compounds. Interestingly, IR data shows features between 950 and 1300 cm^{-1} which were considered to arise from protons, but the absence of these features at corresponding frequencies in the INS data indicate a manganese oxide lattice origin. Notable discoveries of this and related studies include the finding that in undoped spinels 40 % of protons cannot be exchanged and form disordered water, the chemical re-insertion of Li in Li-rich spinel $Li_{1.6}Mn_{1.6}O_4$ removes most of the hydroxyl groups [137], that generally the reversible Li amount is 50 % in both undoped and doped spinels, and that fewer protons are re-exchanged as

　　　　　　　　　　　　　　　　　　　　　　　　　N. Sharma and M. Wagemaker

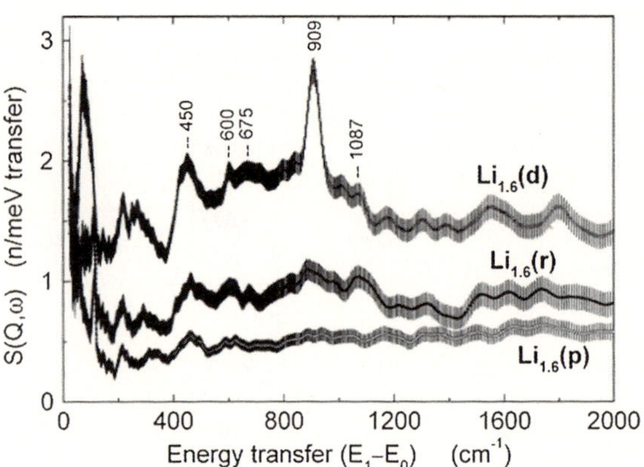

Fig. 7.12 INS spectra of as-synthesized Li$_{1.6}$Mn$_{1.6}$O$_4$ (p), acid-washed (d), and Li chemically re-inserted (r) samples. Reprinted with permission from (M.J. Ariza, D.J. Jones, J. Roziere, R. Chitrakar, K. Ooi, Chem. Mater. **18**, 1885 (2006)) [137]. Copyright (2006) American Chemical Society

the Co concentration increases. The latter is an interesting way to tune the Li-proton exchange capacity of these materials.

Figure 7.12 shows the INS data from a series of Li-rich Li$_{1.6}$Mn$_{1.6}$O$_4$ spinels formed through various methods. The pure sample (bottom of Fig. 7.12) shows some evidence of protons, OH and water, whilst the acid-treated version, where acid results in H–Li exchange, shows strong characteristic peaks for protons and γ(OH) groups. Finally, the acid-treated sample undergoes a chemical Li re-insertion step and results in the loss of the proton and OH signatures. However, the re-inserted material does not replicate the pure sample suggesting some protons remain as structural water and hydroxyl groups [137]. Relative comparisons of the INS intensity can be made between the acid-treated and re-inserted samples, with the 909 cm^{-1} peak showing a larger drop in intensity compared to the 1,087 cm^{-1} peak, which is attributed to an H site being easier to depopulate. A comparison of INS data for two Li-rich variants, Li$_{1.33}$Mn$_{1.67}$O$_4$ and Li$_{1.6}$Mn$_{1.6}$O$_4$, shows that the proton stability is higher in Li$_{1.6}$Mn$_{1.6}$O$_4$ than in Li$_{1.33}$Mn$_{1.67}$O$_4$. This suggests the reason that Li$_{1.6}$Mn$_{1.6}$O$_4$ has a larger Li-ion exchange capacity than Li$_{1.33}$Mn$_{1.67}$O$_4$ concerns the stability of the inserted species (or more specifically the stabilized proton sites).

Studies of λ-MnO$_2$ [135] illustrate subtle differences in INS spectra depending on synthesis precursors, noting that precursors and conditions are both important. This work again highlights the need to focus on the protons (often disordered). A related study investigated proton-exchanged spinels that form λ-MnO$_2$ showing that the proton diffusion was dependent on octahedral Mn vacancies [136]. In this study, certain features in the INS spectrum were found to disappear in the highly crystalline sample, suggesting that motion can be perturbed with crystallinity. Researchers have also looked at the proton and water environments in bare and lithiated MnO$_2$ [132] to demonstrate how lithiation influences the proton and water motions, which can then be used to extract information on lithiation processes. Using neutron total scattering from oxidized and lithiated versions of λ-MnO$_2$ researchers derived models for oxidation and lithiation [138].

Further work on the spinel $LiMn_2O_4$ system investigated the cubic to orthorhombic phase transition near room temperature, which is associated with Mn^{3+}/Mn^{4+} charge ordering [139]. Excess Li was introduced at the $16c$ site to study why the phase transition is suppressed in this situation. QENS was used here, where data were found to be dominated by magnetic contributions rather than that from Li hopping, with the slight narrowing of the elastic line near room temperature leading to the preliminary conclusion that electrons are localized on the Mn. A dynamic transition in Li-rich compounds seems to coincide with the structural transition in the parent. The magnetic properties of $Li_{0.96}Mn_2O_4$ were explored in a related INS study [140] showing that two short-range magnetic transitions are present and related to spin ordering of Mn^{3+} and Mn^{4+}.

7.3 Lithium Diffusion

A critical property of electrode materials is the ability to conduct Li through the host lattice. Li-ion mobility can be directly probed with INS and QENS, however only a few neutron studies report the direct measurement of Li dynamics mostly due to its moderate incoherent neutron-scattering cross section. Typically, each material has to be considered in order to determine whether the signal originates from Li, magnetism or other atoms. Usually, the hopping diffusion in Li-ion insertion electrodes is relatively slow compared to the timescale of INS and QENS in which case the local mobility is observed. As a consequence few studies exist that probe the Li motion directly. A different approach is to probe the Li diffusion in electrode materials with diffraction. Both the anisotropic contribution in the Debye-Waller factor and the deviation from harmonicity due to thermal motions at elevated temperatures can indicate the directionality of Li motion, which in turn may allow the identification of diffusion pathways. Both are illustrated here: anisotropy in the ADPs in combination with a maximum-entropy method (MEM) was used to identify the Li-ion trajectory in the positive-electrode material $LiFePO_4$, and anharmonicity of the ADPs revealed the Li-ion trajectory in the negative electrode material $Li_4Ti_5O_{12}$.

$Li_{12}C_{60}$ fulleride is a good example where Li diffusion was studied using QENS and INS, quantifying the diffusional motion of Li-ions across a phase transition proposing a localized jump-diffusion model in the octahedral voids of the $Li_{12}C_{60}$ structure. This accounts for the changes in the vibrational density of states near the phase transition and results in a model of the dynamical behaviour [141]. Another QENS study revealed the diffusion coefficient of Li in a highly-oriented pyrolytic graphite electrode at high temperatures, deriving an activation energy of 0.35 eV [142]. Interestingly, the diffusion coefficient obtained is similar to that obtained using electrochemical methods despite the diffusion lengths measured by the two techniques differing by a factor of 15,000. Li diffusion is more frequently determined indirectly using neutrons, and an example of this is the studies of anion dynamics to shed light on Li diffusion. Li-containing metal hydride systems have been investigated, such as in $LiBH_4$ and $LiAlH_4$, where translational modes of Li are linked with BH_4 in the high-temperature form of $LiBH_4$ [143]. Additionally, the disappearance of Li-containing lattice vibrations near phase transitions in these compounds is thought to be associated with the delocalisation of Li that enhances diffusivity. In this way, hydrogen-containing group dynamics can provide information on Li dynamics.

One QENS study describes alkali-ion diffusion (including for Li) in alkali-containing silicate melts [144], of interest for cathode materials based on silicates. This study used the decoupling of the incoherent (below 60 ps) and coherent neutron-scattering as a signature for Li-ion diffusion along channels in the immobile Si–O network. The relaxation times for Li were a factor of two smaller than for Na, indicating that Li-ion diffusivity is a factor of two larger, in agreement with conductivity data. QENS experiments have also been performed on single crystals of $^7Li_2MnCl_4$ (an inverse spinel-type structure), revealing a lack of anisotropy in the local Li motion [145]. Li-ions at $8a$ tetrahedral sites were shown to visit neighbouring $16c$ interstitial sites and jump back, but longer-range translational motion was outside the timescale used for the measurement.

Significant insight into Li diffusion can be gained from diffraction. Diffusion pathways can be identified by the anisotropy in ADPs in combination with MEMs. The exceptionally high discharge rate [47] observed in $LiFePO_4$ indicates that ionic mobility in the $LiFePO_4$ matrix is unusually fast. This has raised the question of how this is possible by the small polarons that are strongly localized at Fe sites in phase-separated $LiFePO_4$ and $FePO_4$ [146]. Morgan et al. [27] used the nudged elastic band method in calculations that show high Li ion mobility occurs in tunnels along the [010] direction, but reveal that hopping between tunnels is unlikely, confirmed by calculations of Islam et al. [108]. Fast one-dimensional conduction along the b-axis in the $LiFePO_4$ *Pnma* structure was predicted by atomistic modelling [27, 108] and the first experimental proof of the diffusion trajectory came from Nishimura et al. [147] using NPD in combination with the MEM. To enhance the sensitivity towards Li $^7LiFePO_4$ was prepared using ^{7}Li-enriched Li_2CO_3 as the raw material. In this study the ADPs readily show the direction of the Li-trajectory towards adjacent Li-sites, with green ellipsoids in Fig. 7.13 representing the refined Li vibration (displacement parameters) and indicating preferred diffusion towards the face-shared vacant tetrahedra. This suggests a curved trajectory in the [010] direction, consistent with atomic modelling [27, 39].

To relate further the vibrational motions with diffusion, the material with the overall composition $Li_{0.6}FePO_4$ was heated to approximately 620 K. In this composition $Li_{0.6}FePO_4$ forms a solid solution at a relatively low temperature, ~ 500 K, due to the unusual eutectoid as shown in the phase diagram in Fig. 7.13. This is confirmed as single phase by neutron diffraction. Thereby a large number of Li defects are introduced, that in combination with the higher thermal energy, enhances Li motion. Note that the Li trajectory in the solid solution should represent both end members because the crystal symmetry does not change upon heating and Li insertion. In the refinement of the $Li_{0.6}FePO_4$ structure no reliable solution using harmonically-vibrating Li could be found. To evaluate the dynamic

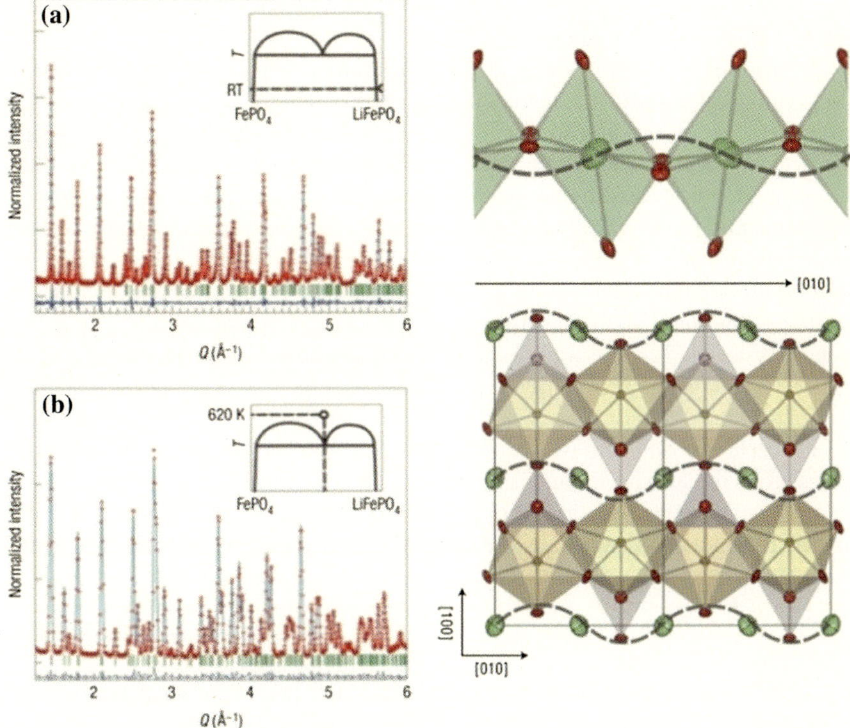

Fig. 7.13 *Left* Neutron diffraction patterns and Rietveld refinement profile of **a** room temperature and **b** 620 K $Li_{0.6}FePO_4$. The specific points of measured composition and temperature are given in the inset phase diagram reported by Delacourt et al. [102] and Dodd et al. [120] *Right* Anisotropic harmonic Li vibration in $LiFePO_4$ shown as *green* ADPs and the expected diffusion path indicated by the *dashed lines*. The ellipsoids were refined by Rietveld/MEM analysis of room-temperature NPD data. Reprinted by permission from (S. Nishimura, G. Kobayashi, K. Ohoyama, R. Kanno, M. Yashima, A. Yamada, Nat. Mater. **7**, 707 (2008)) [147]. Copyright (2008)

disorder of the Li the MEM was used to estimate the nuclear-density distribution from neutron diffraction. By considering the entropy the most probable distributions of nuclear species can be evaluated, making it possible to evaluate not only the missing and overlapping reflections, but also the more complicated nuclear densities. This approach applied to neutron diffraction data of $Li_{0.6}FePO_4$ at 620 K leads to the three-dimensional nuclear distribution of Li (Fig. 7.13). The observed diffusion along the [010] direction is consistent with the shape of the anisotropic thermal motions shown in Fig. 7.13 and atomistic modelling [27, 39]. Note that the Fe, P, and O atoms remain at their normal positions. The data show that the Li-ions move from one octahedral $4a$ site to the next via the intermediate tetrahedral vacant site. Along this trajectory the sites do not face-share with other occupied polyhedra. This is in contrast to, for instance, diffusion along the [001] direction where the intermediate octahedral position shares two faces with PO_4 tetrahedra which will lead to higher activation energies.

Laumann et al. [148] investigated Li migration in commercial spinel $Li_4Ti_5O_{12}$ using variable-temperature neutron diffraction. At 900 °C a marked deviation is observed in the linear dependence of the cell volume, O position, and anisotropic displacement parameters. Refinement of the Li occupancies resulted in almost complete $8a$ site occupation below 900 °C. However, at 900 °C a Li deficiency of approximately 14 % was observed, which was interpreted as the result of anharmonic motions and migration of the Li-ions. Therefore, in the fitting procedure one isotropic anharmonic ADP was refined. Examination of the nuclear density revealed negative scattering-length density peaks next to the $16c$ site. In this way Li-ion occupancy at the $32e$ site was discovered and subsequent refinement of Li at the $32e$ sites results in the probability density shown in Fig. 7.14. This makes it possible to formulate the diffusion pathway. Rather than occupying the $16c$ as an intermediate site between two $8a$ sites, which introduces an unacceptably long Li–O bond, Li passes from the $8a$ site through the face of the surrounding O tetrahedron to the nearby $32e$ site. This is followed by switching to the adjacent $32e$ site where it is bonded to another O atom, and from where it can hop to the next tetrahedral

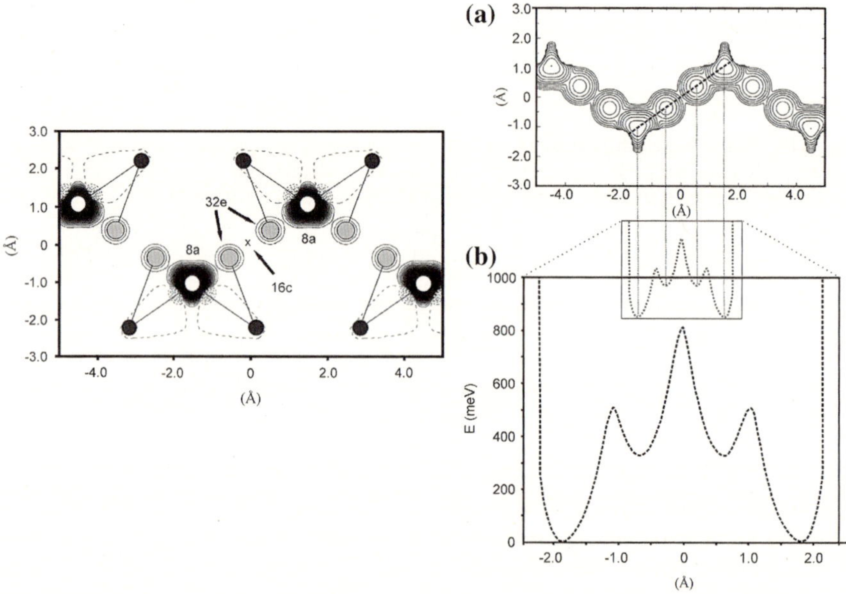

Fig. 7.14 Left: Probability density function derived from the anharmonic ADPs at 900 °C in the (*xxz*) plane through $8a$ and $16c$ sites. The shortest bond distances between Li (*white* at $8a$ and *grey* at $32e$) and O (*black* at $32e$) are indicated. Long dashed lines indicate zero densities and short dashed lines negative densities. Right: One-particle potential of Li at 900 °C in the (*xxz*) plane through $8a$, $32e$, and $16c$ sites (the same section as that in the left figure). Contour lines are in steps of 100 meV. The dotted line shows the linear section along the [111] direction. Reprinted with permission from (A. Laumann, H. Boysen, M. Bremholm, K.T. Fehr, M. Hoelzel, M. Holzapfel, Chem. Mater. **23**, 2753 (2011)) [148]. Copyright (2011) American Chemical Society

$8a$ position. Effectively, this mechanism results in a number of short jumps along the [111] direction between adjacent $8a$ sites. The energy barriers can be approximated by assuming Boltzmann statistics for single-particle motion resulting in the one-particle potential shown in Fig. 7.14. These findings are consistent with nuclear magnetic resonance measurements indicating that the $16c$ site forms the saddle point of the barrier between two $8a$ sites [149]. In this way NPD is able to reveal the details of the three-dimensional long-range diffusion pathway in spinel $Li_4Ti_5O_{12}$.

7.4 Electrolytes

Commercial electrolytes typically contain a Li salt dissolved in an organic solvent and are often composed of two components: one for the dissolution of the salt and another that assists in the formation of a protective layer on the anode to prevent continuous electrolyte-reduction and self-discharge, e.g. ethylene carbonate. These electrolyte systems, being non-aqueous and highly air-sensitive, tend to be flammable and can turn from liquid to gas at elevated temperatures (Fig. 7.15). The electrolyte also determines the cathode and anode materials that can be used by limiting the applicable voltage range which is associated with the HOMO of the cathode and LUMO of the anode [4]. The key factors that determine a good electrolyte are ionic conductivity, flammability and chemical stability, and applicable voltage windows.

To overcome the safety and long-term reliability issues of using organic electrolytes, research has been directed to aqueous electrolyte systems with Li salts. Unfortunately, voltage limitations have hampered significant development of aqueous electrolytes, but these safe electrolyte-systems have found niche use in medical applications. In addition to aqueous electrolytes, liquid electrolytes based on ionic liquids have attracted significant attention.

Fig. 7.15 An example of a LiFePO$_4$‖graphite battery containing 1:1 mol. % ethylene carbonate:dimethyl carbonate heated to 90°C where the dimethyl carbonate (organic solvent) has boiled, expanding the casing of the battery

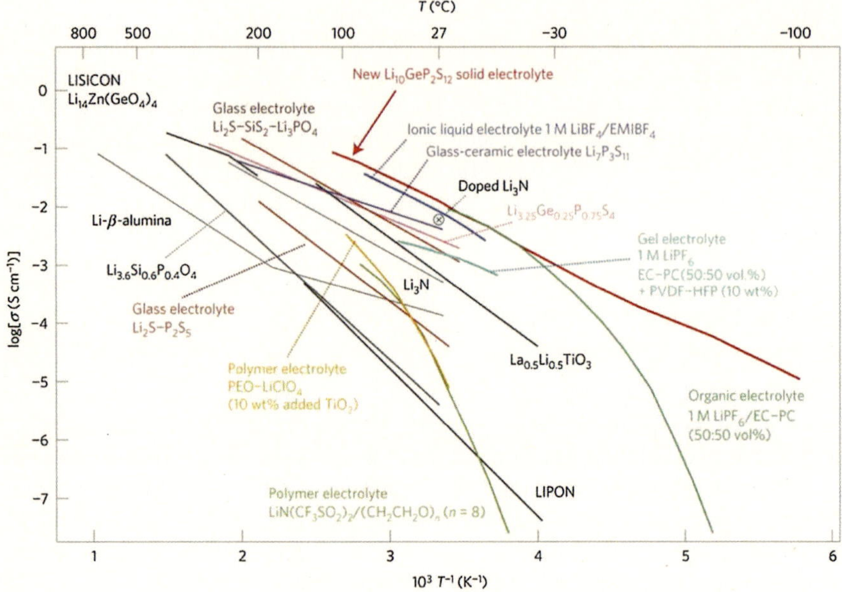

Fig. 7.16 A collection of conductivity data of pertinent electrolytes used for commercial and research-scale Li-ion batteries. Reprinted with permission from (N. Kamaya, K. Homma, Y. Yamakawa, M. Hirayama, R. Kanno, M. Yonemura, T. Kamiyama, Y. Kato, S. Hama, K. Kawamoto, A. Mitsui, Nat. Mater. **10**, 682 (2011)) [150]. Nature Publishing Group

Apart from electrolytes in the liquid state, semi-solid or solid-state electrolytes such as gel and solid polymer electrolytes continue to be a preferred option in overcoming safety and leakage issues. Neutron scattering work has been undertaken on ceramic and glass-ceramic solid state Li-ion conducting electrolytes. Some of these electrolytes feature Li-ion conductivities that can be as good as commercial organic electrolytes as elegantly demonstrated for $Li_{10}GeP_2S_{12}$ [150]. This is the first solid-state electrolyte that shows conductivity that matches that of commercially-available liquid electrolytes (Fig. 7.16).

7.4.1 Structure

Detailed structural analysis has been undertaken with the use of neutron diffraction on a variety of Li ionic conductors as demonstrated by studies of Li argyrodites. Using a combination of simulations and structural refinements against X-ray and NPD data, the structure of the Li-ion conducting argyrodites, Li_6PS_5X where X = Cl, Br, and I, were determined [151]. The Li content and location, along with the mixed occupancies of the X and Li or X and S sites was modelled. For X = Cl and Br, Cl or Br was found on two S sites, whilst I was found on an independent wholly I-containing site. For $X = $ Cl only one Li site was found, while for $X = $ Br and I two Li sites were revealed with a distribution of Li that differs depending on X. This highlighted that the halide plays a critical role in the distribution of atoms, including the location and occupancy of Li. Ionic conductivity was found to be highest for the $X = $ Br samples, suggesting the influence Br has on the atomic distribution to be favourable for this property [151].

These materials were also investigated using variable-temperature neutron diffraction, where the starting reagents were reacted under similar conditions to those used in the laboratory synthesis. This was to determine the optimal reaction temperatures and conditions for favourable properties and whether any intermediate highly-conducting phases were present. The notion was to explore whether synthesis temperatures could be lowered, potentially reducing manufacturing costs, or whether intermediate phases were able to provide superior ionic-conduction properties. Analysis of neutron diffraction data showed that argyrodite formation begins at relatively low temperatures around 100 °C, well below the reported synthesis temperature of 550 °C, but at temperatures around 550 °C the reagents become amorphous or nano-crystalline with all reflections from the sample disappear (Fig. 7.17). Notably, on cooling the desired phase re-condenses and on inspection it is found that the anion ordering, leading to the most conductive phase, is actually found in the re-condensed phase rather than the initially-formed phase [152]. These types of systematic studies on bulk formation shed light on which phases and synthetic routines may provide the best ionic conduction.

Structurally disordered solid-state electrolytes have been investigated using INS. Work [153] exploring low-energy vibrational dynamics of the $^{11}B_2O_3$–$^7Li_2O$

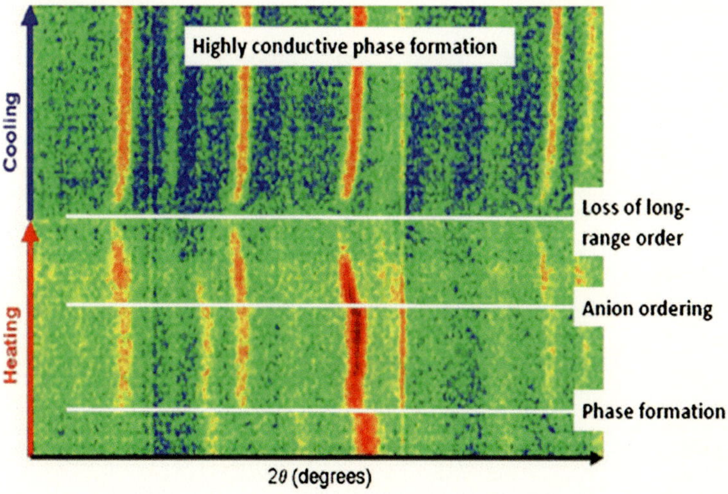

Fig. 7.17 Collated NPD patterns of a heating and cooling sequence applied to Li_6PS_5Br. Although an argyrodite phase forms at relatively low temperatures, it is found to be less conducting than the phase formed after the loss of long-range order

168   N. Sharma and M. Wagemaker

system showed a boson peak between 2 and 10 meV. It was found that with
increasing Li content the intensity and position of the boson peak changed, suggesting the presence of intermediate glass structures. More importantly, this
information can be used to generate a master curve, which suggests a universal
distribution of vibrational density-of-states that is composition independent, even
though the structure changes markedly. Furthermore, increasing the Li content in
these glasses results in chemical structure-induced densification, as fewer low-
density B-containing groups are found. It is argued that the densification may arise
from the same microscopic origin as the boson peak.

7.4.2 Lithium Diffusion

Perovskite structures (ABO_3, where A and B are different-sized cations) have good
ionic conductivities, with $La_{0.5}Li_{0.5}TiO_3$ showing the highest conductivity of these
compounds. Nuclear-density maps derived from neutron diffraction data were used
to illustrate the disordered nature of Li in these systems at high temperatures.
Neutron diffraction data were also used to show the migration pathway of $2c$-$4f$-$2c$
or $2c$-$2d$-$2c$ [154, 155] (Fig. 7.18), with data at lower temperatures showing Li
localized at the $2c$ site. Other work detailed descriptions of TiO_6 tilting, La and
vacancy ordering, and localization of Li. This was achieved using low-temperature
data collection on ^{7}Li-enriched samples in addition to different heat treatments to
determine the origin of the high conductivity in these materials [156]. Notably, it
was shown that quenched samples in the Li-poor $La_{0.18}Li_{0.61}TiO_3$ stabilize into a

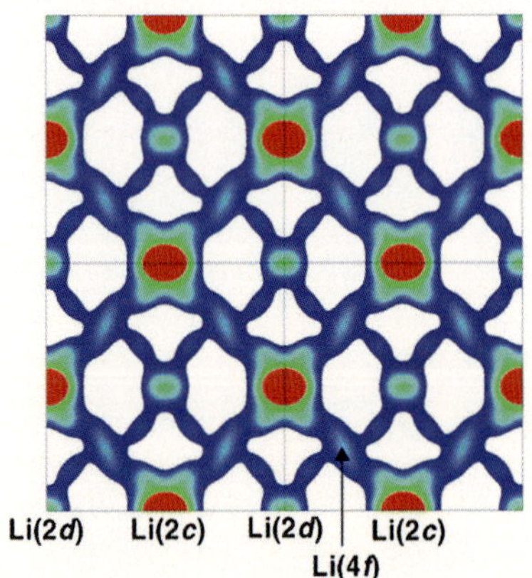

Fig. 7.18 Nuclear-density distribution on the (002) plane of $La_{0.62}Li_{0.16}TiO_3$ at room temperature.
Negative amplitudes indicate Li diffusion paths with *dark ovals* showing maximum scattering
amplitude and *blue* minimum amplitude. Reprinted with permission (M. Yashima, M. Itoh,
Y. Inaguma, Y. Morii, J. Am. Chem. Soc. **127**, 3491 (2005)) [155]. Copyright 2005 American
Chemical Society

structure featuring three-dimensional Li pathways. In contrast, slow-cooled samples only feature two-dimensional pathways and the Li-ions feature unusual four-fold coordination environments. The best conduction is found in samples with vacant A-sites nearby [156]. The crystallographic nature of Li and its conduction pathways are critical in assessing the origin of the high ionic-conductivities found in these compounds, and how to modify the crystal-chemistry to improve it.

Spinels can also have three-dimensional Li diffusion pathways. Recent neutron diffraction work on complex ordered spinels $Li_2(Ni/Zn)Ge_3O_8$ [157] shows that the Ni and Zn compounds have different Li-ion conduction pathways. The Ni-case shows $8c$-$12d$-$8c$ hopping with concerted harmonic displacements of O, whilst the Zn-case shows a complex $4b$-$24e$-$4a$-$24e$-$4b$ pathway. The Zn-case features smaller windows through which Li can pass and this is in part responsible for the lower ionic conduction. The strategy used by the authors involved refining the room and high-temperature structure. In the highest-temperature case the Li anisotropic displacement parameters were kept small and nuclear density maps revealed new Li sites. This process was continued with intermediate temperatures and the number of Li-sites compared to total Li content and thereby showed the pathways for Li conduction [157].

Neutron radiography can also be used to track macroscopic Li diffusion in battery materials. Early work on $Li_{1.33}Ti_{1.67}O_4$ prepared using pure ^7Li and natural Li illustrated and quantified Li-transport [158]. The experimental setup placed two materials, the pure ^7Li-containing material and the natural Li-containing material in contact with each other. Electrodes were applied to the non-adjacent sides and a field applied during heating. Standard samples with controlled $^7Li/^6Li$ contents were also measured to quantify Li content. Figure 7.19 illustrates the experimental setup and some of the results showing how ^7Li moves from the labelled 'smpB' to 'smpA'. Note that the ^7Li near the anode is consumed, as no white regions are left.

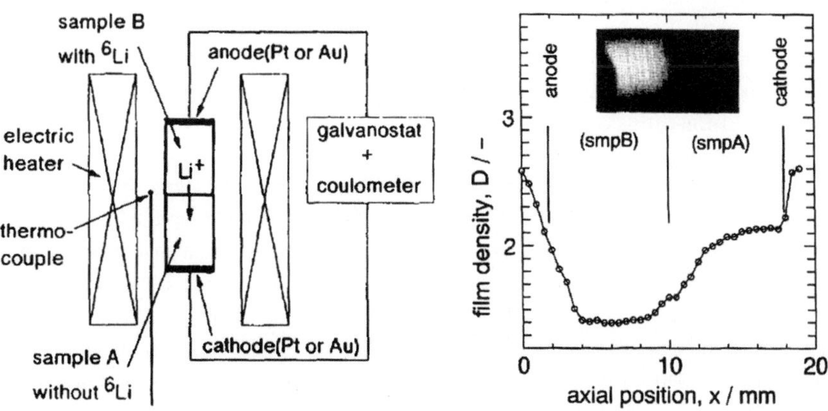

Fig. 7.19 *Left* The experimental setup for the electrolysis experiment. *Right* Neutron radiography image obtained after charge transfer at 200 °C. Reprinted with permission from (M. Kamata, T. Esaka, S. Fujine, K. Yoneda, K. Kanda, Nucl. Instr. Meth. Phys. Res. A **377**, 161 (1996)) [158]. Elsevier

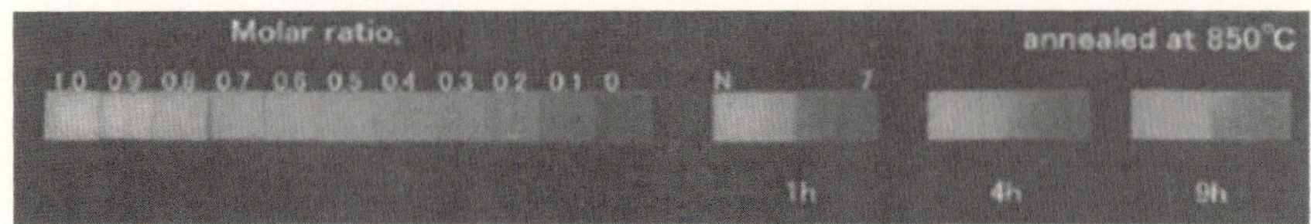

Fig. 7.20 *Left* Neutron radiography images of standard samples varying the Li isotope ratio from ^{N}Li (natural ratio ^{6}Li/^{7}Li) to ^{7}Li, a whiter image representing a higher ^{N}Li concentration. *Right* Annealed samples show increased mixing of ^{N}Li and ^{6}Li species when the annealing time increases illustrating the diffusion of Li in these super ionic conductors. Reprinted with permission from (T. Esaka, Ionics **10**, 358 (2004)) [159]. Springer

An elegant example of neutron imaging to study Li diffusion is given for the lithium zinc germanium oxide as shown in Fig. 7.20. This material is often referred to as a Li super ionic conductor, abbreviated to LISICON. In Fig. 7.20 standard samples with known ^{6}Li/^{7}Li ratios are shown and regions with more ^{6}Li are represented by a lighter colour [159]. The work shows how annealing temperatures influence Li-ion diffusion through the sample. Further studies compared the $Li_{3.5}Zn_{0.25}GeO_4$ material with the analogue $Li_3Zn_{0.5}GeO_4$ and found that ionic-conductivity differences were due to high ion-mobility rather than differences in diffusion paths [160]. The authors obtained tracer diffusion coefficients using neutron radiography data, similar to the work determining the tracer diffusion coefficients in $Li_{1.33}Ti_{1.67}O_4$ [161].

QENS and INS have been used to study diffusion in electrolyte systems. The relationship between ion transport and polymer relaxation in polymer/Li salt systems are considered to be bi-phasic at room temperature [162]. Ionic conduction occurs predominantly in the fluid-like amorphous regions as compared to the crystalline salt-rich regions. The mechanism for conduction is not completely understood. One theory is that Li ions coordinated to O on the polymer move by hopping between the O and the formation and disruption of these coordination bonds influences the local chain structure. Researchers investigated polyethylene oxide (PEO) in two different electrolytes, $LiClO_4$ and LiTFSI (where TFSI = $N(SO_2CF_3)_2$). Each of these electrolytes has salient features that make them good for battery applications. Considering the Li-ion diffusion constant and distance between neighbouring O atoms, the jump time is on the order of nanoseconds (ns). To probe the relaxation and dynamics of the polymer chain in the ns timescale QENS was used. The signal from Li is still insignificant compared to the H atoms in the polymer that make up the largest signal in QENS data. Two processes are found in both electrolytes, a fast process characteristic of rotational dynamics often related to relaxation in polymers, and a not previously-observed slower process showing more translational character. However, in the PEO and LiTFSI electrolyte two fast processes were found (see Fig. 7.21). The fast processes can be related to the fluctuations of the polymer chain segments as Li–O inter and intra-chain coordination bonds form and break during transport, associated with a rapid rotation of the polymer. This can also be visualized as a positive charge approaching and leaving, resulting in rotation of a polymeric region. The slower process appears to arise from translations of polymeric chain segments,

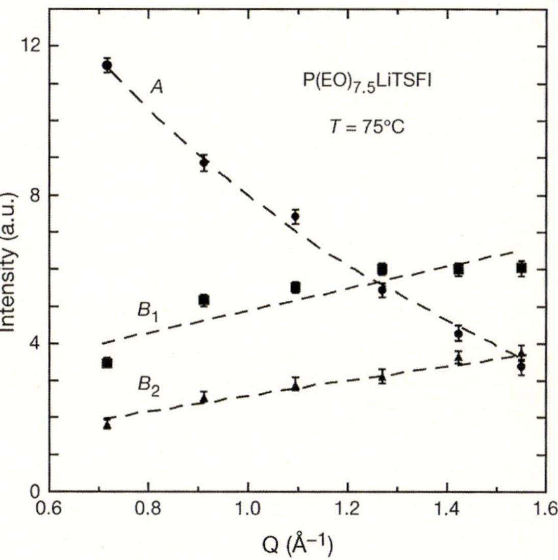

Fig. 7.21 Detail of the fast (*B*) and slow (*A*) processes in the PEO and LiTFSI electrolyte. Reprinted (adapted) with permission from (G. Mao, R.F. Perea, W.S. Howells, D.L. Price, M.-L. Saboungi, Nature **405**, 163 (2000)) [162]. Nature Publishing Group

however, these seem likely to be static on the timescale window used in this study. Overall these findings show two to three kinds of motions in the polymers with ion transport and that the O of the Li–O plays a large role in controlling the Li dynamics.

A second study used QENS to investigate complex polymer-gel electrolytes in which conventional carbonate-based electrolytes were dispersed [163]. There were conflicting ideas in the literature from other techniques that probe different time and length scales. To probe solvent (carbonate) dynamics on an intermediate time and length scale, QENS was used. In order to model the system, a weak solvent and polymer interaction had to occur, which is consistent with better ionic conduction. Stronger interactions would require more complex modelling. The undeuterated carbonates were used as a probe for the deuterated polymer-gel matrix, so that the signal was dominated by the incoherent neutron-scattering from protons in the solvent. The carbonate solution used two carbonates that differed by a methyl group. A model of the Li and carbonate arrangement of the electrolyte was generated using predictions from molecular-dynamics simulations. The interest lay not in the solvent but the complex polymer-gel matrix that was revealed using the carbonate solvent as the probe. Again, two solvent (carbonate) relaxation times were found: A faster rotational relaxation-time on the order of 100 μeV and a slower translational-diffusion signature on the order of 10 μeV. The translational diffusion was similar to that expected in a liquid electrolyte, but the characteristic distance of this process was 5 Å due to the constraints applied by the polymer matrix. This constraint also acts to lower the ionic conductivity relative to a liquid. The constraint is evidenced by a narrowing of the QENS signal at low Q, where without the constraint there should be no Q-dependence in the signal width. The rotational component shows changes in intensity as a function of Q with almost no change in width. The solvent molecules are likely to exhibit both rotational and translational motion simultaneously, so the scattering function is a frequency convolution, which also implies that the rotational peak width/intensity as a function of Q will influence the translational peak width/intensity and vice versa. Fortunately, these processes are well separated in time allowing further assumptions to be made in the model. The model considered how many protons contribute to the signal and what fraction of the protons will take part in the two processes, considered either static or undergoing both process. The possibility of a proton undergoing one process and not the other was excluded. Static protons would be close to the matrix while the free protons further away. Models were designed for the most likely scenario, e.g. uniaxial rotation of the free protons, translational diffusion of the solvent via a microscopic jump-model accounting for the constraint of the polymer matrix by random jump-diffusion between parallel impermeable walls, and longer-range 'free' diffusion. The researchers noted the complexity in deriving an initial model and considered only the main dynamical process (noting also that other processes may occur). However, the insight gained in this study is important to understanding the behaviour of electrolytes in matrices. For example, this study found that the proportion of static protons, which are indicative of a less-mobile solvent, ranges from 8 % at 373 K to 35 % at 253 K [163], which is consistent with the difference between electrolyte liquid solution and polymer matrix systems.

Li-based glasses and glass ceramics have been used as solid-state electrolytes in all solid-state Li-ion batteries. The key limitation in the commercialization of these materials is the low ionic-conductivity under ambient conditions (with minimal electronic-conductivity) and the microstructural aspects associated with solid interfaces and contacts. The first QENS work on solid-state Li-ion conductors was a polarised experiment that determined Li diffusion in single crystal of $^{7}Li_2S$ at elevated temperatures [164]. The polarisation allowed the separation of incoherent and coherent neutron-scattering contributions. High temperatures were required as these phases show higher ionic conduction in these regimes, and fortunately adopt cubic symmetry, simplifying the modelling. Coherent neutron scattering showed minimal changes with temperature, whilst the incoherent neutron scattering changed markedly, which is predominantly from Li. The work shows that an inter-site jump model fits the QENS data, and this corresponds to Li-ions hopping to neighbouring cation sites. More complex models also fit the collected data, but further data are required to verify them, with the diffusion coefficient obtained from the simpler model matching that expected from other analysis techniques.

A related study [165] explored the $^{11}B_2O_3$–Li_2O–LiBr system showing four superimposed oscillating ionic units. It revealed further information such as activation energies for two well-resolved processes: translational jump-diffusion and a localized reorientation motion of ions. The fact that each motion cannot be represented as a fraction but are superimposed dramatically influences the ease of modelling. There was no evidence of Li hopping. Figure 7.22 (left) shows the experimental spectrum and Fig. 7.22 (right) shows the model with the four units used in the modelling. Although it may seem that the choice of four units and their behaviour is arbitrary, comparison with Raman and IR data illustrates that the four units are likely to be Br, BO_3/BO_4, Li, and O. The researchers concluded that the conductivity is due to both jump and reorientation motions [165].

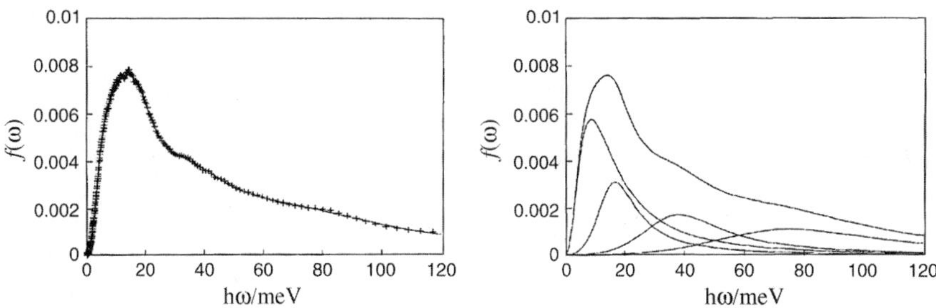

Fig. 7.22 *Left* Experimental frequency spectrum of $B_2O_3 \cdot 0.56Li_2O \cdot 0.45LiBr$ at 573 K and *Right* superposition of the contribution of the four oscillators used in the model. Reprinted (adapted) with permission from (C. Cramer, K. Funke, C. Vortkamp-Ruckert, A.J. Dianoux, Phys. A **191**, 358 (1992)) [165]. Elsevier

In most of the solid-state electrolyte cases, the basic system is Li_2O, which crystallizes with a fluorite arrangement, where phonon-dispersion curves have been measured as a function of temperature and show a sharp decrease approaching the fast-ion conducting phase [166]. Although this material was studied for fusion-reactor applications, it forms the basis of many Li-ion conducting glasses. Li self-diffusion was identified in incoherent QENS and tracer techniques (see references in [166]). Coherent INS data was used to elucidate properties that can be related to those measured physically such as elasticity, which show anomalies near the onset of fast-ion conduction in these compounds. Therefore, a direct relation to dynamics and physical properties could be made. An interesting extension of this work was the study of Li_2O crystals with metallic Li colloids [167] which form after irradiation.

A related QENS study investigated the Li-ion conductor LiI, made in a composite form with mesoporous Al_2O_3 [168]. These materials have potential applications in the nuclear industry but can also be applied as solid-state electrolytes. Notably the activation energies for Li migration derived from QENS were smaller (by a factor of approximately 2) than that determined from electrical conductivity and pulsed-field gradient nuclear-magnetic resonance measurements. This observation was attributed to the migration mechanism in mesoporous structures and possible conductivity enhancement in confined spaces by the increased presence of interfaces in the mesoporous environment. This study highlights the importance of using a reliable model for QENS that is developed specifically for the material under study.

7.5 Interfaces

Acquiring a comprehensive understanding of interfacial reactions in batteries is essential for the design of new materials [4]. However, due to the difficulty in probing the interfaces, relatively little is known about the relevant processes. What is known is that the stability of electrolytes at the electrode interface plays a key role in determining the cycle life and safety of batteries. The most studied interface is between the carbon negative electrode and carbonate organic electrolytes. The instability of carbonate electrolytes with respect to the carbon chemical-potential results in deposition of the electrolyte Li-containing inorganic and organic decomposition products on the electrode surface, otherwise known as the SEI layer. This reduces the amount of active Li available to the cell and degrades the electrolyte. Typically electrolytes contain two components, one for the Li-salt dissolution, and one that assists in the formation of a protective layer on the anode preventing continuous electrolyte-reduction and self-discharge, e.g. ethylene carbonate. This requires the formation of a stable SEI film showing good ionic-conductivity and poor electronic-conductivity. In this way the SEI can passivate against further electrolyte decomposition without severely influencing battery performance. The SEI formation and maintenance during further cycling is expected to play an essential role in the cycle life and stability of batteries, however, the growth mechanism under variable battery-conditions is largely unexplored. In addition, the charge-transfer reaction at the interface, most likely influenced by the SEI development [169], is an essential parameter that in many cases limits the power of batteries. Also, interfaces within the electrode material can establish upon phase transitions during (de)lithiation. From an applications point of view these transitions have the favourable property of being associated with a constant potential that is independent of the composition, but the disadvantage of being associated with volumetric changes that may restrict the cycle life. Probing such interfaces under in situ conditions is possible using neutron reflectometry.

One of the first reported neutron-reflectometry studies on a Li-ion battery system determined the Li insertion and extraction mechanism in thin film anatase TiO_2, a negative electrode material operating around 1.7 V versus Li/Li^+. Lithiation of the tetragonal anatase TiO_2 leads, via a first-order phase transition, to the orthorhombic $Li_{0.5}TiO_2$ Li-titanate phase. The aim of the neutron-reflectometry study was to discover the phase-evolution scheme in this electrode material, which is of more general value for electrode materials undergoing first-order phase transitions. Previous studies suggested the establishment and movement of a diffusion-controlled phase boundary, parallel to the electrode surface, between the Li-rich Li-titanate and the Li-poor anatase phase [170]. This is in contrast to, for instance, a percolation scheme where the Li-titanate phase would penetrate the original anatase layer only at certain regions of the thin film. Further intercalation would increase both the perpendicular and the lateral dimension of these percolation paths, eventually leading to a homogeneously-intercalated film. Van de Krol et al. suggested a specific scheme in order to explain the more facile Li-ion extraction rate compared to the insertion rate [170]. Based on the assumed faster Li-diffusion in the Li-anatase phase [171] one might expect fast depletion of Li in the near-surface region of the Li-titanate phase containing electrode, which is in contact with the electrolyte. As a result, during Li extraction, the Li-anatase phase starts to grow at the electrolyte surface into the layer at the expense of the Li-titanate phase.

The contrast difference between the $Li_{0.5}TiO_2$ Li-titanate phase and the TiO_2 anatase phase for neutrons should make it possible for neutron reflectometry to determine the phase-evolution scheme both during Li insertion and extraction.

An approximately 25 nm smooth anatase TiO_2 electrode was deposited on a thin ~ 20 nm Au current collector on a 10 mm thick single-crystal quartz block that served as the medium for the incoming and reflected neutron beam. The latter is practically transparent to thermal neutrons allowing approximately 70 % transmission over 10 cm path length. The TiO_2 electrode is exposed to a 1 M solution of $LiClO_4$ in propylene carbonate electrolyte using Li metal both as counter and as reference electrode. Li was galvanostatically inserted in two steps and extracted in two steps using 10 mA (C/5) in the same voltage window. Neutron-reflection experiments were performed after each step when a constant equilibrium-potential was achieved. The results, including the fit and the associated scattering-length density (SLD) profiles are shown in Fig. 7.23. The profound change observed in the neutron reflection from the virgin state and the state after the electrochemistry can be explained by the formation of a SEI layer on the TiO_2 surface.

For the half lithiated state the best fit of the neutron reflection data was achieved by assuming a Li-rich Li-titanate phase ($Li_{0.5}TiO_2$) in contact with the electrolyte. As a result of the negative coherent neutron-scattering length of Li the SLD of lithiated TiO_2 being is smaller than that of pure TiO_2. Neutron reflectometry proved the

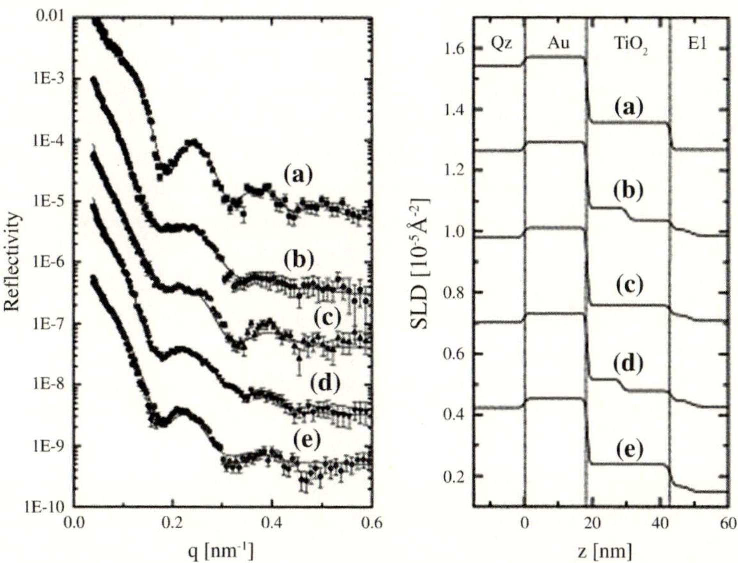

Fig. 7.23 *Left* Neutron reflectometry results measured at different stages in the intercalation cycle, including the best fit that corresponds to the model described in the text. **a** Virgin state, before any electrochemistry is performed, **b** approximately half-way in the intercalation, **c** fully intercalated state, **d** approximately half way in the de-intercalation, and **e** empty state after the de-intercalation. *Right* SLD profiles (**a**–**e**) as in the left figure corresponding to the fits of the neutron reflectivity data also shown left. "Qz" refers to the quartz which is the incoming and outgoing medium for the neutrons, "El" refers to the electrolyte, 1 M $LiClO_4$ in propylene carbonate. Reprinted with permission from (M. Wagemaker, R. van de Krol, A.A. van Well, Phys. B. **336**, 124 (2003)) [220]. Elsevier

establishment of a phase boundary parallel to the electrode surface rather than a percolation model, which would lead to a homogeneous change of the SLD. The fully-lithiated state was fitted with a single electrode layer with the SLD corresponding to the composition $Li_{0.52}TiO_2$. The neutron reflection data after half delithiation indicated that the phase front moves back via the way in which it came, with the Li-titanate phase being in contact with the electrolyte. This is in contrast to the expected Li depletion during extraction, which should lead to TiO_2 formation at the electrolyte interface. This symmetric phase-front movement does not immediately explain the difference in insertion and extraction rate. However, nuclear magnetic resonance experiments show the diffusion over the phase boundary to be the rate-limiting step [172, 173], giving a rationale for the more sluggish lithiation of TiO_2, which as opposed to delithiation, requires diffusion over the phase boundary.

Information related to the structure and composition of the SEI layers is mostly based on ex situ spectroscopic and microscopic studies [174, 175], but because of the reactive and delicate nature of these layers, in situ analysis is essential to improve our understanding. Being relatively sensitive to the light organic and inorganic species present in the SEI and to the surface layers ranging from a few to hundreds of nanometers, neutron reflectometry is an exceptionally suitable technique for in situ studies of the growth, composition, and the structure of the SEI.

Owejan et al. [176] used neutron reflectometry to study the formation and structure of the SEI layer in a Li battery. A requirement for neutron reflectometry is a flat and smooth surface as it probes the average in-plane SLD profile. A Li half-cell was configured with Cu as the 'counter' electrode to prevent Li reaction with the electrode, so that all electrochemical charge can be attributed to decomposition of the electrolyte and SEI layer formation. The use of a non-intercalating electrode, such as Cu, as model electrode for electrolyte decomposition appears to be justified by the similarity of the SEI layers formed by C materials at low potentials in Li-salt containing electrolytes [46, 47]. Additionally, the thermodynamics of electrolyte reduction appear to be governed by the cation that is used in the electrolyte [43]. The scattering contrast of the electrolyte was increased by preparation of a 1 M $LiPF_6$ solution in a 1:2 (v/v) ratio of deuterated ethylene carbonate and isotopically-normal diethyl carbonate. The deuterated ethylene carbonate also offers the opportunity to identify the possible preferential decomposition of cyclic (ethylene) over acyclic (diethyl) carbonates. By deuterating selected components in the electrolyte solution researchers can access which component contributes or forms the SEI layer.

In Fig. 7.24 the neutron reflectivity versus Q is shown for the pristine Cu electrode immersed in the electrolyte at the open-cell potential. This electrode underwent 10 cyclic voltammogram sweeps between 0.05 and 3 V at a 10 mVs^{-1} rate, followed by holding the potential at 0.25 V versus Li/Li$^+$ (potentiostatic reducing conditions). A clear difference between the peak amplitudes and oscillations (positions) in the reflectivity of the fresh and electrochemically-cycled electrode is observed. Initially, at the Cu-electrolyte interface, copper carbonate/hydroxide ligand-containing layers appear to be present which are removed after the cyclic voltammetry. Interestingly, after 10 cyclic voltammetry sweeps and the potential-hold step under reducing conditions, a 4.0 nm thick SEI layer at the

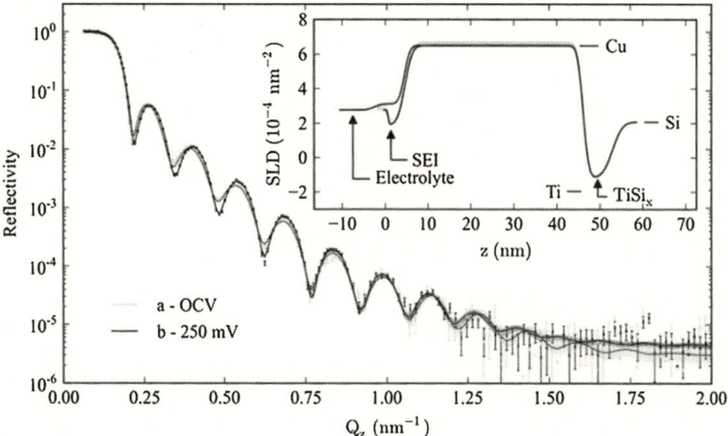

Fig. 7.24 Neutron reflectivity versus Q shown for the sample at the open-cell voltage (OCV) and after 10 cyclic voltammetry cycles during a hold at 250 mV. The *solid lines* are the best fit to the two data sets. *Inset* SLD of Si, Cu, and Ti are indicated and electrolyte, SEI and $TiSi_x$ layers are identified. For both parts, the darker and lighter shaded regions are the 68 and 95 % confidence intervals, respectively. Reprinted with permission from (J.E. Owejan, J.P. Owejan, S.C. DeCaluwe, J.A. Dura, Chem. Mater. **24**, 2133 (2012)) [176]. Copyright (2012) American Chemical Society

interface had developed with a SLD much lower than that of the electrolyte. A further 10 additional cyclic voltammetry sweeps led to only a small growth of the SEI layer. For completeness, the authors then took a number of data sets at different potentials by slowly ramping the potential at $10~mVs^{-1}$ to the next potential value and holding the potential during neutron reflectometry data collection. The results

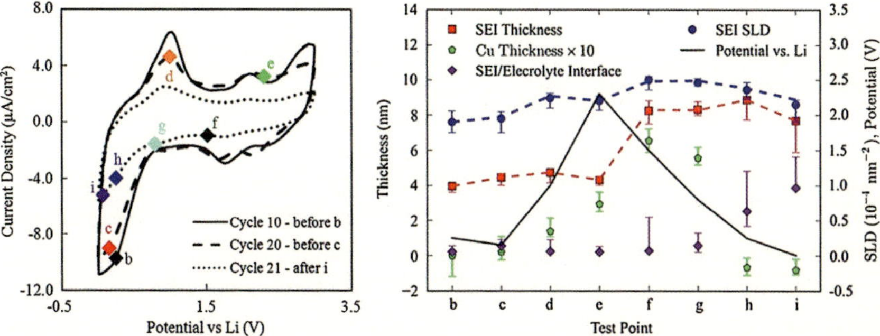

Fig. 7.25 *Left* Cyclic voltammograms for selected scans. The points *b-i* denote the location of potentiostatic holds. *Right* Selected fitting parameters for test points. For reference, the hold potential versus Li is also shown. The *dashed lines* are the total thickness and SLD from composition modelling, matching well with measured values. Reprinted with permission from (J. E. Owejan, J.P. Owejan, S.C. DeCaluwe, J.A. Dura, Chem. Mater. **24**, 2133 (2012)) [176]. Copyright (2012) American Chemical Society

are summarized in Fig. 7.25. During the first two points *d* and *e*, using an oxidation current, a small decrease in the SEI layer thickness was observed and the SLD suggests a shrinking of the SEI layer due to solubility of SEI components. However, at the next point, holding at a reducing current the SEI layer grows significantly up to 8.9 nm. Most of the neutron reflection measurements indicate rather homogeneous SLD profiles with little roughness, in contrast to proposed structures in literature. After point *f* even the lowest potentials do not lead to further SEI layer growth, illustrating the passivating nature of the SEI layer. The systematic decrease of the SLD at lower potentials indicates that the SEI is increasingly composed of low SLD elements, which indicate Li-rich molecules.

Further insight into the composition of the SEI layer was obtained by combining X-ray photoelectron spectroscopy-derived compositions with the neutron reflectometry results for the SEI layer. This indicated an increase in LiOH and LiF molecules, and the decrease of lithium alkyl carbonates at the lower reducing-potentials. This study demonstrated the advantage of neutron reflectometry in giving direct insight into the growth and composition of the SEI layer and its relationship to the electrochemical conditions.

7.6 Battery Function

The progress of Li-ion batteries is severely hindered by the complexity of the chemical and physical processes and most importantly, the difficulty of observing these processes in situ during operation. Direct observation of Li ions in a battery in a non-destructive way is not possible by any conventional material analysis technique. The consequence is that a large number of practical and fundamental questions remain including: how do Li-ion battery degradation mechanisms depend on battery conditions? How do the structural changes and electrochemical processes depend on the (dis)charge rate, and what actually determines the (dis)charge rate? An understanding of the interplay between structure, electrochemistry, and reaction mechanisms is required for battery design.

To answer these questions we require time-resolved and non-destructive structural information including Li-ion positions and the Li distribution under operando conditions. These possibilities are offered by in situ neutron techniques that have been realized by the recent developments in neutron sources, detectors, and analysis methods. In situ neutron diffraction enables researchers to follow the structural changes and Li-positions upon all possible electrochemical manipulations in both the positive and negative electrodes. In situ neutron depth profiling determines Li-ion concentrations with high resolution in flat electrodes giving direct insight into the cylindrical batteries. Finally in situ neutron imaging allows a full three-dimensional picture of Li distribution in the battery to be determined. Recent advances in these three techniques will be discussed, including one of the major challenges, the cell design.

7.6.1 In Situ Neutron Powder Diffraction

Historically, in situ NPD has seen comparatively fewer applications to Li-ion battery research relative to in situ X-ray or synchrotron diffraction (see Ref. Brant et al. [177] for further details regarding in situ X-ray-based studies). This is due to a variety of factors, including the inherent complexities of the measurement technique, sample requirements, and the number of neutron diffractometers available for such experiments. However, in recent years neutron diffractometers and research trends have overcome the perceived difficulty of such complex experiments.

Unlike conventional neutron measurements where, in most cases, only the material under study is in the beam, with in situ methods, everything comprising the device can be in the beam, and thus contribute to the observed signal. For diffraction, H can be particularly problematic in the analysis of batteries [178] as the separator (e.g. polyethylene), electrolyte solutions, and the binder are often H-rich. Adding further to the background signal is the liquid or paste-like electrolyte. Therefore, attention has been devoted to custom-made cells for in situ neutron diffraction studies.

Commercial batteries are often produced with minimal quantities of electrolyte to maximise lifetime and avoid the issue of electrolyte leakage. Additionally, the electrodes are often coated on both sides of current collectors, and the overall of quantity of electrodes is significantly larger than that achieved in custom-made batteries. Furthermore, these batteries can be cycled at relatively high rates and are used in "real-world" applications. These considerations can outweigh the detrimental contribution to the background that H-containing components make and yield significant information on the evolution of electrode structure.

As the challenges in battery design and construction are investigated, we also look toward the best instruments for this task by considering the neutron flux, detector, and acquisition time.

7.6.2 Commercial Batteries

In situ neutron diffraction on commercial batteries has been used to provide structural information of electrodes at various states of charge [179, 180], under overcharge (or overdischarge) conditions [181], with fresh and fatigued or used batteries [182, 183], and at different temperatures and electrochemical conditions (applied currents) [180, 183–187]. In situ NPD allows structural snapshots of electrodes within a battery to be obtained, and depending on the diffractometer, these snapshots can be extremely fast such that the time-resolved structural evolution can be captured. Notably, new aspects of the graphitic anode and $LiCoO_2$ cathode that were commercialised in the 1990s are being discovered with this probe. Such insights include the existence of a small quantity of a spinel phase in the previously-thought layered $LiCoO_2$ cathode, an apparent lack of staged Li insertion into graphite to form LiC_{12} with low current rates [180], and the transformation of the graphite anode to a wholly LiC_6 anode with voltages around 4.5 V—above and beyond the recommend limits applied by manufacturers [181]. Figure 7.26 shows $LiCoO_2$ and Li_xC_6 reflections and their evolution as a function of time. Structural changes are a function of the applied charge/discharge rates, with faster structural evolution occurring at higher rates. Importantly, higher rates produce a lower capacity which is directly related to a lower quantity of the LiC_6 (charged anode phase) being formed.

This information explains the processes in electrodes that are well known and in situ neutron diffraction can be used to explore a wide variety of battery-function parameters ranging from current, voltage, and lifetime. Kinetic processes in batteries can be probed with time-resolved data, where rates of structural changes are determined for electrode materials and related to the applied current. In most cases [184, 188, 189] the rate of structural change is directly proportional to the applied current. The rate of lattice expansion and contraction can be used to determine the

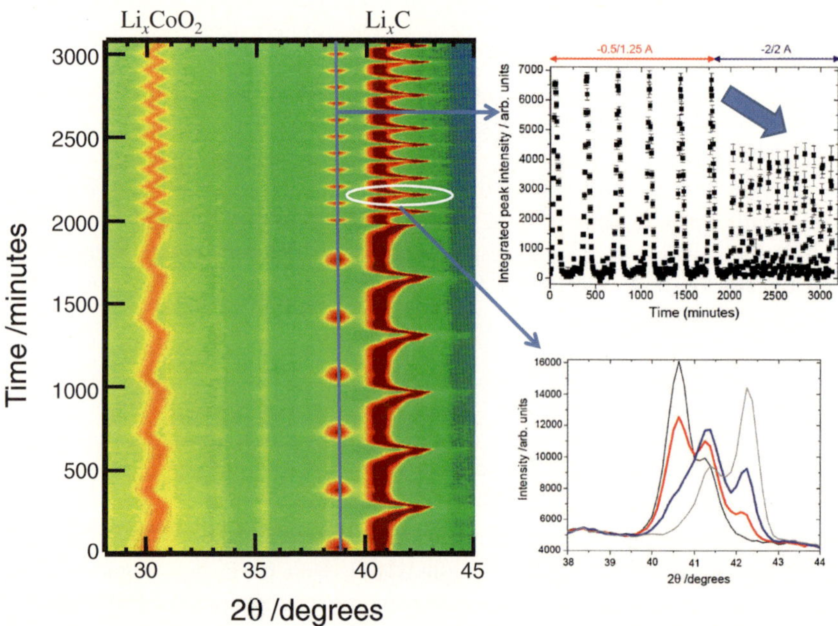

Fig. 7.26 *Left* In situ NPD data collected in consecutive 5-minute intervals. The *red* regions represent significant intensity, while the *yellow/green* regions are the background. Reflections arising from the Li_xCoO_2 cathode and Li_xC_6 anode are labelled. *Right* The integrated intensity of the LiC_6 anode reflection (*top*) corresponding to the line marked on the figure left. As the current is increased the integrated intensity drops, indicating a smaller quantity of LiC_6 forming. The composition of the anode displayed as individual neutron diffraction patterns during the high current experiment (*bottom*). Reprinted (adapted) from (L. Cai, K. An, Z. Feng, C. Liang, S.J. Harris, J. Power Sources **236**, 163 (2013)) [183]

viability of electrode materials for higher power applications. However, the relationship between kinetic structural parameters and the electrochemical capabilities of the battery are yet to be explored in detail. This is an active research area that may yield valuable information with more advanced experiments.

7.6.3 Custom-Made Batteries

Maximising the signal from materials of interest and maintaining acceptable electrochemical performance has been the overriding factor in designing neutron-friendly batteries. Initial designs were plagued by the need for large quantities of electrode materials and the associated requirement to use low current to ensure the reaction of the bulk of the electrode, for example studies of $LiMn_2O_4$ electrodes used 5 g of material as shown in Fig. 7.27 (left) [190, 191]. This design has evolved to designs shown in Fig. 7.27 (right) [178] which increasingly resemble their commercial equivalents, allowing high current to be used and a more direct comparison with commercial performance. For most of these examples the polyethylene separator is replaced with a separator containing a smaller amount of H, e.g. polyvinylidene difluoride, and the electrolyte solution is replaced with deuterated equivalents. By using the design in Fig. 7.27 it was possible to show the loss of long-range order of the MoS_2 anode during its first discharge [178], the composite nature of the $TiO_2/Li_4Ti_5O_{12}$ anode [192], relaxation phenomena in $LiCo_{0.16}Mn_{1.84}O_4$ cathodes [193], evolution of $LiMn_2O_4$ structure [190, 191], and the reaction mechanism evolution of $LiFePO_4$ [188, 194].

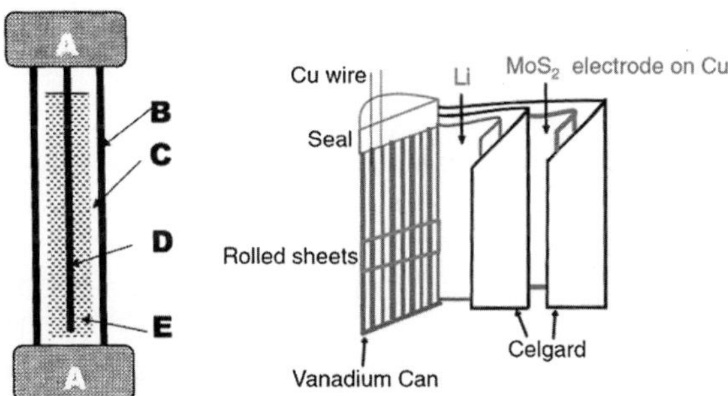

Fig. 7.27 *Left* One of the first batteries developed for in situ neutron diffraction, where A are brass plugs, B is a Pyrex® tube lined with Li foil, C is the separator soaked in H-containing electrolyte, D is the stainless-steel current collector, and E is the active material mixed with C black and binder. *Right* A more recent in situ neutron diffraction battery design with components as labelled. Reprinted (adapted) from (N. Sharma, G. Du, A.J. Studer, Z. Guo, V.K. Peterson, Solid State Ionics **199–200**, 37 (2011)) [178]

Alternate designs have been developed for in situ neutron diffraction experiments and these include coin-type cells [195–197] which still feature relatively-thick electrodes but have been used to successfully investigate $Li_4Ti_5O_{12}$, graphite, and $LiFePO_4$. Similarly, pouch-type cells with alternate layers of cathode and anode-coated current collectors are applicable for investigating full cells, as opposed to the use of Li metal in the majority of the previous examples. Studies have been conducted on $Li[Ni_{1/3}Mn_{1/3}Co_{1/3}]O_2$‖graphite, $Li[Li_{0.2}Ni_{0.18}Mn_{0.53}Co_{0.1}]O_2$‖graphite [198], and $LiNi_{0.5}Mn_{1.5}O_4$‖$Li_4Ti_5O_{12}$ [199] full cells.

The motivation for designing these neutron-friendly cells is that any electrode material can be tested in situ in a real cell. Effectively, some of these designs can be manufactured using relatively-small electrode sizes (0.5–1 g) allowing a variety of materials to be investigated, and the interplay between structure, electrochemistry, and reaction mechanism can be elucidated. This information can then be used to direct the choice of future electrode-materials.

Some of these cells have been used to extract time-dependent information which reveals the rate of reactions as a function current applied, relating structural perturbations to electrochemical factors [188, 193]. Of particular note has been the study of the reaction mechanism of $LiFePO_4$ [188]. The evolution of $LiFePO_4$, either by a single-phase solid-solution reaction, or a two-phase reaction, during charge/discharge has been extensively discussed in the literature (see [188]). Some parameters that lead to a particular type of reaction mechanism being favoured have been detailed. However, there was a lack of time-resolved information concerning bulk-electrode behaviour in a commercially-equivalent cell to definitively establish the working mechanism of $LiFePO_4$. Time-resolved in situ NPD data showed the evolution of the reaction mechanism of $LiFePO_4$ during charge/discharge processes. This is significant because the experiment probed the material under real working-conditions at the bulk-electrode scale. It should be noted that the $LiFePO_4$ sample used was expected to have only two-phase behaviour, and this work revealed a solid-solution reaction mechanism region during charge/discharge which is followed by a two-phase reaction mechanism. Moreover, the transition between the 'competing' reaction mechanisms was identified and characterized to be a gradual transition with solid-solution reactions persisting into the two-phase reaction region, rather than an abrupt transition. Figure 7.28 details this evolution and the co-existing reaction mechanism region.

Therefore, in situ NPD not only provides information on the evolution of electrode structure, but also on the evolution of the (de)lithiation reaction mechanisms of the electrode. This information can be time-dependent and as a function of the electrochemical process, and can be used to design alternative electrodes that avoid, or undergo, certain reaction mechanisms to enhance battery performance.

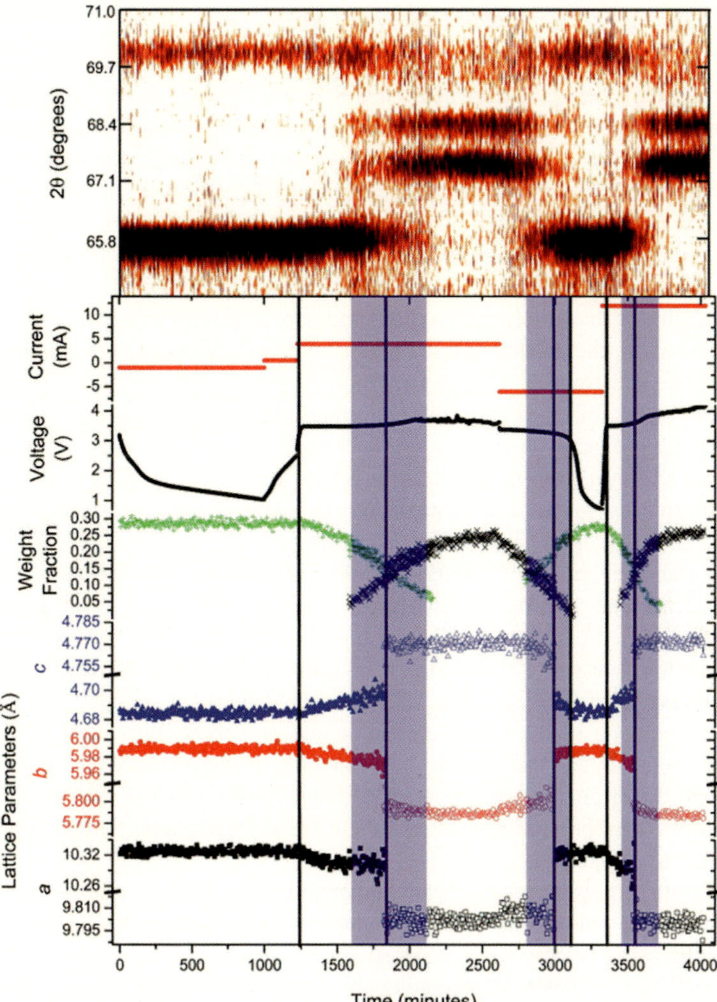

Fig. 7.28 In situ NPD data of the Li‖LiFePO$_4$ battery (*top*) with scaled intensity highlighting the LiFePO$_4$ and FePO$_4$ 221 and 202 reflections. *Bottom* The applied current is the *red line* and the measured voltage is the *black line*. Parameters derived from the neutron data are shown including the phase fraction of LiFePO$_4$ (*green crosses*), the phase fraction of FePO$_4$ (*black crosses*), and the lattice parameters, where a is *black*, b is *red*, and c is *blue*. The lattice parameters for LiFePO$_4$ are solid symbols and those for FePO$_4$ are open symbols. *Vertical black lines* represent the onset of the solid-solution reaction and *vertical purple lines* indicate the chronological transition from a composition that is predominantly Li$_{1-y}$FePO$_4$ to predominantly Li$_x$FePO$_4$, where $x \approx 0.03$ and $y \approx 0.04$. Shaded regions indicate the coexistence of solid solution and two-phase reactions. Reprinted from (N. Sharma, X. Guo, G. Du, Z. Guo, J. Wang, Z. Wang, V.K. Peterson, J. Am. Chem. Soc.**134**, 7867 (2012)) [188]

7.6.4 Kinetics of Lithium Distribution

Ultimately, the goal of in situ neutron diffraction is to track the Li content and location in crystalline electrode materials as a function of charge and discharge. This is a difficult task, that until recently was only demonstrated at limited battery states of charge [191] with long collection times and effectively under equilibrium conditions. Other studies have inferred Li content via electrochemical approximations (amount of charge transferred) and the evolution of reflection intensities [196]. If the Li composition of an electrode can be reliably determined as a function of discharge/charge this would give a direct measure of the capacity of the battery, or more accurately, the capacity of the battery that is stored in the crystalline component of the electrode.

Recently, the ability to track the Li location and content as a function of time (and charge/discharge) has been demonstrated using commercial $Li_{1+y}Mn_2O_4$ cathodes [189]. Arguably, this represents the most Li-centric view of a Li-ion battery during operation. The Li evolution is found to differ at a structural level during charge/discharge (Fig. 7.29) accounting for the ease of discharging these types of cathodes, relative to charging. Additionally, the Li evolution is shown to progress from one to two crystallographic sites during the charge/discharge processes. The lattice parameter follows a linear relationship with Li content during single Li site processes (Vegard's law) and during processes involving two Li sites the relationship between the lattice parameter and the Li occupancy and site is a linear combination of the individual single site processes (Fig. 7.30). This work provided unparalleled insight into the function of the cathode and is used to understand the origins of how the electrode functions. Further studies on structural permutations may provide insight on how these electrodes can be improved from the perspective of the Li.

$LiNi_{0.5}Mn_{1.5}O_4$ is attracting significant attention for cathode applications due to the high-voltage redox couple during battery function, and $Li_4Ti_5O_{12}$ is attracting attention for anode applications due its small volume change during Li insertion and extraction. By specifically constructing a neutron-friendly cell made of this electrode-combination it was possible to study the structural evolution of these materials using time-resolved in situ neutron diffraction [199]. This highlights another advantage of using custom-made cells for in situ neutron diffraction experiments, where research is not limited to commercially-available materials. In this case, it was possible to determine the evolution of Li occupation in the cathode and indirectly infer the Li occupation in the anode (Fig. 7.31) in addition to determining the reaction mechanism evolution for the electrodes. It was found that a solid-solution reaction occurred at the cathode with the Ni^{2+}/Ni^{3+} redox couple at ~ 3.1 V and a two-phase reaction with the Ni^{3+}/Ni^{4+} redox couple at ~ 3.2 V. Thus, the extraction of Li from the cathode and insertion of Li into the anode during charge was directly determined, again in real-time. This opens up a way to evaluate a range of materials used as electrodes in Li-ion batteries, where how Li is extracted and inserted while a battery functions can be determined.

Fig. 7.29 **a** The evolution of Li at the $8a$ (*black*) and $16c$ sites and in a formula unit of $Li_{1+y}Mn_2O_4$ (lower plot) during charge/discharge. The discharge process shows an increase in Li content, whilst charge shows a decrease. **b** A representation of the sites described in (**a**) for $Li_{1+y}Mn_2O_4$. Reprinted from (N. Sharma, D. Yu, Y. Zhu, Y. Wu, V.K. Peterson, Chem. Mater. **25**, 754 (2013)) [189]

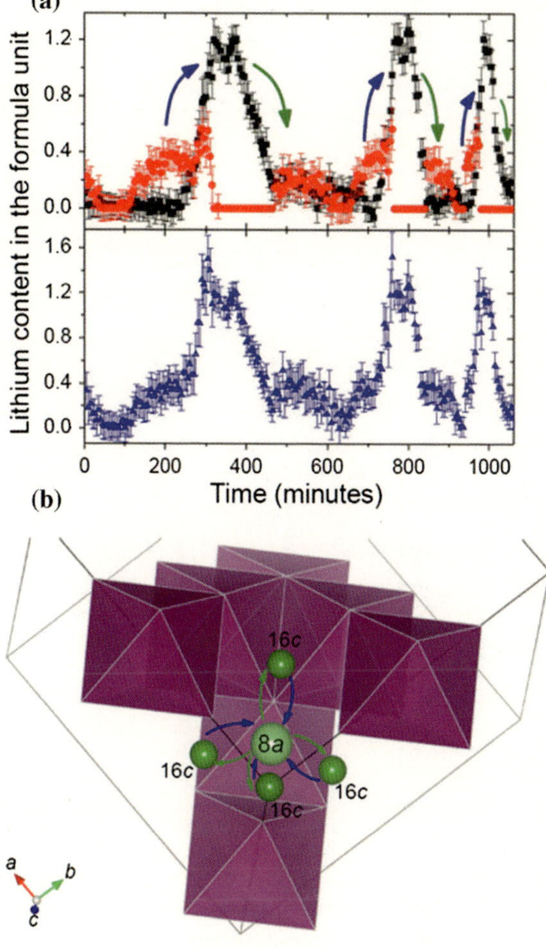

Fig. 7.30 The relationship between Li site occupancies and the lattice parameters of $Li_{1+y}Mn_2O_4$. Single-site regions are shown in black and red and mixed-site regions are in blue and purple. Reprinted from (N. Sharma, D. Yu, Y. Zhu, Y. Wu, V.K. Peterson, Chem. Mater. **25**, 754 (2013)) [189]

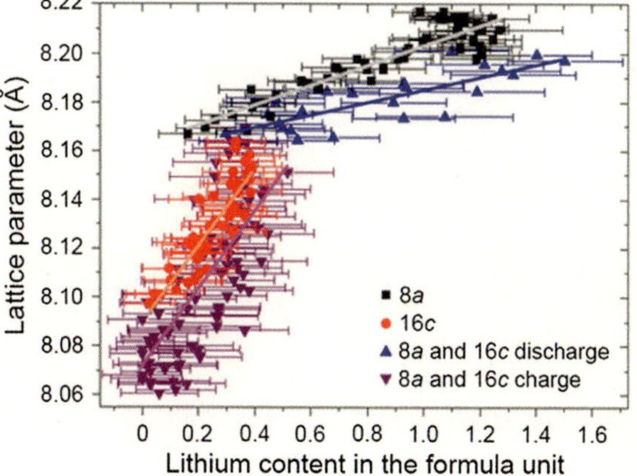

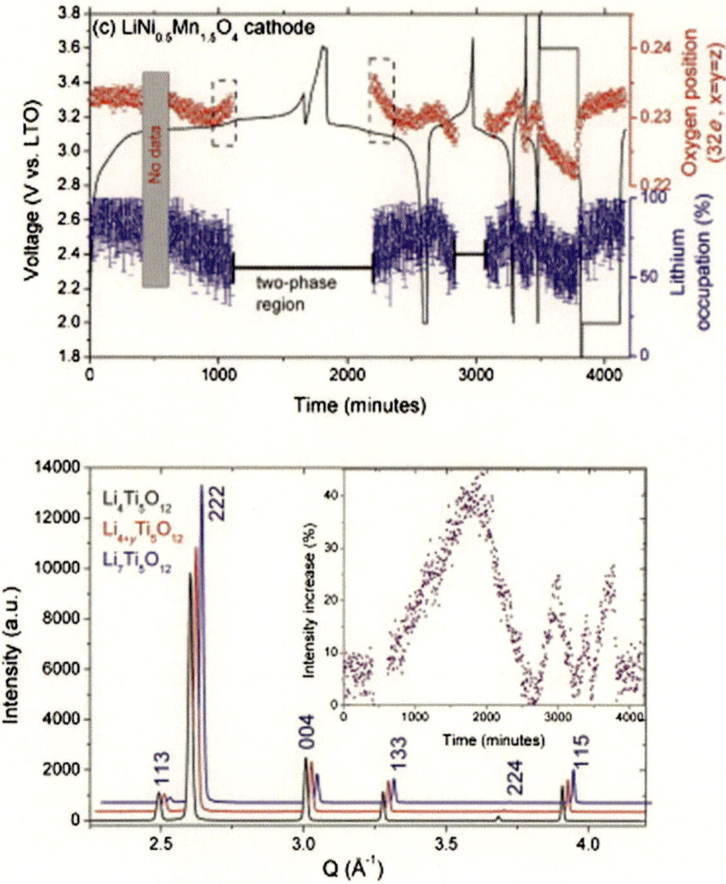

Fig. 7.31 *Top* Changes in the Li occupation and O positional parameter extracted from in situ neutron diffraction data of the $LiNi_{0.5}Mn_{1.5}O_4$ cathode. The battery operation is shown by the potential curve in black. *Bottom*: Simulated patterns of the $Li_{4+y}Ti_5O_{12}$ anode slightly offset in Q to show the differences in reflection intensity. The inset shows the evolution of the $Li_{4+y}Ti_5O_{12}$ 222 reflection which coincides with the expected variation with (de)lithiation from simulations. Reprinted from (W.K. Pang, N. Sharma, V.K. Peterson, J.-J. Shiu, S.H. Wu, J. Power Sources **246**, 464–472(2014)) [199]

7.6.5 Neutron Depth Profiling

A unique method to "see" Li-ion concentration profiles is provided by neutron depth profiling (NDP), Fig. 7.32. Previously it has been shown that NDP is capable of determining Li concentration gradients in optical waveguides [200], electrochromic devices [201], and under ex situ conditions in thin film battery electrodes and electrolytes [202]. NDP uses a neutron-capture reaction for ^6Li resulting in:

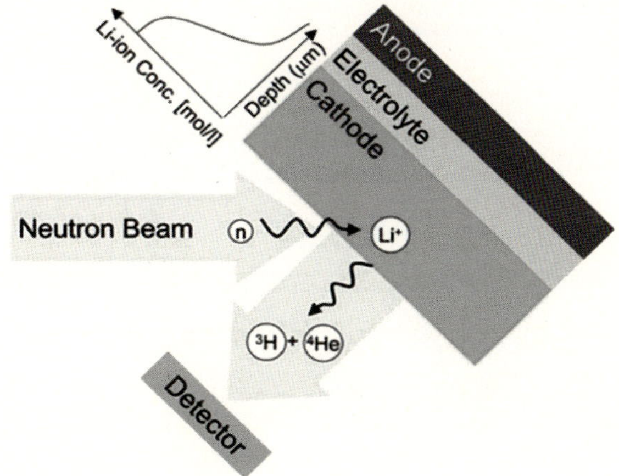

Fig. 7.32 Schematic principle of NDP applied to Li-ion battery systems

$$^{6}\text{Li} + n_{\text{thermal}} \rightarrow\, ^{4}\text{He}\ (2.06\,\text{MeV}) +\, ^{3}\text{H}(2.73\,\text{MeV}) \tag{7.1}$$

The kinetic energy of the products due to $\Delta E = \Delta mc^2$, where E is the energy, m is the rest mass of the particles and c is the speed of light, is distributed over the tritium (^{3}H) and the alpha particle (^{4}He), while the incoming thermal energy of the neutron at $\sim$25 meV, is negligible. Due to the small particle flux and the inherently-low interaction of neutrons with matter, NDP is a totally non-destructive technique. When such a capture reaction takes place in a Li-ion battery electrode, the particles produced (^{4}He and ^{3}H) lose part of their kinetic energy due to the scattering by the electrode material, referred to as stopping power. The stopping power is directly related to the composition and density of the electrode and hence is a known quantity. Therefore, by measuring the energy of the ^{4}He and ^{3}H ions when they exit from the electrode, the depth of the capture reaction can be reconstructed. Typically, the spatial resolution of NDP for well-defined homogeneous layers is on the order of tens of nano-meters, Currently, the main restrictions of the NDP technique is the maximum depth that can be probed and the time resolution which, depending on the material investigated and the in situ cell design, are approximately 5–50 microns and 10–20 min, respectively. Ideal solid-state batteries can be designed with high spatial homogeneity for initial experiments, before proceeding to more complicated systems.

NPD has only been applied occasionally to Li-ion battery research, however, in these ex situ studies [202–205] NDP is very powerful in identifying Li-ion transport and aging mechanisms. The possibilities of NDP in Li-ion battery research is demonstrated with the first in situ study on thin film solid-state batteries probing the kinetic processes in these Li-ion batteries.

Oudenhoven et al. brought NDP one step further, demonstrating that Li depth profiles can be measured in situ in an all solid-state micro battery system during (dis)charging [206]. The Li-ion distribution was studied in a thin film solid state battery stack containing a monocrystalline Si substrate with a 200 nm Pt current

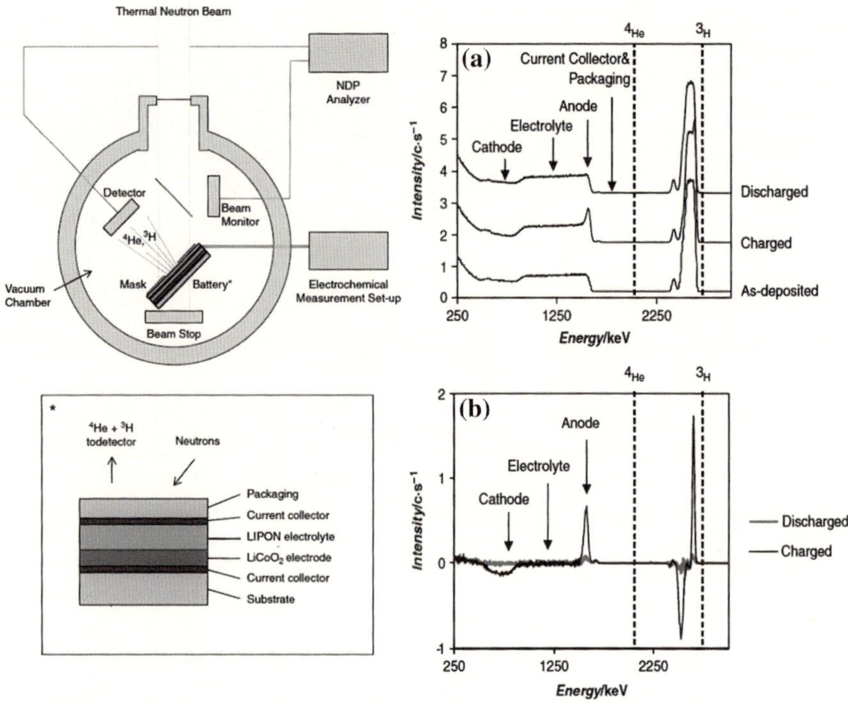

Fig. 7.33 *Left* Schematic representation of the NDP set-up. The inset below shows the orientation of the battery inside the NDP measurement chamber. *Right* **a** Overview of the NDP spectrum of the as-deposited battery and the battery after the first charge and discharge. An offset is applied to distinguish the various spectra. Based on the ^{4}He and ^{3}H reference energies (indicated by the *dashed lines*) the packaging/top current collector, the anode, the electrolyte and the cathode can be clearly distinguished. **b** When the spectrum of the as-deposited state is subtracted from the spectra of the charged and discharged states, the amount of Li moved during use of the battery can straightforwardly be determined. Reprinted with permission from (J.F.M. Oudenhoven, F. Labohm, M. Mulder, R.A.H. Niessen, F.M. Mulder, P.H.L. Notten, Adv. Mater. **23**, 4103 (2011)) [221]. Wiley

collector, 500 nm $LiCoO_2$ positive electrode, 1.5 μm N-doped Li_3PO_4 (LiPON) electrolyte, and a 150 nm Cu top current-collector. The basic setup of the experiment is shown in Fig. 7.33. By subtracting the NDP spectrum of the as-prepared electrode from the charged and discharged spectra, the changes in Li-distributions can be observed directly, see Fig. 7.33. Upon charging, Li in the positive $LiCoO_2$ electrode is depleted and increased at the negative Cu current collector.

The development of large concentration-gradients in both the $LiCoO_2$ electrode and the LiPON solid electrolyte, Fig. 7.34, reveals that in this system ionic transport in both electrolyte and electrode limit the overall charge-rate. The cathode was enriched with ^{6}Li to highlight the redistribution of ^{6}Li and the natural abundance of ^{6,7}Li in the electrolyte during time-dependent experiments. In this case, the NDP intensity increases by approximately a factor of 13. Apart from being able to observe

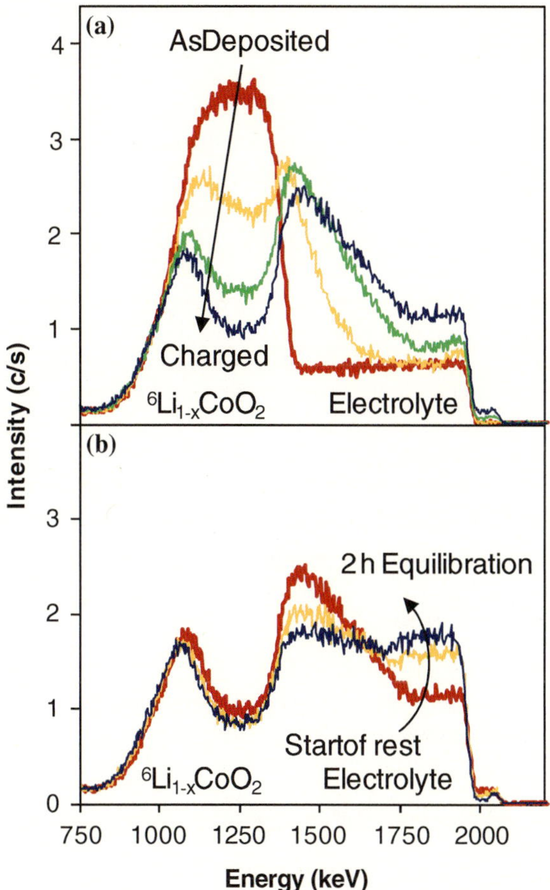

Fig. 7.34 First in situ NDP spectra representing the Li concentration depth profiles of a battery during operation. **a** Enriched $^6LiCoO_2$ cathode and naturally abundant $^{6,7}LiPON$ electrolyte during several stages of the charging process showing the removal of ^6Li from the electrode and large ^6Li concentration-gradients in both electrode and electrolyte. **b** Equilibration in the charged state after 0.1 and 2 h showing the disappearance of the ^6Li concentration-gradient in the LiPON electrolyte and remaining concentration gradients in the cathode indicating two-phase separation into a Li rich and Li poor phase. Since in the cathode area the red curve is higher than the blue, some equilibration takes place between cathode and electrode. Reprinted with permission from (J.F.M. Oudenhoven, F. Labohm, M. Mulder, R.A.H. Niessen, F.M. Mulder, P.H.L. Notten, Adv. Mater. **23**, 4103 (2011)) [221]. Wiley

where ^6Li is going, the expected diffusive equilibration of the ^6Li concentration within the electrolyte was observed during a 2 h equilibration period. Interestingly, the enriched Li remains in the $LiCoO_2$ electrode, even though the exchange current that establishes the dynamic equilibrium would be expected to redistribute the ^6Li equally throughout the $LiCoO_2$ electrode and the LiPON electrolyte. The absence of vacancies at the initial stage probably makes the exchange-current extremely small.

As the battery is charged at 0.5 °C, this results in a large decrease in the Li-ion signal of the LiCoO$_2$ electrode, as shown in Fig. 7.34.

The stronger decrease in Li-ion signal near the interface with the electrolyte suggests an inhomogeneous Li-ion distribution in the electrode. Although this may be the case, a redistribution of the ^{6}Li ions due to exchange with the electrolyte will lead to a lower Li-ion signal in the electrode. That this is indeed part of the explanation is clear from the almost 70 % decrease in Li-ion signal. This decrease is more than would be expected under the mild electrochemical conditions that should lead to Li$_{0.5}$–CoO$_2$, and hence at most a factor of two decrease in the Li-ion signal is to be expected. However, the inhomogeneous signal from the electrode indicated that the exchange does not reach the back part of the electrode that is closest to the current collector. The inhomogeneous distribution of the Li-ion signal originating from the electrolyte indicates the presence of an inhomogeneous ^{6}Li and Li-ion distribution. The evolution of this non-equilibrium situation was investigated by relaxing the system after charging during a period of 2 h and taking NDP spectra, shown in Fig. 7.34. After 2 h the ^{6}Li gradient almost vanishes in the electrolyte, whereas it remains in the LiCoO$_2$ electrode. Clearly, Li-ions are much more mobile in the electrolyte compared to in the electrode. The work of Oudenhoven et al. shows for the first time that the evolution of the Li distribution and gradient under dynamic conditions can be studied.

7.6.6 Neutron Imaging

Neutron imaging (radiography) is becoming increasingly important in the study of Li-ion batteries as the spatial and temporal resolution of the detectors continually improve, and more advanced computational methods allow tomographic and/or three-dimensional rendering [207]. Neutron radiography (NR) is used to show macroscopic information concerning the Li distribution within a Li-ion battery, and in some cases while a process is occurring or at different states-of-charge [208]. Additionally, the H distribution in the electrolyte can be probed [209]. Examples of such studies include the Li distribution at the charged state versus the discharged state, during high-temperature battery operation, during fast charge/discharge cycling, and during overcharging [207, 208, 210, 211].

Neutron imaging has also been used to study alkaline [212, 213] and Li-air batteries [214]. The future for neutron radiography relies on new instruments with improved spatial resolution, but also temporal resolution to allow time-resolved in situ experiments. Another method under considerable investigation is the combination of diffraction and imaging, which requires the definition of a gauge volume which is imaged and from which diffraction data can also be collected. This has been demonstrated for physically-larger batteries such as Na metal halide batteries, which usually have larger electrodes [215]. However, to be pertinent for Li-ion battery research, the gauge volume has to be reduced to become comparable to the thickness of electrode layers.

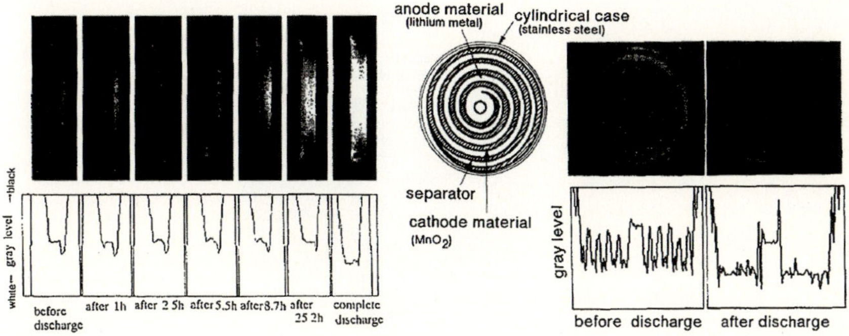

Fig. 7.35 Neutron imaging studies of coin (*left*) and cylindrical (*right*) cells. The dark images correspond to images at different states of battery charge, with *white* regions representing high Li concentration. The graphs below the images show the integrated intensity highlighting the evolution of the Li distribution. The cylindrical battery construction is also shown (*right*). **a** Variation of neutron radiography images of CR 1220 with discharge. **b** The NR images of CR1/3-1H before and after discharge. Reprinted (adapted) with permission from (M. Kamata, T. Esaka, S. Fujine, K. Yoneda, K. Kanda, Nucl. Instr. Meth. Phys. Res. A **377**, 161 (1996)) [158] and (M. Kamata, T. Esaka, S. Fujine, K. Yoneda, K. Kanda, J. Power Sources **68**, 459 (1997)) [216]. Elsevier

Early work on imaging Li-ion batteries explored the different types of battery construction, e.g. coin, prismatic, and cylindrical cells, and the distribution of Li at various battery states or during charge/discharge [216]. Figure 7.35 (left) shows images of a coin cell (CR1220 from Panasonic) from the charged to the discharged state, where lighter (white) regions at the charged state correspond to the Li anode (Li metal) and electrolyte (arrow). Over the course of discharge the Li-ions move towards the cathode (MnO_2) resulting in an even distribution of white regions. The authors comment that if standard components and standardised cells are constructed then a more quantitative description of the Li distribution can be made. They also explored charging rates and other constructions, some of which had further experimental difficulties due to the internal structure of the batteries and the need to account for absorption by various layers. An example of the same electrode chemistry in the cylindrical case (CR1/3–1H) is shown in Fig. 7.35 (*right*) before and after discharge [158] showing similar Li distributions at the charged and discharged states.

In commercial batteries overcharging can be a potentially-devastating failure mechanism, and imaging studies on commercial graphite//$LiNi_{0.8}Co_{0.15}Al_{0.05}O_2$ (NCA) batteries show what is deposited on the graphite anode during overcharge [208]. By performing an in situ measurement the deposition of a material on the graphite anode was studied (Fig. 7.36) and later determined to be Li. In addition, the authors were able to characterize 'where' the Li deposits during battery processes. Another work explored 'fresh' and 'fatigued' batteries, where batteries that had been cycled 200 times. The 18650 cylindrical batteries showed no differences at the macroscopic level in the neutron images of between fatigued and fresh batteries [182], even though neutron diffraction data indicated less Li insertion in fatigued graphite.

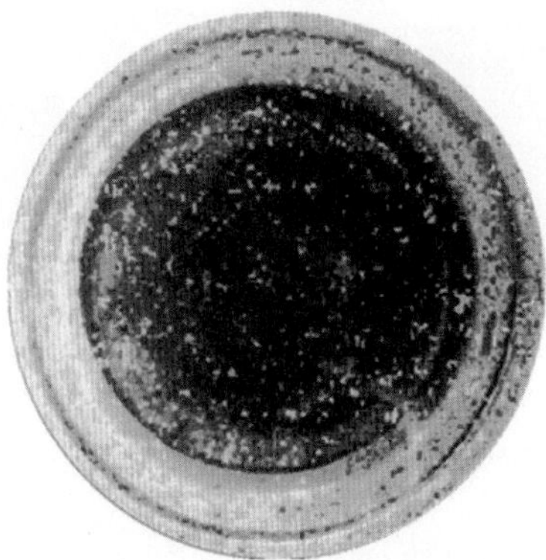

Fig. 7.36 Neutron image from a coin cell at 4.8 V with the black region showing the anode. Other *shaded regions* represent regions of high neutron-attenuation (likely to be Li-containing). Reprinted (adapted) with permission from (A. Same, V. Battaglia, H.-Y. Tang, J.W. Park, J. Appl. Electrochem. **42**, 1 (2012)) [208]. Springer

Another study illustrated that a 14 μm spatial resolution is attainable for battery samples using neutron imaging [210]. This work used a purpose-built graphite-containing cell to quantify Li content during charge/discharge and the residual Li content after each cycle, showing quantification over several cycles (e.g. capacity loss). Figure 7.37 shows the evolution of Li content and its distribution in graphite during the first discharge. The Li distribution was compared during cycling and between cycles. A slight difference in Li content between the separator and current collector was found. Further work investigated $LiFePO_4$‖graphite pouch-cells and revealed Li concentration gradients across electrodes and in their bent regions [211]. Figure 7.38 shows the distribution of Li in the layers of the pouch cell at various states-of-charge. The authors used the Beer-Lambert law to correlate colour gradients, shown in Fig. 7.38, to the Li concentration. One advantage of using a pouch cell is that one image contains many layers, so an increase in electrode thickness can be seen in multiple layers verifying the result (as can Li concentration gradients). Clearly, this information can direct the development of better per-forming electrodes.

More recent work used cold neutrons rather than thermal neutrons, harnessing the stronger interaction of colder neutrons with matter, to visualize Li-ion distributions in Li–I batteries used in pacemakers [217]. This work also used three-dimensional imaging (tomography) and discussed methods to improve the signal-to-noise ratio in the images. The authors collected 50 images at 0.3 s for each angular step (rotation) of 0.91° which were then used to construct the three-dimemsional image. Figure 7.39 is an example of a cross section of a neutron tomography image of the fresh battery (left) and after a certain period of discharge (right). The battery is made of plates of Li and I. An unexpected change in the Li distribution (white) was observed in this study, where the smooth distribution became highly irregular.

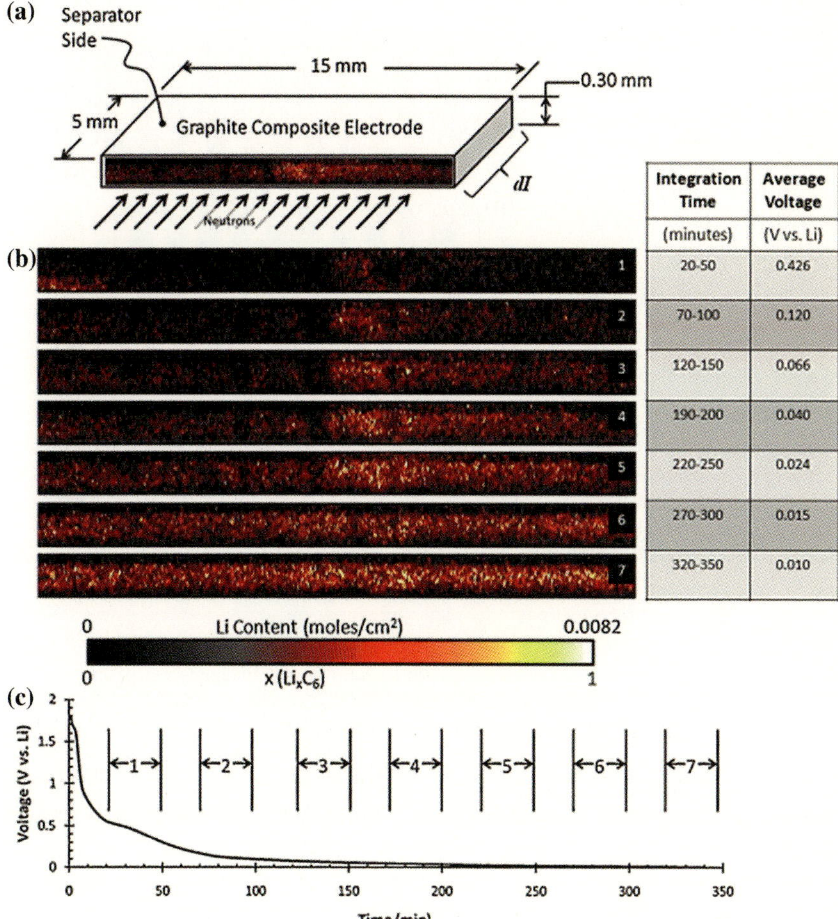

Integration Time	Average Voltage
(minutes)	(V vs. Li)
20-50	0.426
70-100	0.120
120-150	0.066
190-200	0.040
220-250	0.024
270-300	0.015
320-350	0.010

Fig. 7.37 Li distribution in a graphite electrode during first discharge **a** showing the geometry of the experiment, **b** time-resolved radiographs and parameters and **c** the potential profile. Reprinted (adapted) with permission from (J.P. Owejana, J.J. Gagliardo, S.J. Harris, H. Wang, D.S. Hussey, D.L. Jacobson, Electrochim. Acta **66**, 94 (2012)) [210]. Elsevier

The irregularity of the Li distribution after discharge is extracted in the three-dimensional image shown in Fig. 7.39, where the Li formations are clearly seen.

Further experimentation was undertaken on a Li-ion polymer battery using monochromatic imaging with cold neutrons, specifically targeting the anode and the processes that occur within it [218]. The LiC$_6$ compound, but no evidence of the staging phenomenon often observed in LixC$_6$ anodes was seen. This work was the first real-time in situ imaging of a commercial Li-ion battery, with some results shown in Fig. 7.40.

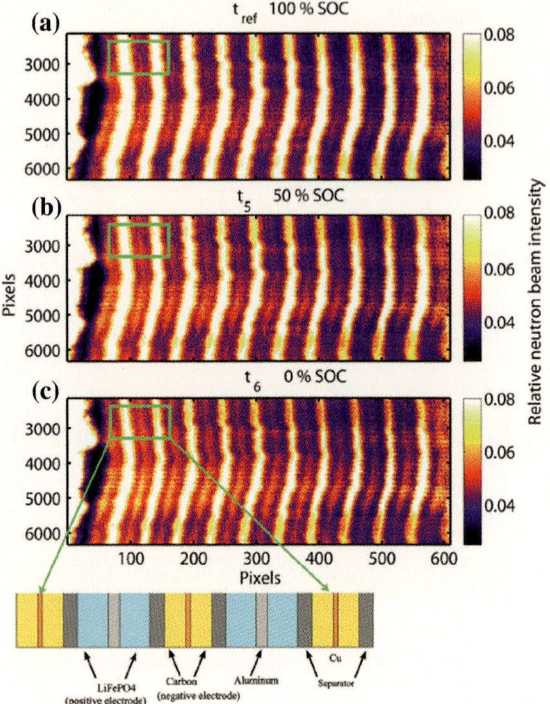

Fig. 7.38 Images from a LiFePO$_4$‖graphite pouch cell during charge/discharge. Reprinted with permission from (J.B. Siegel, X. Lin, A.G. Stefanopoulou, D.S. Hussey, D.L. Jacobson, D. Gorsich, J. Electrochem. Soc. **159**, A523 (2011)) [211], Copyright (2011). The Electrochemical Society

A subset of radiography research using commercial batteries, and in some cases custom-made batteries, is the study of gas evolution [209]. One in situ study showed how excess electrolyte present in batteries is consumed in the first charge-cycle, resulting in the formation of the SEI layer and some volume expansion. Additionally, gases were found to be evolved during the first charge. Figure 7.41 shows the consumption of excess electrolyte in these cells. The authors were also able to approximate the amount of expansion and contraction of the electrodes indirectly.

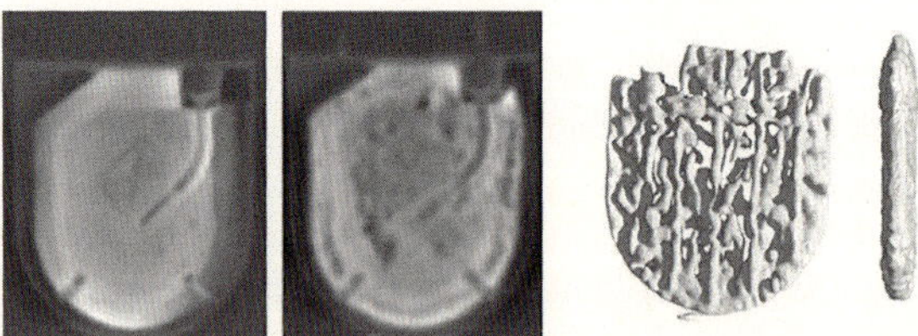

Fig. 7.39 *Left* Cross-sections of a tomography image before (*left-most*) and after discharge. High neutron-attenuation regions are white. *Right* A three-dimensional rendering with only the Li contributions (other components removed due to the large contrast available). Reprinted (adapted) with permission from (N. Kardjilov, A. Hilger, I. Manke, M. Strobl, W. Treimera, J. Banhart, Nucl. Instr. Meth. Phys. Res. A **542**, 16 (2005)) [217]. Elsevier

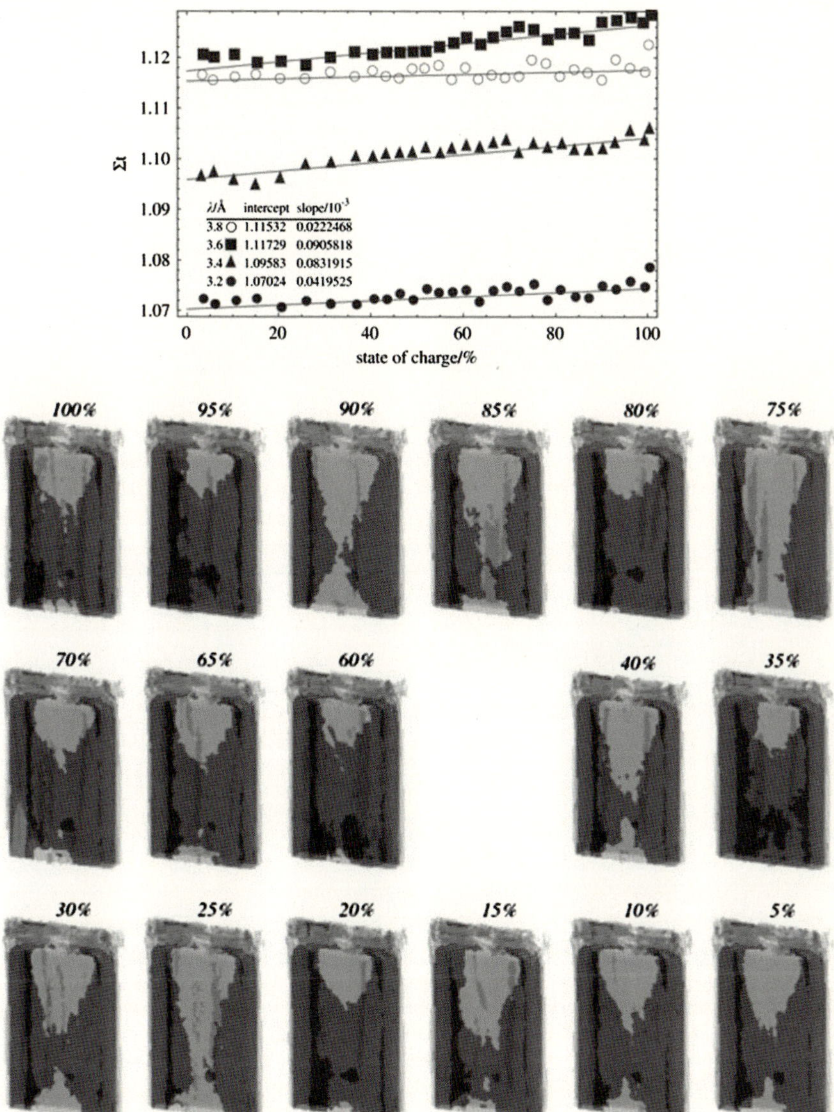

Fig. 7.40 *Left* Selected attenuations (corresponding to LiC_6) at different wavelengths plotted as a function of charge. The 3.6 Å attenuation seems to show the largest response to the formation of LiC_6. *Right* Real-time tomography of a commercial Li-ion battery. Reprinted (adapted) with permission from (L.G. Butler, B. Schillinger, K. Ham, T.A. Dobbins, P. Liu, J.J. Vajo, Nucl. Instr. Meth. Phys. Res. A **651**, 320 (2011)) [218]. Elsevier

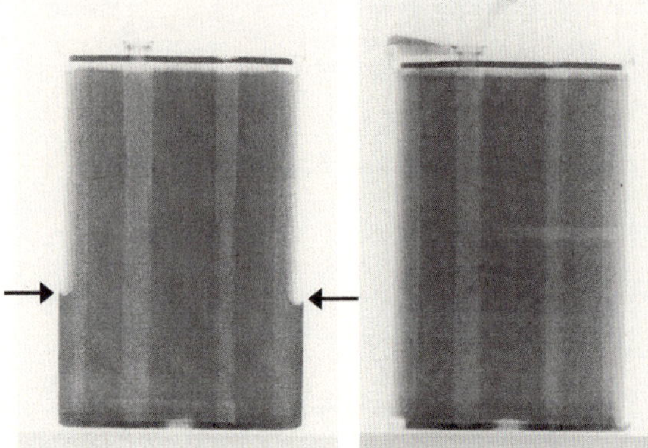

Fig. 7.41 Neutron radiography image of a fresh Li-ion battery (*left*) and a battery cycled 70 times (*right*). The arrow indicates the excess electrolyte level. Reprinted (adapted) with permission from (M. Lanz, E. Lehmann, R. Imhof, I. Exnar, P. Novak, J. Power Sources **101**, 177 (2001)) [209]. Elsevier

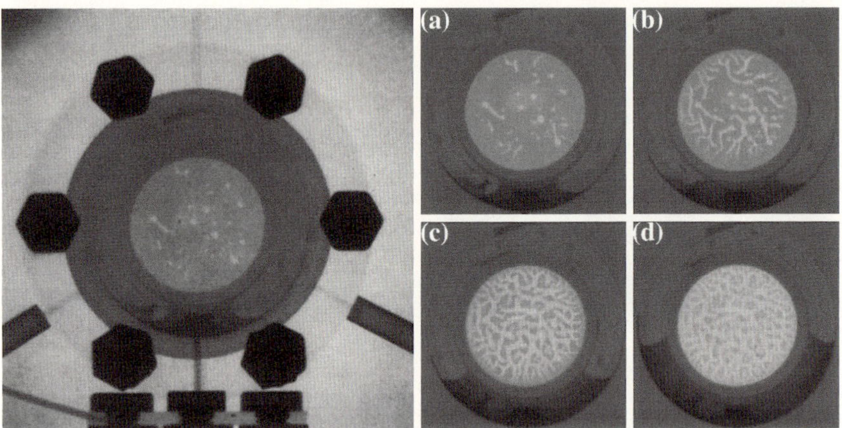

Fig. 7.42 *Left* A neutron radiography image of the test cell prior to electrochemical cycling. *Right* Images at progressive states of cycling from (a) to (d). Reprinted (adapted) with permission from (D. Goers, M. Holzapfel, W. Scheifele, E. Lehmann, P. Vontobel, P. Novak, J. Power Sources **130**, 221 (2004)) [219]. Elsevier

In situ neutron radiography has been extensively used to study the interface between graphite and a range of gel-based electrolytes [219]. By using this technique, the generation of gas bubbles in the first charge can be visualized and quantified. This information allows the best electrolyte to be proposed, noting that the generation of gas bubbles, particularly on graphite surfaces, leads to performance degradation. This measurement also provided information on the spatial distribution and kinetic evolution of gas bubbles, as well as the electrolyte displacement and volume expansion in graphite. In order to undertake these measurements, specialised cells were developed. A neutron image of the cell is shown in Fig. 7.42. For the in situ experiment the exposure time for each image was 20 s and an image was recorded every 2 min. Figure 7.42 shows how channels of gases are formed seemingly-randomly in the cell and their evolution at different times. It was found that LiC_6 is formed only where gas emission is absent, illustrating some heterogeneities in the charge distribution and electrode composition. The gel-based electrolytes tested in this study show less gas evolution (3 %) compared to liquid-based electrolytes (60 %) and this was related to the smaller amount of gas evolution during the first cycle.

7.7 Perspectives

This chapter has aimed at demonstrating how neutron-scattering methods allow researchers to elucidate crucial structural and kinetic properties of electrodes, electrolytes, and complete batteries. Neutron-scattering techniques play a key role in the development of new materials by relating structure to functional properties. Future battery research and development will in particular profit from the advances in in situ neutron-scattering techniques, probing complete battery systems. This gives the opportunity to relate battery performance to material and electrode structural, morphological, and dynamic properties under non-equilibrium and ageing conditions, which is vital information for the design of future batteries.

References

1. J.M. Tarascon, M. Armand, Nature **414**, 359 (2001)
2. M. Armand, J.M. Tarascon, Nature **451**, 652 (2008)
3. M. Winter, J.O. Besenhard, M.E. Spahr, P. Novak, Adv. Mater. **10**, 725 (1998)
4. J.B. Goodenough, Y. Kim, Chem. Mater. **22**, 587 (2010)
5. B.L. Ellis, K.T. Lee, L.F. Nazar, Chem. Mater. **22**, 691 (2010)
6. M.S. Whittingham, MRS Bull. **33**, 411 (2008)
7. M. Gaberscek, J. Jamnik, Solid State Ionics **177**, 2647 (2006)
8. J. Jamnik, R. Dominko, B. Erjavec, M. Remskar, A. Pintar, M. Gaberscek, Adv Mater **21**, 2715 (2009)
9. C. Fongy, S. Jouanneau, D. Guyomard, J.C. Badot, B. Lestriez, J. Electrochem. Soc. **157**, A1347 (2010)
10. C. Fongy, A.C. Gaillot, S. Jouanneau, D. Guyomard, B. Lestriez, J. Electrochem. Soc. **157**, A885 (2010)
11. D.P. Singh, F.M. Mulder, A.M. Abdelkader, M. Wagemaker, Adv Energy Mater **3**, 572 (2013)
12. D.P. Singh, F.M. Mulder, M. Wagemaker, Electrochem. Commun. **35**, 124 (2013)
13. M.M. Thackeray, J. Electrochem. Soc. **142**, 2558 (1995)
14. A.K. Padhi, K.S. Nanjundaswamy, J.B. Goodenough, J. Electrochem. Soc. **144**, 1188 (1997)
15. U. Kasavajjula, C.S. Wang, A.J. Appleby, J. Power Sources **163**, 1003 (2007)
16. J.R. Szczech, S. Jin, Energy Environ. Sci. **4**, 56 (2011)
17. B.L. Ellis, K. Town, L.F. Nazar, Electrochim. Acta **84**, 145 (2012)
18. T. Ohzuku, Y. Makimura, Chem. Lett. **7**, 642 (2001)

19. H. Kawai, M. Nagata, H. Tukamoto, A.R. West, J. Power Sources **81**, 67 (1999)
20. K. Amine, H. Tukamoto, H. Yasuda, Y. Fujita, J. Power Sources **68**, 604 (1997)
21. K. Amine, H. Tukamoto, H. Yasuda, Y. Fujita, J. Electrochem. Soc. **143**, 1607 (1996)
22. Y. Xia, Q. Zhang, H. Wang, H. Nakamura, H. Noguchi, M. Yoshio, Electrochim. Acta **52**, 4708 (2007)
23. N. Ravet, J.B. Goodenough, S. Besner, M. Simoneau, P. Hovington, M. Armand, in *Improved iron based cathode material. Abstract No. 127*, in *Electrochemical Society Fall Meeting, Honolulu, Hawaii* (Electrochemical Society, Pennington, New Jersey, 1999)
24. H. Huang, S.C. Yin, L.F. Nazar, Electrochem. Solid State Lett. **4**, A170 (2001)
25. C. Delacourt, P. Poizot, S. Levasseur, C. Masquelier, Electrochem. Solid State Lett. **9**, A352 (2006)
26. F. Zhou, T. Maxisch, G. Ceder, Phys. Rev. Lett. **97**, 15 (2006)
27. D. Morgan, A. Van der Ven, G. Ceder, Electrochem. Solid State Lett. **7**, A30 (2004)
28. V. Srinivasan, J. Newman, J. Electrochem. Soc. **151**, A1517 (2004)
29. B.L. Ellis, K.T. Lee, L.F. Nazar, Chem. Mater. **22**, 691 (2010)
30. J.R. Dahn, R. Fong, M.J. Spoon, Phys. Rev. B **42**, 6424 (1990)
31. M. Winter, J.O. Besenhard, Electrochim. Acta **45**, 31 (1999)
32. O. Mao, R.A. Dunlap, J.R. Dahn, J. Electrochem. Soc. **146**, 405 (1999)
33. J. Graetz, C.C. Ahn, R. Yazami, B. Fultz, Electrochem. Solid State Lett. **6**, A194 (2003)
34. J. Yang, B.F. Wang, K. Wang, Y. Liu, J.Y. Xie, Z.S. Wen, Electrochem. Solid State Lett. **6**, A154 (2003)
35. E. Ferg, R.J. Gummow, A. Dekock, M.M. Thackeray, J. Electrochem. Soc. **141**, L147 (1994)
36. D. Deng, M.G. Kim, J.Y. Lee, J. Cho, Energy Environ. Sci. **2**, 818 (2009)
37. M. Wagemaker, F.M. Mulder, Acc. Chem. Res. **46**, 1206 (2013)
38. A.R. Armstrong, C. Lyness, P.M. Panchmatia, M.S. Islam, P.G. Bruce, Nat. Mater. **10**, 223 (2011)
39. J. Maier, Nat. Mater. **4**, 805 (2005)
40. A.S. Arico, P. Bruce, B. Scrosati, J.M. Tarascon, W. Van Schalkwijk, Nat. Mater. **4**, 366 (2005)
41. N. Meethong, H.Y.S. Huang, W.C. Carter, Y.M. Chiang, Electrochem. Solid State Lett. **10**, A134 (2007)
42. G. Sudant, E. Baudrin, D. Larcher, J.M. Tarascon, J. Mater. Chem. **15**, 1263 (2005)
43. W.J.H. Borghols, M. Wagemaker, U. Lafont, E.M. Kelder, F.M. Mulder, J. Am. Chem. Soc. **131**, 17786 (2009)
44. G. Kobayashi, S.I. Nishimura, M.S. Park, R. Kanno, M. Yashima, T. Ida, A. Yamada, Adv. Funct. Mater. **19**, 395 (2009)
45. M. Wagemaker, W.J.H. Borghols, F.M. Mulder, J. Am. Chem. Soc. **129**, 4323 (2007)
46. M. Wagemaker, D.P. Singh, W.J.H. Borghols, U. Lafont, L. Haverkate, V.K. Peterson, F.M. Mulder, J. Am. Chem. Soc. **133**, 10222 (2011)
47. B. Kang, G. Ceder, Nature **458**, 190 (2009)
48. Y.S. Hu, L. Kienle, Y.G. Guo, J. Maier, Adv. Mater. **18**, 1421 (2006)
49. X.P. Gao, Y. Lan, H.Y. Zhu, J.W. Liu, Y.P. Ge, F. Wu, D.Y. Song, Electrochem. Solid State Lett. **8**, A26 (2005)
50. A.R. Armstrong, G. Armstrong, J. Canales, R. Garcia, P.G. Bruce, Adv. Mater. **17**, 862 (2005)
51. P. Balaya, H. Li, L. Kienle, J. Maier, Adv. Funct. Mater. **13**, 621 (2003)
52. J.R. Dahn, Phys. Rev. B **44**, 9170 (1991)
53. O. Dolotko, A. Senyshyn, M.J. Muehlbauer, K. Nikolowski, F. Scheiba, H. Ehrenberg, J. Electrochem. Soc. **159**, A2082 (2012)
54. A. Senyshyn, O. Dolotko, M.J. Muehlbauer, K. Nikolowski, H. Fuess, H. Ehrenberg, J. Electrochem. Soc. **160**, A3198 (2013)
55. R.J. Cava, D.W. Murphy, S. Zahurak, A. Santoro, R.S. Roth, J. Solid State Chem. **53**, 64 (1984)

56. M. Wagemaker, G.J. Kearley, A.A. van Well, H. Mutka, F.M. Mulder, J. Am. Chem. Soc. **125**, 840 (2003)
57. D.W. Murphy, M. Greenblatt, S.M. Zahurak, R.J. Cava, J.V. Waszczak, G.W. Hull, R.S. Hutton, Rev. Chim. Miner. **19**, 441 (1982)
58. W.J.H. Borghols, M. Wagemaker, U. Lafont, E.M. Kelder, F.M. Mulder, Chem. Mater. **20**, 2949 (2008)
59. A.R. Armstrong, C. Arrouvel, V. Gentili, S.C. Parker, M.S. Islam, P.G. Bruce, Chem. Mater. **22**, 6426 (2010)
60. A. Deschanv, B. Raveau, Z. Sekkal, Mater. Res. Bull. **6**, 699 (1971)
61. D.C. Johnston, J. Low Temp. Phys. **25**, 145 (1976)
62. K.M. Colbow, J.R. Dahn, R.R. Haering, J. Power Sources **26**, 397 (1989)
63. T. Ohzuku, A. Ueda, N. Yamamoto, J. Electrochem. Soc. **142**, 1431 (1995)
64. A. Deschanv, B. Raveau, Z. Sekkal, Mater. Res. Bull. **6**, 699 (1971)
65. S. Scharner, W. Weppner, P. Schmid-Beurmann, J. Electrochem. Soc. **146**, 857 (1999)
66. G. Du, N. Sharma, V.K. Peterson, J.A. Kimpton, D. Jia, Z. Guo, Adv. Funct. Mater. **21**, 3990 (2011)
67. L. Kavan, M. Kalbac, M. Zukalova, I. Exnar, V. Lorenzen, R. Nesper, M. Graetzel, Chem. Mater. **16**, 477 (2004)
68. M.A. Reddy, M.S. Kishore, V. Pralong, U.V. Varadaraju, B. Raveau, Electrochem. Solid State Lett. **10**, A29 (2007)
69. M. Zukalova, M. Kalbac, L. Kavan, I. Exnar, M. Graetzel, Chem. Mater. **17**, 1248 (2005)
70. V. Gentili, S. Brutti, L.J. Hardwick, A.R. Armstrong, S. Panero, P.G. Bruce, Chem. Mater. **24**, 4468 (2012)
71. W.J.H. Borghols, M. Wagemaker, U. Laffont, E.M. Kelder, F.M. Mulder, Chem. Mater. **20**, 2949 (2008)
72. N.-S. Choi, J.-S. Kim, R.-Z. Yin, S.-S. Kim, Mater. Chem. Phys. **116**, 603 (2009)
73. R.J. Gummow, M.M. Thackeray, Solid State Ionics **53**, 681 (1992)
74. R.J. Gummow, M.M. Thackeray, W.I.F. David, S. Hull, Mater. Res. Bull. **27**, 327 (1992)
75. M. Carewska, S. Scaccia, F. Croce, S. Arumugam, Y. Wang, S. Greenbaum, Solid State Ionics **93**, 227 (1997)
76. S. Levasseur, M. Menetrier, E. Suard, C. Delmas, Solid State Ionics **128**, 11 (2000)
77. S. Levasseur, M. Menetrier, Y. Shao-Horn, L. Gautier, A. Audemer, G. Demazeau, A. Largeteau, C. Delmas, Chem. Mater. **15**, 348 (2003)
78. J.M. Kim, H.T. Chung, Electrochim. Acta **49**, 937 (2004)
79. P.S. Whitfield, I.J. Davidson, L.M.D. Cranswick, I.P. Swainson, P.W. Stephens, Solid State Ionics **176**, 463 (2005)
80. Y. Idemoto, T. Matsui, Solid State Ionics **179**, 625 (2008)
81. C. Fong, B.J. Kennedy, M.M. Elcombe, Zeitschrift Fur Kristallographie **209**, 941 (1994)
82. A.S. Wills, N.P. Raju, J.E. Greedan, Chem. Mater. **11**, 1510 (1999)
83. J. Rodriguez-Carvajal, G. Rousse, C. Masquelier, M. Hervieu, Phys. Rev. Lett. **81**, 4660 (1998)
84. R. Benedek, M.M. Thackeray, Electrochem. Solid State Lett. **9**, A265 (2006)
85. H. Berg, E.M. Kelder, J.O. Thomas, J. Mater. Chem. **9**, 427 (1999)
86. H. Berg, K. Goransson, B. Nolang, J.O. Thomas, J. Mater. Chem. **10**, 1437 (2000)
87. M.V. Reddy, M.J. Silvister Raju, N. Sharma, P.Y. Quan, S.H. Nowshad, H.E.-C. Emmanuel, V.K. Peterson, B.V.R. Chowdari, J. Electrochem. Soc. **158**, A1231 (2011)
88. Y. Liu, H. Kabbour, C.M. Brown, D.A. Neumann, C.C. Ahn, Langmuir **24**, 4772 (2008)
89. M. Yonemura, A. Yamada, H. Kobayashi, M. Tabuchi, T. Kamiyama, Y. Kawamoto, R. Kanno, J. Mater. Chem. **14**, 1948 (2004)
90. D. Gryffroy, R.E. Vandenberghe, J. Phys. Chem. Solids **53**, 777 (1992)
91. F.G.B. Ooms, M. Wagemaker, A.A. van Well, F.M. Mulder, E.M. Kelder, J. Schoonman, J. Appl. Phys. A **74**, S1089 (2002)
92. M. Wagemaker, F.G.B. Ooms, E.M. Kelder, J. Schoonman, G.J. Kearley, F.M. Mulder, J. Am. Chem. Soc. **126**, 13526 (2004)

93. A. Yamada, Y. Takei, H. Koizumi, N. Sonoyama, R. Kanno, K. Itoh, M. Yonemura, T. Kamiyama, Chem. Mater. **18**, 804 (2006)
94. A. Nyten, J.O. Thomas, Solid State Ionics **177**, 1327 (2006)
95. J. Xu, S.-L. Chou, M. Avdeev, M. Sale, H.-K. Liu, S.-X. Dou, Electrochim. Acta **88**, 865 (2013)
96. C.-Y. Chiang, H.-C. Su, P.-J. Wu, H.-J. Liu, C.-W. Hu, N. Sharma, V.K. Peterson, H.-W. Hsieh, Y.-F. Lin, W.-C. Chou, C.-H. Lee, J.-F. Lee, B.-Y. Shew, J. Phys. Chem. C **116**, 24424 (2012)
97. K.L. Harrison, C.A. Bridges, M.P. Paranthaman, C.U. Segre, J. Katsoudas, V.A. Maroni, J.C. Idrobo, J.B. Goodenough, A. Manthiram, Chem. Mater. **25**, 768 (2013)
98. N. Marx, L. Croguennec, D. Carlier, A. Wattiaux, F. Le Cras, E. Suard, C. Delmas, Dalton Trans. **39**, 5108 (2010)
99. B.L. Ellis, T.N. Ramesh, W.N. Rowan-Weetaluktuk, D.H. Ryan, L.F. Nazar, J. Mater. Chem. **22**, 4759 (2012)
100. H. Kim, S. Lee, Y.-U. Park, H. Kim, J. Kim, S. Jeon, K. Kang, Chem. Mater. **23**, 3930 (2011)
101. G. Rousse, J. Rodriguez-Carvajal, S. Patoux, C. Masquelier, Chem. Mater. **15**, 4082 (2003)
102. C. Delacourt, P. Poizot, J.M. Tarascon, C. Masquelier, Nat. Mater. **4**, 254 (2005)
103. C. Delacourt, J. Rodriguez-Carvajal, B. Schmitt, J.M. Tarascon, C. Masquelier, Solid State Sci. **7**, 1506 (2005)
104. Y. Orikasa, T. Maeda, Y. Koyama, H. Murayama, K. Fukuda, H. Tanida, H. Arai, E. Matsubara, Y. Uchimoto, Z. Ogumi, J. Am. Chem. Soc. **135**, 5497 (2013)
105. A. Yamada, H. Koizumi, S.I. Nishimura, N. Sonoyama, R. Kanno, M. Yonemura, T. Nakamura, Y. Kobayashi, Nat. Mater. **5**, 357 (2006)
106. S.Y. Chung, J.T. Bloking, Y.M. Chiang, Nat. Mater. **1**, 123 (2002)
107. M. Wagemaker, B.L. Ellis, D. Luetzenkirchen-Hecht, F.M. Mulder, L.F. Nazar, Chem. Mater. **20**, 6313 (2008)
108. M.S. Islam, D.J. Driscoll, C.A.J. Fisher, P.R. Slater, Chem. Mater. **17**, 5085 (2005)
109. J.J. Chen, M.S. Whittingham, Electrochem. Commun. **8**, 855 (2006)
110. S.Y. Chung, S.Y. Choi, T. Yamamoto, Y. Ikuhara, Phys. Rev. Lett. **100**, 12 (2008)
111. K.M.O. Jensen, M. Christensen, H.P. Gunnlaugsson, N. Lock, E.D. Bojesen, T. Proffen, B.B. Iversen, Chem. Mater. **25**, 2282 (2013)
112. P. Gibot, M. Casas-Cabanas, L. Laffont, S. Levasseur, P. Carlach, S. Hamelet, J.M. Tarascon, C. Masquelier, Nat. Mater. **7**, 741 (2008)
113. S.-P. Badi, M. Wagemaker, B.L. Ellis, D.P. Singh, W.J.H. Borghols, W.H. Kan, D.H. Ryan, F.M. Mulder, L.F. Nazar, J. Mater. Chem. **21**, 10085 (2011)
114. S.-Y. Chung, S.-Y. Choi, S. Lee, Y. Ikuhara, Phys. Rev. Lett. **108**, 19 (2012)
115. F. Zhou, C.A. Marianetti, M. Cococcioni, D. Morgan, G. Ceder, Phys. Rev. B **69**, 20 (2004)
116. N. Meethong, H.Y.S. Huang, S.A. Speakman, W.C. Carter, Y.M. Chiang, Adv. Funct. Mater. **17**, 1115 (2007)
117. A. Yamada, H. Koizumi, S.I. Nishimura, N. Sonoyama, R. Kanno, M. Yonemura, T. Nakamura, Y. Kobayashi, Nat. Mater. **5**, 357 (2006)
118. D. Burch, M.Z. Bazant, Nano Lett. **9**, 3795 (2009)
119. M. Wagemaker, F.M. Mulder, A. van der Ven, Adv. Mater. **21**, 1 (2009)
120. J.L. Dodd, R. Yazami, B. Fultz, Electrochem. Solid State Lett. **9**, A151 (2006)
121. R. Stevens, J.L. Dodd, M.G. Kresch, R. Yazami, B. Fultz, B. Ellis, L.F. Nazar, J. Phys. Chem. B **110**, 22732 (2006)
122. D.A. Cogswell, M.Z. Bazant, ACS Nano **6**, 2215 (2012)
123. P. Papanek, W.A. Kamitakahara, P. Zhou, J.E. Fischer, J. Phys. Condens. Matter **13**, 8287 (2001)
124. P. Zhou, P. Papanek, R. Lee, J.E. Fischer, W.A. Kamitakahara, J. Electrochem. Soc. **144**, 1744 (1997)
125. G.W. Neilson, R.D. Broadbent, I. Howell, R.H. Tromp, J. Chem. Soc. Faraday Trans. **89**, 2927 (1993)

126. P. Zhou, P. Papanek, C. Bindra, R. Lee, J.E. Fischer, J. Power Sources **68**, 296 (1997)
127. Y. Reynier, R. Yazami, B. Fultz, J. Power Sources **119–121**, 850 (2003)
128. Y. Reynier, J. Graetz, T. Swan-Wood, P. Rez, R. Yazami, B. Fultz, Phys. Rev. B **70**, 174304 (2004)
129. J. Li, V.O. Garlea, J.L. Zarestky, D. Vaknin, Phys. Rev. B **73**, 024410 (2006)
130. J. Li, T.B.S. Jensen, N.H. Andersen, J.L. Zarestky, W. McCallum, J.-H. Chung, J.W. Lynn, D. Vaknin, Phys. Rev. B **79**, 174435 (2009)
131. J. Li, W. Tian, Y. Chen, J.L. Zarestky, J.W. Lynn, D. Vaknin, Phys. Rev. B **79**, 144410 (2009)
132. C.S. Johnson, M.M. Thackeray, J.C. Nipko, C.-K. Loong, Phys. B **241–243**, 1252 (1998)
133. P. Aitchison, B. Ammundsen, J. Roziere, G.R. Burns, D.J. Jones, Solid State Ionics **176**, 813 (2005)
134. P. Aitchison, B. Ammundsen, T. Bell, D. Jones, J. Roziere, G. Burns, H. Berg, R. Tellgren, J. Thomas, Phys. B **276–278**, 847 (2000)
135. B. Ammundsen, D.J. Jones, J. Roziere, G.R. Burns, Chem. Mater. **7**, 2151 (1995)
136. B. Ammundsen, P.B. Aitchison, G.R. Burns, D.J. Jones, J. Roziere, Solid State Ionics **97**, 269 (1997)
137. M.J. Ariza, D.J. Jones, J. Roziere, R. Chitrakar, K. Ooi, Chem. Mater. **18**, 1885 (2006)
138. W.I. Jung, M. Nagao, C. Pitteloud, K. Itoh, A. Yamada, R. Kanno, J. Mater. Chem. **19**, 800 (2009)
139. A.J.M. Schmets, G.J. Kearley, H. Mutka, I.M. de Schepper, Phys. B **350**, e991 (2004)
140. C.C. Yang, S.Y. Wu, W.-H. Li, K.C. Lee, J.W. Lynn, R.S. Liu, C.H. Shen, Mater. Sci. Eng. B **95**, 162 (2002)
141. L. Cristofolini, P. Facci, M.P. Fontana, G. Cicognani, A.J. Dianoux, Phys. Rev. B **61**, 3404 (2000)
142. M.D. Levi, E. Markevich, D. Aurbach, J. Phys. Chem. B **109**, 7420 (2005)
143. A. Borgschulte, A. Jain, A.J. Ramirez-Cuesta, P. Martelli, A. Remhof, O. Friedrichs, R. Gremaud, A. Zuttel, Faraday Discuss **151**, 213 (2011)
144. F. Kargl, A. Meyer, M.M. Koza, H. Schober, Phys. Rev. B **74**, 014304 (2006)
145. G.A. Eckstein, G. Eckold, W. Schmidt, H.J. Steiner, H.D. Lutz, Solid State Ionics **111**, 283 (1998)
146. T. Maxisch, F. Zhou, G. Ceder, Phys. Rev. B **73**, 10 (2006)
147. S. Nishimura, G. Kobayashi, K. Ohoyama, R. Kanno, M. Yashima, A. Yamada, Nat. Mater. **7**, 707 (2008)
148. A. Laumann, H. Boysen, M. Bremholm, K.T. Fehr, M. Hoelzel, M. Holzapfel, Chem. Mater. **23**, 2753 (2011)
149. M. Wagemaker, E.R.H. van Eck, A.P.M. Kentgens, F.M. Mulder, J. Phys. Chem. B **113**, 224 (2009)
150. N. Kamaya, K. Homma, Y. Yamakawa, M. Hirayama, R. Kanno, M. Yonemura, T. Kamiyama, Y. Kato, S. Hama, K. Kawamoto, A. Mitsui, Nature Mater. **10**, 682 (2011)
151. P.R. Rayavarapu, N. Sharma, V.K. Peterson, S. Adams, J. Solid State Electrochem. **16**, 1807 (2012)
152. P.R. Rayavarapu, N. Sharma, V.K. Peterson, S. Adams, Solid State Ionics **230**, 72 (2013)
153. Y. Matsuda, M. Kawashima, Y. Moriya, T. Yamada, O. Yamamuro, S. Kojima, J. Phys. Soc. Jpn. **79**, 033801 (2010)
154. M. Yashima, J. Ceram. Soc. Jpn. **117**, 1055 (2009)
155. M. Yashima, M. Itoh, Y. Inaguma, Y. Morii, J. Am. Chem. Soc. **127**, 3491 (2005)
156. A. Attia, M. Zukalova, L. Pospisil, L. Kavan, J. Solid State Electrochem. **11**, 1163 (2007)
157. N. Reeves-McLaren, R.I. Smith, A.R. West, Chem. Mater. **23**, 3556 (2011)
158. M. Kamata, T. Esaka, S. Fujine, K. Yoneda, K. Kanda, Nucl. Instr. Meth. Phys. Res. A **377**, 161 (1996)
159. T. Esaka, Ionics **10**, 358 (2004)
160. S. Takai, K. Kurihara, K. Yoneda, S. Fujine, Y. Kawabata, T. Esaka, Solid State Ionics **171**, 107 (2004)

161. S. Takai, M. Kamata, S. Fujine, K. Yoneda, K. Kanda, T. Esaka, Solid State Ionics **123**, 165 (1999)
162. G. Mao, R.F. Perea, W.S. Howells, D.L. Price, M.-L. Saboungi, Nature **405**, 163 (2000)
163. D. Andersson, D. Engberg, J. Swenson, C. Svanberg, W.S. Howells, L. Borjesson, J. Chem. Phys. **122**, 234905 (2005)
164. F. Altorfer, W. Buhrer, I. Anderson, O. Scharpf, H. Bill, P.L. Carron, H.G. Smith, Physica B, **180–181**, 795 (1992)
165. C. Cramer, K. Funke, C. Vortkamp-Ruckert, A.J. Dianoux, Phys. A **191**, 358 (1992)
166. S. Hull, T.W.D. Farlwy, W. Hayes, M.T. Hutchings, J. Nucl. Mater. **160**, 125 (1988)
167. P. Vajda, F. Beuneu, G. Krexner, M. Prem, O. Blaschko, C. Maier, Nucl. Instr. Meth. Phys. Res. B **166–167**, 275 (2000)
168. H. Maekawa, Y. Fujimaki, H. Shen, J. Kawamura, T. Yamamura, Solid State Ionics **177**, 2711 (2006)
169. S. Ganapathy, E.R.H. van Eck, A.P.M. Kentgens, F.M. Mulder, M. Wagemaker, Chem. Eur. J. **17**, 14811 (2011)
170. R. van de Krol, A. Goossens, J. Schoonman, J. Phys. Chem. B **103**, 7151 (1999)
171. H. Lindstrom, S. Sodergren, A. Solbrand, H. Rensmo, J. Hjelm, A. Hagfeldt, S.E. Lindquist, J. Phys. Chem. B **101**, 7710 (1997)
172. M. Wagemaker, A.P.M. Kentgens, F.M. Mulder, Nature **418**, 397 (2002)
173. M. Wagemaker, R. van de Krol, A.P.M. Kentgens, A.A. van Well, F.M. Mulder, J. Am. Chem. Soc. **123**, 11454 (2001)
174. K. Xu, Chem. Rev. **104**, 4303 (2004)
175. P. Verma, P. Maire, P. Novak, Electrochem. Acta **55**, 6332 (2010)
176. J.E. Owejan, J.P. Owejan, S.C. DeCaluwe, J.A. Dura, Chem. Mater. **24**, 2133 (2012)
177. W.R. Brant, S. Schmid, G. Du, Q. Gu, N. Sharma, J. Power Sources **244**, 109 (2013)
178. N. Sharma, G. Du, A.J. Studer, Z. Guo, V.K. Peterson, Solid State Ionics **199–200**, 37 (2011)
179. X.-L. Wang, K. An, L. Cai, Z. Feng, S.E. Nagler, C. Daniel, K.J. Rhodes, A.D. Stoica, H.D. Skorpenske, C. Liang, W. Zhang, J. Kim, Y. Qi, S.J. Harris, Sci. Rep. **2**, 1 (2012)
180. N. Sharma, V.K. Peterson, M.M. Elcombe, M. Avdeev, A.J. Studer, N. Blagojevic, R. Yusoff, N. Kamarulzaman, J. Power Sources **195**, 8258 (2010)
181. N. Sharma, V.K. Peterson, J. Power Sources **244**, 695 (2013)
182. A. Senyshyn, M.J. Muhlbauer, K. Nikolowski, T. Pirling, H. Ehrenberg, J. Power Sources **203**, 126 (2012)
183. L. Cai, K. An, Z. Feng, C. Liang, S.J. Harris, J. Power Sources **236**, 163 (2013)
184. N. Sharma, V.K. Peterson, Electrochim. Acta **101**, 79 (2013)
185. C.-W. Hu, N. Sharma, C.-Y. Chiang, H.-C. Su, V.K. Peterson, H.-W. Hsieh, Y.-F. Lin, W.-C. Chou, B.-Y. Shew, C.-H. Lee, J. Power Sources **244**, 158 (2013)
186. M.A. Rodriguez, D. Ingersoll, S.C. Vogel, D.J. Williams, Electrochem. Solid-State Lett. **7**, A8 (2004)
187. M.A. Rodriguez, M.H. Van Benthem, D. Ingersoll, S.C. Vogel, H.M. Reiche, Powder Diffr. **25**, 143 (2010)
188. N. Sharma, X. Guo, G. Du, Z. Guo, J. Wang, Z. Wang, V.K. Peterson, J. Am. Chem. Soc. **134**, 7867 (2012)
189. N. Sharma, D. Yu, Y. Zhu, Y. Wu, V.K. Peterson, Chem. Mater. **25**, 754 (2013)
190. O. Bergstom, A.M. Andersson, K. Edstrom, T. Gustafsson, J. Appl. Cryst. **31**, 823 (1998)
191. H. Berg, H. Rundlov, J.O. Thomas, Solid State Ionics **144**, 65 (2001)
192. G. Du, N. Sharma, J.A. Kimpton, D. Jia, V.K. Peterson, Z. Guo, Adv. Funct. Mater. **21**, 3990 (2011)
193. N. Sharma, M.V. Reddy, G. Du, S. Adams, B.V.R. Chowdari, Z. Guo, V.K. Peterson, J. Phys. Chem. C **115**, 21473 (2011)
194. M. Roberts, J.J. Biendicho, S. Hull, P. Beran, T. Gustafsson, G. Svensson, K. Edstrom, J. Power Sources **226**, 249 (2013)
195. F. Rosciano, M. Holzapfel, W. Scheifele, P. Novak, J. Appl. Cryst. **41**, 690 (2008)
196. J.-F. Colin, V. Godbole, P. Novák, Electrochem. Commun. **12**, 804 (2010)

197. V.A. Godbole, M. Hess, C. Villevieille, H. Kaiser, J.F. Colin, P. Novak, RSC Adv. **3**, 757 (2013)
198. H. Liu, C.R. Fell, K. An, L. Cai, Y.S. Meng, J. Power Sources **240**, 772 (2013)
199. W.K. Pang, N. Sharma, V.K. Peterson, J.-J. Shiu, S.H. Wu, J. Power Sources **246**, 464 (2014)
200. P. Kolarova, J. Vacik, J. Spirkova-Hradilova, J. Cervena, Nucl. Instrum. Methods Phys. Res. Section B-Beam Interact. Mater. Atoms **141**, 498 (1998)
201. L.H.M. Krings, Y. Tamminga, J. van Berkum, F. Labohm, A. van Veen, W.M. Arnoldbik, J. Vac. Sci. Technol. Vac. Surfaces Films **17**, 198 (1999)
202. G.P. Lamaze, H.H. Chen-Mayer, D.A. Becker, F. Vereda, R.B. Goldner, T. Haas, P. Zerigian, J. Power Sources **119**, 680 (2003)
203. S.M. Whitney, S.R.F. Biegalski, G. Downing, J. Radioanal. Nucl. Chem. **282**, 173 (2009)
204. S. Whitney, S.R. Biegalski, Y.H. Huang, J.B. Goodenough, J. Electrochem. Soc. **156**, A886 (2009)
205. S.C. Nagpure, R.G. Downing, B. Bhushan, S.S. Babu, L. Cao, Electrochim. Acta **56**, 4735 (2011)
206. J.F.M. Oudenhoven, F. Labohm, M. Mulder, R.A.H. Niessen, F.M. Mulder, P.H.L. Notten, Adv. Mater. **23**, 4103 (2012)
207. N. Kardjilov, I. Manke, A. Hilger, M. Strobl, J. Banhart, Mater. Today **14**, 248 (2011)
208. A. Same, V. Battaglia, H.-Y. Tang, J.W. Park, J. Appl. Electrochem. **42**, 1 (2012)
209. M. Lanz, E. Lehmann, R. Imhof, I. Exnar, P. Novak, J. Power Sources **101**, 177 (2001)
210. J.P. Owejana, J.J. Gagliardo, S.J. Harris, H. Wang, D.S. Hussey, D.L. Jacobson, Electrochim. Acta **66**, 94 (2012)
211. J.B. Siegel, X. Lin, A.G. Stefanopoulou, D.S. Hussey, D.L. Jacobson, D. Gorsich, J. Electrochem. Soc. **159**, A523 (2011)
212. I. Manke, J. Banhart, A. Haibel, A. Rack, S. Zabler, N. Kardjilov, A. Hilger, A. Melzer, H. Riesemeier, Appl. Phys. Lett. **90**, 214102 (2007)
213. G.V. Riley, D.S. Hussey, D.L. Jacobson, ECS Trans. **25**, 75 (2010)
214. J. Nanda, H. Bilheux, S. Voisin, G.M. Veith, R. Archibald, L. Walker, A. Allu, N.J. Dudney, S. Pannala, J. Phys. Chem. C **116**, 8401 (2012)
215. M. Hofmann, R. Gilles, Y. Gao, J.T. Rijssenbeek, M.J. Muhlbauer, J. Electrochem. Soc. **159**, A1827 (2012)
216. M. Kamata, T. Esaka, S. Fujine, K. Yoneda, K. Kanda, J. Power Sources **68**, 459 (1997)
217. N. Kardjilov, A. Hilger, I. Manke, M. Strobl, W. Treimera, J. Banhart, Nucl. Instr. Meth. Phys. Res. A **542**, 16 (2005)
218. L.G. Butler, B. Schillinger, K. Ham, T.A. Dobbins, P. Liu, J.J. Vajo, Nucl. Instr. Meth. Phys. Res. A **651**, 320 (2011)
219. D. Goers, M. Holzapfel, W. Scheifele, E. Lehmann, P. Vontobel, P. Novak, J. Power Sources **130**, 221 (2004)
220. M. Wagemaker, R. van de Krol, A.A. van Well, Phys. B. **336**, 124 (2003)
221. J.F.M. Oudenhoven, F. Labohm, M. Mulder, R.A.H. Niessen, F.M. Mulder, P.H.L. Notten, Adv. Mater. **23**, 4103 (2011)

Chapter 8
Hydrogen Storage Materials

Juergen Eckert and Wiebke Lohstroh

Abstract An eventual realization of a Hydrogen Economy requires working solutions in three fundamental areas, namely hydrogen production, hydrogen storage, and fuel cells, in addition to the development of an extensive, new infrastructure. While neutron scattering experiments and the associated techniques of analysis have been of utility in all three of these research areas, they have had by far the most significant impact on the development and understanding of materials for hydrogen storage applications. This chapter examines some of these contributions.

8.1 Hydrogen Storage for Mobile Applications

Hydrogen can readily be stored in large quantities as a cryogenic liquid for stationary applications, e.g. for use as a rocket fuel, since the size and weight of the storage tanks are not limited in practice. This is, however, not the case for mobile applications, for which the US Department of Energy (DOE) has established some well-known criteria for a working system that would provide a driving range of 300 miles (480 km) for a hydrogen fuelled car. These include both gravimetric and volumetric storage capacities, operating conditions, and several other important factors. The requirements for a hydrogen storage system for mobile applications can be summarized in a qualitative way as follows:

(i) Appropriate thermodynamics (favourable enthalpies of hydrogen absorption and desorption).
(ii) Fast kinetics (quick uptake and release).

J. Eckert (✉)
Department of Chemistry, University of South Florida, Tampa, FL, USA
e-mail: juergen@usf.edu

W. Lohstroh
Heinz Maier-Leibnitz Zentrum (MLZ), Technische Universität München,
Garching, Germany
e-mail: wiebke.lohstroh@frm2.tum.de

© Springer International Publishing Switzerland 2015
G.J. Kearley and V.K. Peterson (eds.), *Neutron Applications in Materials for Energy*, Neutron Scattering Applications and Techniques,
DOI 10.1007/978-3-319-06656-1_8

   (iii) High storage capacity.
   (iv) Effective heat transfer.
   (v) High gravimetric and volumetric densities (light in weight and conservative in use of space).
   (vi) Long cycle lifetime for hydrogen absorption/desorption.
   (vii) High mechanical strength and durability of material and containers.
   (viii) Safety under normal use and acceptable risk under abnormal conditions.

Most of the considerable effort to realize such a working system has been focused on increasing the capacities of storage materials in terms of wt.%. The ultimate requirement specified by the US DOE (updated to 7 wt.% for the entire system, not just the storage medium) requires that the hydrogen molecules be packed much more closely than they are in liquid H_2. This can, of course, more readily be accomplished if hydrogen is stored in atomic form, which occupies much less space. In these cases hydrogen is rather strongly bound, as, for example, in chemical compounds, which does, however, make desorption kinetics and reversibility more difficult if not impossible.

The types of materials that are candidate hydrogen-storage systems (Fig. 8.1) may briefly be summarized as follows:

(1) Hydrogen in metals, such as $FeTiH_x$ and $LaNi_5H_6$, that were actively investigated in the early 1980 and tested for vehicular use. Here H_2 dissociates at the metal surface and forms a solid solution with the metal. The hydrogen gas can be released at elevated temperatures. These materials can have very high volumetric capacities, but their gravimetric capacities are low because of the high densities of the metal hosts.

(2) In so-called complex or light metal hydrides, where hydrogen is essentially covalently bonded to metals such as Al or Li in the form of a chemical compound. Elevated temperatures and/or a catalyst are required for a release of hydrogen, and on-board regeneration is not readily possible.

(3) Chemical hydrides, such as hydrocarbons, ammonia, or amino compounds, which have the highest hydrogen content (both volumetric and gravimetric), but where the release of hydrogen is only by a chemical reaction, and not all of the hydrogen content is available under reasonable conditions. On-board regeneration does not seem feasible.

(4) Molecular hydrogen adsorbed on surfaces, such as carbons of various types, or inside large pore materials, such as zeolites, clathrates, and metal-organic frameworks (MOFs). High volumetric and gravimetric capacities have been achieved in cases where the surface area is very large, but at operating conditions (low temperature and elevated pressures) that are not currently considered to be suitable for vehicular applications, on account of the weak interaction (physisorption) with the host material.

(5) Molecular hydrogen stored as a gas or liquid. These are straightforward, established technologies, and therefore are typically used in cars used to test hydrogen fuel-cells, or simply for combustion in a normal engine.

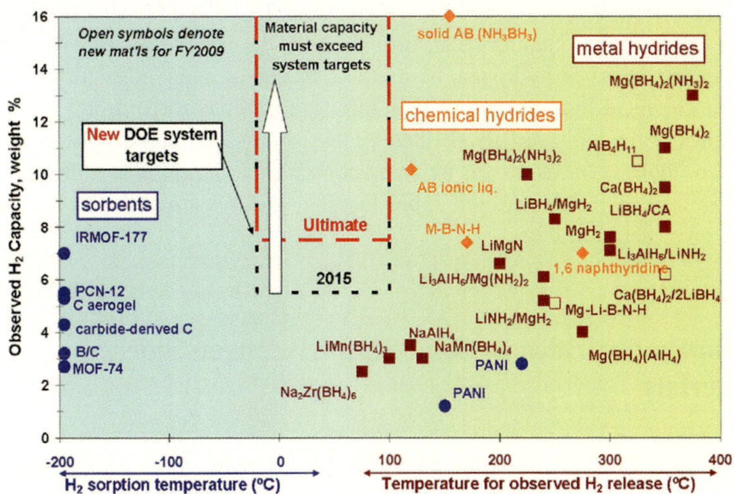

DOE: G. Thomas (2007), G. Sandrock (2008), B. Bowman (2009)

Fig. 8.1 Capacities and operating conditions for various proposed hydrogen-storage systems

Targets for the capacities of a hydrogen storage system have been somewhat modified by the US DOE from those of the original 'Freedom Car' program. The emphasis has shifted towards system targets that must be met, i.e. the weights of the containers, and all associated plumbing, valves etc. must be included. This can easily be a factor of two, so that a material needs to have an intrinsic capacity of 12 wt.% recoverable H_2 to meet the system target (for 2015) of 6 wt.%.

Prospects for a practical hydrogen storage system based on the DOE guidelines basically depend on two critical, but very difficult developments (Fig. 8.1): (1) in the case of sorption-based systems binding energies of hydrogen to the host must be substantially increased so that the necessary capacities can be achieved at room temperature and modest pressures, while (2) for chemical or light metal hydrides desorption conditions must be improved to the point where all available hydrogen is released at or below 100 °C, and, more importantly, on-board regeneration becomes facile. DOE guidelines, for example, also specify a refueling rate of 2 kg H_2/min in addition to the better-known targets on gravimetric and volumetric capacities. Progress in these areas ultimately depends on materials synthesis, namely the ability to design and to produce (on a large scale, at low cost) the storage material, which overcomes the current limitations. While much progress in synthesis can be made by applying chemical principles in conjunction with standard thermodynamic measurements (i.e. adsorption isotherms), the use of advanced, molecular-level characterization methods should provide additional essential details on the interactions of hydrogen with the host material, and this is the type of information which serves both as input and all-important validation of the extensive computational studies, which are also being employed in the search for suitable hydrogen storage media.

Perhaps the most powerful experimental methods for molecular-level studies of hydrogen in materials involve the use of neutrons because of their outstanding sensitivity to hydrogen. Neutrons can provide both the rotational and vibrational dynamics of the adsorbed hydrogen, be it H_2, H, or one of many other forms in the complex or chemical hydrides, as well as diffusive motions through the material. The dynamics of the molecule are affected by its surroundings and can therefore be taken as an indirect measure of its interaction. The microscopic diffusion of hydrogen is critical for the desired rapid loading, and release, but has not been studied in great detail to date.

8.2 Complex Hydrides as Solid-State Hydrogen Storage Materials

The discovery of reversible hydrogen uptake and release from sodium alanate ($NaAlH_4$) doped with titanium by Bogdanovic and Schwickardi [1] in 1997, brought about a considerable increase in research activities in the field of hydrogen storage in complex hydrides. Many of the fundamental properties of complex hydrides had not been adequately investigated up that point, and a great number of important questions remain to be answered to this date. Neutron scattering investigations should be an indispensable tool in this effort because of their great sensitivity to hydrogen. The interest in complex hydrides mainly stems from the high hydrogen content of light complex hydrides (e.g. 18.4 wt.% H_2 in $LiBH_4$ [2, 3]), which make them attractive as potential solid-state hydrogen storage materials. Complex hydrides may simply be viewed as salt-like compounds of the type $A^{n+}[XH_M]_n^-$ where A is either an alkaline, alkaline earth, or early transition metal and X typically is B, Al, or N. The bond between hydrogen and X has covalent character and the resulting complex is a rather stable entity which fulfills the "18 electron" rule, i.e. the hydrogen containing entity such as $[AlH_4]^-$ has a closed-shell electronic structure. While complex hydrides do have high hydrogen-content, only a few systems exhibit reversible hydrogen uptake and release, and the thermodynamic and kinetic factors which govern the hydrogen exchange reaction are not yet fully understood. Hydrogen release and uptake is typically accompanied by a solid-state reaction which involves long-range diffusion of atoms heavier than hydrogen, which potentially limits fast reaction-rates. The covalent bond of hydrogen with the transition metal also appears to be very stable, so that the mechanisms for bond splitting (and reformation during hydrogen charging) still need to be fully understood.

Neutron scattering experiments can address virtually all fundamental atomic-level questions on the function of complex hydrides, such as the crystallographic and dynamic properties of compounds which are often not well characterized, materials development such as the mapping of the reaction pathway during hydrogen exchange, or technology development such as using imaging techniques to monitor the H uptake in prototype tanks. The focus of most investigations has been on structure determination using neutron powder diffraction and characterization of the H dynamics using vibrational spectroscopy and quasielastic neutron scattering (QENS). The latter process gives access to stochastic motions of H, i.e. translational and rotational diffusion inside the crystal lattice, while vibrational spectroscopy can be used to determine the H density of states and has become an essential tool to benchmark first-principles calculations. In the following sections we will describe a few examples of neutron scattering experiments for the characterization of complex hydrides and their development as solid-state H storage materials. Apart from fundamental insights into these materials, which in many cases lack adequate characterization, phase transitions during hydrogen absorption and desorption can be monitored and thus giving insight into the reaction mechanism, intermediate reaction steps, and kinetic limitations.

8.2.1 Alanates: NaAlH$_4$

The archetypical example NaAlH$_4$ decomposes (and reabsorbs) hydrogen in two distinct reaction steps as follows:

$$3NaAlH_4 \leftrightarrow Na_3AlH_6 + 2Al + 3H_2$$
$$3.7 \text{ wt.\%}, \Delta H = 38 \text{ kJmol}^{-1} \tag{8.1}$$

$$Na_3AlH_6 \leftrightarrow 3NaH + Al + 1.5 H_2$$
$$1.7 \text{ wt.\%}, \Delta H = 47 \text{ kJmol}^{-1} \tag{8.2}$$

Additives such as TiCl$_3$, ScCl$_3$, or CeCl$_3$ enable nearly complete reversible conversion and rehydrogenation at 373 K and 100 bar H$_2$. Undoped NaAlH$_4$, on the other hand, exhibits negligible reversibility under these conditions. The function of the catalyst is still being debated, and while numerous different materials have been identified to be able to catalyze these reactions [4–7], a unified picture for their activity is still lacking.

Neutron powder diffraction measurements have identified the structure of NaAlD$_4$ as body-centred tetragonal $I4_1/a$ (NaAlH$_4$) [8]. The intermediate decomposition product, Na$_3$AlH$_6$ occurs in two modifications, i.e. low temperature monoclinic α-Na$_3$AlH$_6$ ($P2_1/n$) [9, 10], and β-Na$_3$AlH$_6$ with a cubic structure ($Fm\,\overline{3}m$) [11] is observed above 525 K. The reversible hydrogen exchange reaction is greatly facilitated by Ti additives and numerous attempts have been made to localize the additive and to unravel its catalytic mode of operation. Among the suggested mechanism are: (a) Ti acts as a surface catalyst that facilitates the splitting of the AlH$_4$/AlH$_3$ bonds [12, 13], (b) Ti initiates vacancy formation and hence promotes H diffusion [11, 14], (c) Ti weakens the Al–H bond [15], or (d) Ti acts as a grain refiner [16]. While neutron powder diffraction measurements did not show any indication of a solid solution with Ti-based additives directly after ball milling, the formation of $Al_{1-x}Ti_x$ species were observed upon cycling [17, 18]. Synchrotron X-ray diffraction studies of Ti-doped $NaAlH_4$ suggested that Ti could substitute in Al-sites [19] while density functional theory (DFT) calculations show that both the substitution of Na or Al in the $NaAlH_4$ structure should be possible in the bulk [20–23] or small clusters [24]. The influence of Ti-doping on the native vacancy formation was also investigated by DFT methods. Ti-doping in $NaAlH_4$ can yield the formation of hydrogen vacancies and interstitials [25]. In contrast, the same paper suggests that in Ti-doped $LiBH_4$ only a reorientation of the BH_4 units occurs. Since structural characterization of the Ti sites proved to be difficult, the localization of the additive and/or its induced vacancy diffusion mechanisms was studied by neutron spectroscopy techniques. QENS data on Ti-doped $NaAlH_4$ at temperatures below 350 K indicate that in both $NaAlH_4$ and Na_3AlH_6, the hydrogen mobility is vacancy mediated [26, 27] but the relative amount of mobile hydrogen at these temperatures is less than 1 % in $NaAlH_4$ even at 390 K and there is no significant difference on the bulk diffusion induced by $TiCl_3$ additions. Similarly, neutron vibrational spectroscopy showed no distinct differences for pure $NaAlH_4$ and Ti-doped $NaAlH_4$ [24, 28] while significant changes of the vibrational density-of-states were observed after thermal treatment of both $NaAlH_4$ and Na_3AlH_6 [29]. These were attributed to the presence of different ionic species resulting from the partial decomposition of the samples. Neutron spectroscopy data (Fig. 8.2) suggest the formation of AlH_3 and its oligomers $(AlH_3)_n$ [28] during reabsorption of the H depleted mixture NaH/Al (after 0.5 h at 140 bar H_2, 403 K).

Here, the measured inelastic neutron scattering (INS) spectra are compared with calculated spectra of AlH_3 and of Al_4H_{12}. Volatile species of this kind had been suggested to be an intermediate species responsible for mass-transfer processes during the hydrogen exchange reaction [30] and neutron spectroscopy is a unique tool to detect these intermediate species.

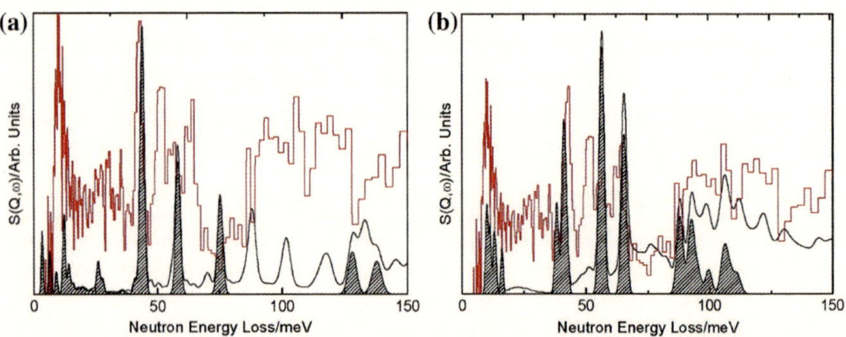

Fig. 8.2 **a** Calculated spectrum for AlH_3 adsorbed on Al metal (*bottom line*) and experimental data of NaH/Al (after 0.5 h at 140 bar, 403 K) (*top*). **b** Calculated spectrum for the Al_4H_{12} (crown) molecule. The shaded area represents the contribution from fundamental vibrations, and the total spectrum contains the overtones and combinations. Reprinted with permission from (Q.J. Fu, A. Ramirez-Cuesta, S.C. Tsang, J. Phys. Chem. B **110**, 711 (2006)) [28]. Copyright (2006) American Chemical Society

8.2.2 Borohydrides

Borohydrides are also under investigation as potential hydrogen storage materials, particularly those with high hydrogen contents. For instance, $LiBH_4$ can be reformed from the elements at 973 K and 150 bar H_2 [31]. Conditions are slightly milder when LiB_x precursors such as LiB_3 or Li_7B_6 intermetallics are used. Other systems with limited reversibility include e.g. $Mg(BH_4)_2$ or $Ca(BH_4)_2$ which can be (partially) rehydrogenated at high pressures (90 MPa) and high temperatures (673 K) [32, 33]. $Mg(BH_4)_2$ would be an attractive storage material since it has a gravimetric storage capacity of 14.9 wt.% H_2 and the reported enthalpy of reaction is 39 kJmol^{-1} for the reaction $Mg(BH_4)_2 \rightarrow MgH_2 + B + 3H_2$ [34], so that the estimated equilibrium pressure at RT is 1 bar H_2.

The structure of $LiBH_4$ at low temperature is orthorhombic (LT) and changes near 380 K to a hexagonal high-temperature (HT) structure [35]. The four H atoms are bound covalently to the central B atom. Early structural studies found a strong asymmetry of the BH_4-tetrahedron [36] in the LT phase, while later reports describe the geometry of the BH_4-entities to be close to that of an ideal tetrahedron both below and above the structural phase transition [35]. The B–H distance was estimated to be $d_{B-H} = 1.16–1.26$ Å in the HT phase from synchrotron X-ray diffraction [35], while the B-D distances were determined to be $d_{B-D} = 1.18–1.20$ Å at 302 K in neutron diffraction experiments [37]. The structural phase transition is accompanied by increased thermal motion of the BH_4-complexes, which is indicative of an order-disorder transition. The nature of the reorientations of the BH_4-tetrahedron was investigated using quasielastic neutron scattering (QENS) for both the high temperature and low temperature phase. The type of reorientation was identified from the elastic incoherent structure-factor (EISF), which is proportional to the Fourier transform of the probability density of finding a hydrogen atom at a given position. The hydrogen mobility in both phases can be described by rotational jumps of the BH_4-tetrahedra. The mean residence times τ between successive jumps are on the order of a few picoseconds at temperatures near the phase transition [38]. While 120° rotations around one C_3 symmetry axis of the tetrahedron could be identified in the LT phase [38], but there are also clear indications for two different dynamic processes with distinct activation energies both from NMR data [39], as well as from inelastic fixed window scans in QENS experiments [40]. The reported activation energies are 162 ± 2 and 232 ± 11 meV, respectively. The BH_4 dynamics becomes even faster above the phase transition [38] and the QENS data can then best be described by an orbit exchange-model [41] where three out of the four H atoms rotate fairly freely around a C_3 axis while the axial H displays only occasional jumps as illustrated in Fig. 8.3.

The EISF obtained for HT $LiBH_4$ may be compared (Fig. 8.3) with theoretical models for several reorientation mechanisms, such as 120° rotations around the C_3 axis, tetragonal or cubic tumbling and the orbit exchange-model. It is apparent that it is essential to monitor as much of the reciprocal space as possible in the QENS experiments for the identification of the true nature of the motions, since a clear distinction

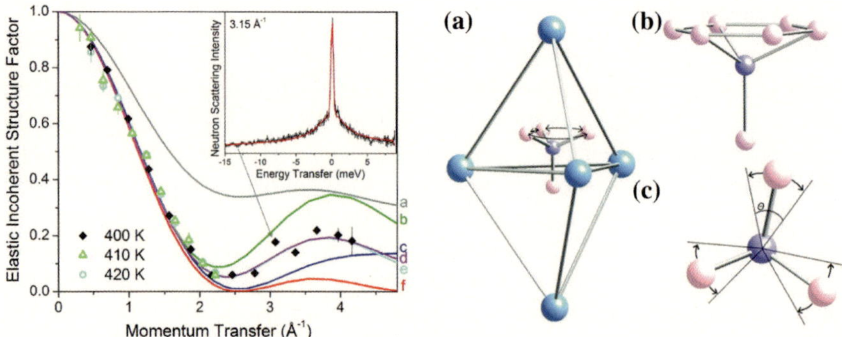

Fig. 8.3 EISF data (*left*) derived from QENS measurements for LiBH$_4$ at 400 K (*black diamonds*), 410 K (*green triangles*), and 420 K (*cyan circles*) compared with calculated curves for various reorientation models: **a** continuous rotation around the trigonal axis with a fixed axial H (*grey*), **b** tetrahedral tumbling (*green*), **c** cubic tumbling (*blue*), **d** the high-temperature model for $N = 6$ (*magenta*), **e** the high-temperature model for $N \geq 12$ (*cyan*), and **f** isotropic rotational diffusion (*red*). The inset exemplifies a quasielastic spectrum (*black data*), collected at 400 K at 3.15 Å^{-1}, and the corresponding fit (*red line*). Vertical error bars denote ±1σ. The lower part illustrates local environment of the BH$_4$-tetrahedron in HT LIBH$_4$ and the proposed reorientation mechanism. Reprinted with permission from (N. Verdal, T.J. Udovic, J.J. Rush, J. Phys. Chem. C **116**, 1614 (2012)) [41]. Copyright (2012) American Chemical Society

between different models can only be made above Q = 2.5 Å^{-1}. The best agreement between the prediction of a given reorientation model and the experimental data is obtained for the orbit exchange-model where frequent H jumps occur between sites on a circle around the C_3 axis while the axial H only occasionally jumps (see Fig. 8.3, lines d and e). The model was calculated using either $N = 6$ or 12 equivalent positions on the circle, however already $N = 6$ was sufficient to describe the data.

While reorientation of the BH$_4$ - tetrahedra has been observed at temperatures as low as 175 K, translational diffusion of BH$_4$ could only be seen above the melting transition [42] at 553 K. Exchange of atomic hydrogen between adjacent BH$_4$ tetrahedra remains slow even at these temperatures and transport of H takes mainly place by way of the BH$_4^-$ units. Li$^+$ mobility, on the other hand, becomes much faster when going from the LT to the HT phase, which is indicated by a an increase of three orders of magnitude of the Li$^+$ conductivity from $10^{-6} - 10^{-8}$ Scm^{-1} in the LT phase to 10^{-3} Scm^{-1} in the HT phase [43]. Substitution of the BH$_4^-$ units by I in LiBH$_4$/LiI mixtures results in a stabilization of the HT phase at room temperature (RT). QENS measurements accordingly show significantly-enhanced dynamics of BH$_4$ at RT compared to the LT phase (at RT) and a similar reorientation mechanism as in the pure LiBH$_4$ HT phase [44]. The stabilization of the HT phase by I substitution also preserves the high Li$^+$-conductivity at RT. It has been suggested that Li$^+$ conductivity is supported by the BH$_4^-$ movements [43] in a paddle-wheel like mechanism.

The alkaline borohydrides NaBH$_4$ and KBH$_4$ exhibit a structural phase-transition associated with an order–disorder transition of the BH$_4$ subunits, from a tetragonal LT phase to a cubic HT phase (at ~ 190 K for $NaBH_4$ [45], and 65–70 K for KBH_4 [46]). QENS data on $NaBH_4$ suggest reorientational jumps around the C_2 and C_3 axis, which is in agreement with the crystal symmetry. Reorientation of the BH_4 entities around a C_4 axis have been reported for both $NaBH_4$ and KBH_4, so that the four hydrogens of the tetrahedron occupy the eight corners of a cube but with half occupancy [47]. Thermodynamic measurements provided evidence for an order–disorder transition in $RbBH_4$ at 44 K and in $CsBH_4$ at 27 K [48] but this could not be discerned from neutron powder diffraction down to low temperature [46]. However, vibrational spectroscopy of the torsion band indicates that the heavier alkali borohydrides also exhibit a similar order-disorder transition although the ordering of the BH_4 seems to occur on a shorter length scale [47].

The variety of structural phases is especially rich for the alkaline earth borohydrides $Mg(BH_4)_2$ and $Ca(BH_4)_2$. Both of these can be synthesized in solvent free form, but the crystal structure depends strongly on the preparation conditions. Two polymorphs were initially identified for $Mg(BH_4)_2$: an α-phase, which transforms irreversibly into β-$Mg(BH_4)_2$ at temperatures above 490 K. Both polymorphs have unexpectedly-complex crystal structures which differ from numerous theoretical predictions [49–52]. Crystal structure determination from powder data proved difficult especially concerning the hydrogen positions but eventually, the structure of the α-phase was identified as hexagonal with space group $P6_122$ [50]. Adjacent BH_4-groups are coordinated by a central Mg via two opposite edges of the tetrahedron, but the resulting $Mg–H_2BH_2–Mg$ bridges are not exactly planar. α-$Mg(BH_4)_2$ has 6.4 % unoccupied voids (amounting to 37 $Å^2$ per unit cell) whereas β-$Mg(BH_4)_2$ which has an orthorhombic structure *(Fddd)* and is much more dense with no unoccupied voids. The α and β-polymorphs are built up of a corner sharing network of $Mg^{2+}[BH_4^-]_4$ tetrahedra. A highly porous form, γ-$Mg(BH_4)_2$, was synthesized more recently, and found to possess a cubic structure $(Id\bar{3}a)$ with a network of interconnected channels and 33 % porosity, which could be utilized for the adsorption of molecular hydrogen (or nitrogen) at low temperatures [52]. Application of pressure (1−1.6 GPa, diamond anvil cell) causes the structure to collapse and results in tetragonal δ-$Mg(BH_4)_2$ [52]. Synchrotron X-ray diffraction studies identified yet another polymorph, ε-$Mg(BH_4)_2$, during the decomposition γ-$Mg(BH_4)_2$ [53]. DFT calculations find that many structural models for $Mg(BH_4)_2$ are nearly degenerate in energy, which can explain difficulties with the correct prediction of crystal structures and decomposition pathways [54, 55]. Moreover it is likely that more polymorphs of $Mg(BH_4)_2$ exist depending on preparation conditions (and/or impurity content).

The reorientational motion of the BH_4-tetrahedra in β-$Mg(BH_4)_2$ have been studied using QENS on two instruments with different energy resolution and therefore different timescales. Two thermally-activated reorientation processes have been observed on these timescales in the temperature range from 120−437 K, and have been identified as rotations around the $C_{2\parallel}$-axis (which connects the two Mg atoms in the $Mg–H_2BH_2–Mg$ bridge) or around the C_3-axis of the tetrahedron [56] (Fig. 8.4).

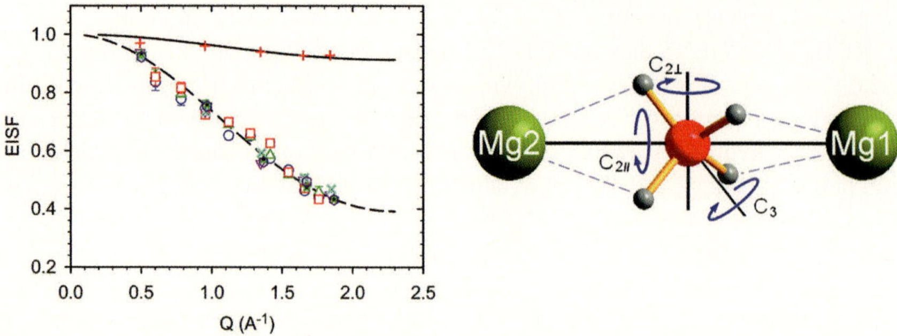

Fig. 8.4 *Left* The measured and modelled EISFs for β-Mg(BH$_4$)$_2$: Using the mica analyser high resolution time-of-flight backscattering spectrometer (MARS) at the Paul Scherrer Institute: (*red plus*) 120 K, (*blue x*) 150 K, (*red triangle*) 180 K, (*green asterisk*) 210 K, and (*black hexagon*) 240 K. Using the spectrometer for high energy resolution (SPHERES) at the Heinz Maier-Leibnitz Zentrum: (*red square*) 318 K, (*green triangle*) 365 K, and (*blue circle*) 473 K. The calculated EISF are for the reorientation models: (—) $C_{2\parallel}$ (MARS data) and C_3 (SPHERES data) rotational diffusion-model and hindered rotation diffusion-model (—). Right: BH$_4$ reorientation as hindered rotation around $C_{2\parallel}$. Reprinted with permission from (P. Martelli, D. Blanchard, J.B. Maronsson, M.D. Riktor, J. Kheres, D. Sveinbj, E.G. Bardají, J. Phys. Chem. C **116**, 2013 (2013)) [56]. Copyright (2013) American Chemical Society

The activation energies were determined to be 39 ± 0.5 meV and 76 ± 5 meV for the $C_{2\parallel}$ rotation and 214 ± 4 meV for the C_3 axis rotation, respectively [55]. Low-frequency vibrational spectroscopy using Raman scattering revealed a complex structure of both internal and external modes in a mixed α/β-Mg(BH$_4$)$_2$ sample which is considerably modified when passing through the phase transition [57]. Neutron spectroscopy revealed an additional band which was attributed to BH$_4$ librations [56] but more spectroscopic work along with DFT calculations is needed for a more comprehensive understanding of Mg(BH$_4$)$_2$.

Ca(BH$_4$)$_2$ is another interesting candidate for storage application with a gravimetric hydrogen storage density of 11.6 wt.% H$_2$ or 130 kgm^{-3} of H$_2$. There are four known polymorphs: α, α', and γ-phase, which are considered LT phases while β-Ca (BH$_4$)$_2$ is an HT phase. The α and γ modification have an orthorhombic structure (*Fddd* [58, 59] or *F2dd* [60] for α-Ca(BH$_4$)$_2$ and *Pbca* [58] for γ-Ca(BH$_4$)$_2$) whereas the α'-modification is tetragonal (*I* $\overline{4}$*2d* [59]). Above 400 K, a tetragonal (*P*$\overline{4}$) high temperature phase β-Ca(BH$_4$)$_2$ is observed [57, 58]. Each Ca is surrounded by 6 boron atoms in the α, β, and γ-phases forming a CaB$_6$ octahedron [58] and the polymorphism results from different arrangements and connections of the CaB$_6$ octahedra. The energy differences of the different polymorphs are small [57, 61] which explains why frequently more than one crystalline phase is found in the same sample batch. Low energy spectroscopic measurements indicate that the phase stability is linked to the librations of the BH$_4$- units [62] and it was suggested that entropic contributions are driving the phase transitions. Moreover, the different polymorphs exhibit different decomposition properties [63], hence a thorough understanding of the bare materials is essential for their potential application in solid-state hydrogen storage. The hydrogen dynamics in (predominantly) β-Ca(BH$_4$)$_2$ have been investigated by QENS [64]. Two different thermally-activated rotational-reorientation processes have been identified in the temperature range from 100 to 260 K, which were attributed to rotations around the C_2 and C_3 axis, respectively. Translational diffusion with a jump length of 2.5 Å was also observed, which the authors attributed to the H$_2$ diffusion of trapped impurities, while no long-range diffusion of the BH$_4$ units was observed at these temperatures.

Besides the pure alkaline and alkaline-earth metal borohydrides, mixtures combining different cations have been investigated [65–67] in an attempt to modify the reaction enthalpy for hydrogen desorption. The decomposition temperature for borohydrides correlates with the electronegativity of the cation [68, 69] and thus dual cation borohydrides could be ideal candidates for tuning the thermodynamic properties.

8.2.3 Amides

Lithium amides have also been studied in some detail for hydrogen-storage applications following the discovery of Chen et al. that mixtures of LiNH$_2$ + 2 LiH can be reversibly cycled according to the following reactions [70]:

$$LiNH_2 + LiH \leftrightarrow Li_2NH + H_2$$
$$6.5 \text{ wt.\% } H_2, \Delta H = 45 \text{ kJmol}^{-1}$$

(8.3)

$$Li_2NH + LiH \leftrightarrow Li_3N + H_2$$
$$5.5 \text{ wt.\%} H_2, \Delta H = 161 \text{ kJmol}^{-1}$$

(8.4)

Experimentally 11.5 wt.% H$_2$ (relative to the hydrogen-free Li$_3$N) are reversibly stored in the mixture, and temperatures of 473 K and 593 K are required for the first and second reaction step, respectively. Pure LiNH$_2$ decomposes by evolving ammonia in contrast to the reaction system LiNH$_2$ + LiH. Some of the Li can be substituted by Mg which leads to lower desorption-temperatures and a lower reaction enthalpy [71]. During the first desorption 2 LiNH$_2$ + MgH$_2$ undergoes a metathesis reaction [72–74] and the resulting reversible reaction is modified to:

$$Mg(NH_2)_2 + 2LiH \leftrightarrow Li_2Mg(NH)_2 + 2H_2$$
$$5.6 \text{ wt.\% } H_2, \Delta H = 39 \text{ kJmol}^{-1}$$

(8.5)

The reaction takes place at 473 K and in total 5.6 wt.% H$_2$ can be reversibly exchanged, but hydrogen release takes place in distinct reaction steps. The high reaction temperature suggests the presence of kinetic barriers hampering the hydrogen exchange reactions and research efforts are focused on the understanding of the reaction steps and the mitigation of the kinetic limitations. Neutron powder

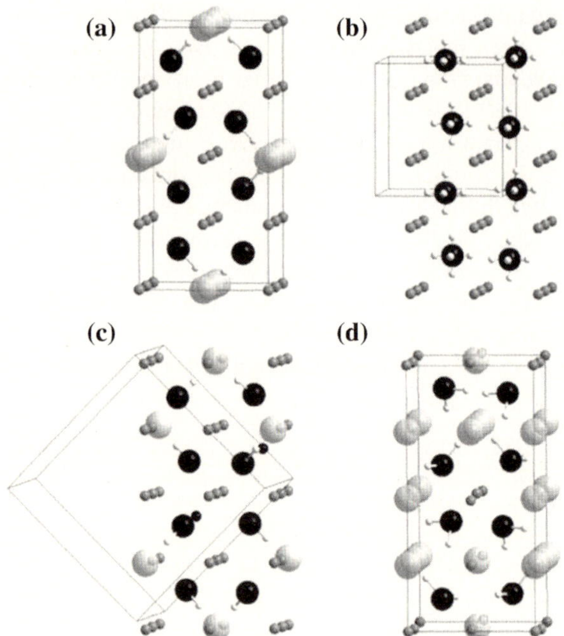

Fig. 8.5 Comparison of the structures of (**a**) alpha-$Li_2Mg(NH)_2$, (**b**) high temperature cubic Li_2NH, (**c**) low temperature orthorhombic Li_2NH and (**d**) $LiNH_2$. Nitrogen atoms (dark grey), protons (white), lithium and/or magnesium (small grey) and vacancies (large transparent grey) are shown. Reprinted figure with permission from (Y. Wang, M. Chou, Phys. Rev. B **76**, 014116 (2007)) [81]. Copyright (2007) by the American Physical Society

diffraction was used to determine the structure of the hydrogen-rich components $LiNH_2$ [70, 75], $Mg(NH_2)_2$ [76] as well as of the hydrogen-depleted imide forms Li_2NH [77, 78], α-$Li_2Mg(NH)_2$, β-$Li_2Mg(NH)_2$ [79], and $MgNH$ [80], all of which are observed as intermediate decomposition-products. With the exception of $MgNH$, the experimentally-observed crystal structures exhibit characteristic similarities, i.e. they are related to the antifluorite structure where the N atoms build an approximate face-centred cubic lattice with the Li or Mg ions (and vacancies) residing in tetrahedral interstitials (see Fig. 8.5). Substitution of Li^+ with Mg^{2+} leads to vacancy ordering on the tetrahedral sites, which results in an orthorhombic structure of α-$Li_2Mg(NH)_2$ with an approximately doubled unit cell relative to cubic Li_2NH [73]. First-principles calculations identified the hydrogen and cation local arrangement as the structural building blocks of the amide/imide structure [81] and their energetics is consistent with the high degree of disorder that is observed in the mixed imide $Li_2Mg(NH)_2$, especially when going from the low temperature α-modification to the high temperature β-phase.

The structural phase evolution during hydrogen absorption and desorption of $Mg(NH_2)_2 + 2LiH$ has been monitored in situ using neutron powder diffraction (Fig. 8.6). Several studies found that the reaction passes through an intermediate

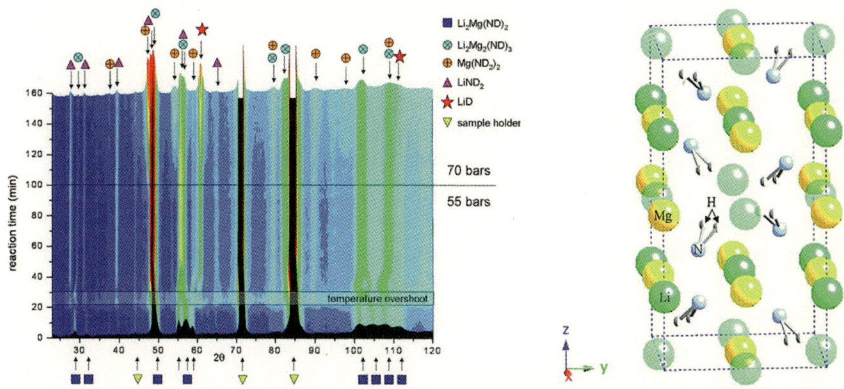

Fig. 8.6 *Left* In situ neutron diffraction of the rehydrogenation of LiN$_2$MgNH$_2$. *Right* Structure of the proposed intermediate Li$_2$Mg(NH$_2$)$_3$. From Ref [82]

reaction step, however, the composition of the intermediate phase was not solved unambiguously. Weidner et al. suggested an intermediate phase of composition Li$_2$Mg$_2$(NH)$_3$ [82] which was subsequently also identified in systems with LiH excess [83].

Aoki et al. [84] reported an intermediate phase with composition Li$_4$Mg$_3$(NH$_2$)$_2$(NH)$_4$ for the system Mg(NH$_2$)$_2$ + 6LiH which upon further hydrogen release exhibits a continuous transition towards Li$_2$Mg(NH)$_2$ by way of solid-solution-like compounds of the form Li$_{4+x}$Mg$_3$(NH$_2$)$_{2-x}$(NH)$_4$. Intermediate phases with non-stoichiometric composition have also been observed experimentally during the decomposition of LiNH$_2$ + LiH [85, 86] and possible intermediates have been investigated using first-principles calculations [87, 88]. The defect structure and vacancy ordering for both systems, LiNH$_2$ + 2LiH and Mg(NH$_2$)$_2$ + 2LiH, has a significant influence on energy landscapes of the systems. Further efforts on theory and experiment are needed, however, to fully comprehend this system.

Additions of LiBH$_4$ to the Mg(NH$_2$)$_2$ + 2LiH system have been reported to improve the hydrogen-exchange reaction because of the intermittent formation of Li$_4$(BH$_4$)(NH$_2$)$_3$ [89]. Neutron powder diffraction on annealed samples of 1LiBD$_4$ + 3LiNH$_2$ was used to characterize the structure.

The beneficial effect of the LiBH$_4$ on the desorption properties is thought to stem from a combination of factors, including an altered reaction-pathway from the removal of intermittent LiNH$_2$ from the mixture, improved recrystallization of Mg (NH$_2$)$_2$, and the exothermic heat-effect during the formation of the quaternary compound Li$_4$(BH$_4$)(NH$_2$)$_3$ which has body-centred cubic symmetry [90].

Mixtures of LiNH$_2$/LiBH$_4$ with varying composition have also been investigated as hydrogen storage materials. More than 10 wt.% H$_2$ is released in an exothermic process [91–93] for the stoichiometric composition (LiNH$_2$)$_{0.67}$(LiBH$_4$)$_{0.33}$.

8.3 Adsorbents

Adsorption of hydrogen on graphite, or in zeolites, has been investigated at a molecular level for primarily fundamental reasons long before this became an urgent practical issue for the ultimate realization of a hydrogen economy. The use of neutron scattering techniques in both of these examples illustrates two basic requirements that can make neutrons a surface-sensitive probe, the first is that the adsorbate needs to have a significantly greater neutron scattering cross-section (e.g. H) than the host material (e.g. C or silica), and the second is that adsorbent must have a very large surface-area so that the total number of surface-bound molecules is sufficiently large. This is the case for most carbon materials, such as exfoliated graphite, which was used [94] to determine the two-dimensional structural phase diagram of monolayers of deuterium. The highest coverage was found for an incommensurate structure of 0.078 molecules $\mathring{A}^{-2}$ which corresponds to about 1.3×10^{-5} mol.m^{-2}. This number indicates a limit of the storage capacity for simple physisorption on carbons, namely that a material with a maximum surface area of 1,500 m^{2}g^{-1} will bind close to 0.02 mol of H_2 (0.04 g, 4 wt.%), which still falls short of what is required. Surface areas for the interior pores in zeolites are typically considerably less than that because of the high density of the silicate framework.

It has therefore long been recognized that surface areas have to be considerably increased, and this has been accomplished with the synthesis of a large number of metal-organic framework (MOF) materials with as much as 5,000 m^{2}g^{-1} or more. Capacities around 8 wt.% have been achieved in these materials albeit under unfavourable conditions (77 K and pressures of approximately 60 bar), but the amount of hydrogen stored in MOFs at ambient temperature is far below those required for practical goals. The reason for this is, of course, that the binding energies for hydrogen provided by physisorption are too low (of the order of 5 kJmol^{-1}) and hence vapour pressures are too high for the operation of such a storage system at ambient conditions.

Materials therefore need to be designed to not only have very high surface-areas but also binding energies for hydrogen at least three times as high as that for normal physisorption (e.g. on carbons) in order for such a system to operate at room temperature and modest pressures. Heats of adsorption must also be approximately constant over the entire range of hydrogen loading, i.e. of similar magnitude for all accessible binding sites.

An enormous effort to design and synthesize new and better porous materials guided by chemical or structural principles, including the use of combinatorial methods, has resulted in a huge variety of metal-organic materials not only for applications as hydrogen storage materials, but also for the capture of CO_2, gas separation, and catalysis (see Chaps. 2 and 3). Binding energies of hydrogen molecules have more than doubled from that of simple physisorption in some of these materials, but are still quite far away from what is needed for a practical storage medium operating at ambient conditions.

The use of neutrons makes it possible to employ a systematic, molecular-level approach to the development of improved hydrogen storage materials, particularly when combined with advanced computational analysis. Such experiments can provide the location of the adsorbed hydrogen molecules at different sites (neutron diffraction) and give some measure of the strength of the interaction with the host material at each site (inelastic neutron scattering), whereas conventional thermo-dynamic measurements (adsorption isotherms) only give average properties for all sites that are occupied at a given loading. Results of such experiments should then serve as the basis for detailed computational analysis, which can rationalize both the locations of the hydrogen molecules and their binding strength in terms of the interatomic guest-interactions. This type of information should then in principle be useful in the design and synthesis of improved storage materials.

A detailed knowledge of the binding sites of hydrogen molecules in crystalline porous materials can be obtained by neutron diffraction methods. Adsorbed hydrogen molecules cannot be "seen" in single crystal X-ray diffraction experiments even at very low temperature, but this method has been used to obtain a very detailed picture of the adsorption and binding sites of Ar and N_2, i.e. molecules with many electrons, in the prototypical metal-organic material $Zn_4O(bdc)_3$ (where bdc = 1,4-benzenedicarboxylate), also known as MOF-5. The primary binding site at 30 K for both types adsorbates was found to be in the pocket of the Zn_4O cluster in the corner of the framework (α-site) [95]. A number of other binding sites were clearly identified, including ones that are triply or doubly bridging the carboxylate oxygens (β and γ sites, respectively), which connect the cluster the organic linker. Interaction with the benzene ring leads to two binding sites, one near the two aromatic hydrogens in a side-on configuration, and the other on top of the ring.

Ideally a similar approach with single-crystal neutron diffraction should provide similar detail for hydrogen binding sites. This has, however, been accomplished just once, namely for hydrogen in MOF-5 [96], since the sizes of MOF single crystals are usually too small for the relatively-low intensities in neutron diffraction experiments. This experiment was carried out on a crystal that barely was of adequate size. Nonetheless, the most favourable binding site for H_2 in MOF-5 at 5 K was clearly identified to be in the pocket of the Zn_4O cluster (α-site) with one end of the molecule on the three-fold axis pointing at the central oxygen atom. The H–H bond of the molecule is at an angle relative to the three-fold axis of the crystal so that the other end is disordered over three equivalent orientations (Fig. 8.7) in a similar manner as N_2. This site is essentially fully occupied at 5 K. Hydrogen was also located on the β-site, but could not be refined as two separate atoms because of rather large atomic displacement parameters.

Because of the aforementioned difficulties in obtaining sufficiently-large crystals of MOFs, a number of studies to characterize hydrogen binding sites have been carried out by neutron powder diffraction, including also on MOF-5 [97]. For neutron powder diffraction it is highly advantageous, although not always neces-sary, to have all Hs in the system replaced by deuterium (D) in the synthesis of the material, which can, of course, be rather difficult and expensive, and to use D_2 as the adsorbate. The strong background of the incoherent scattering by H can thereby

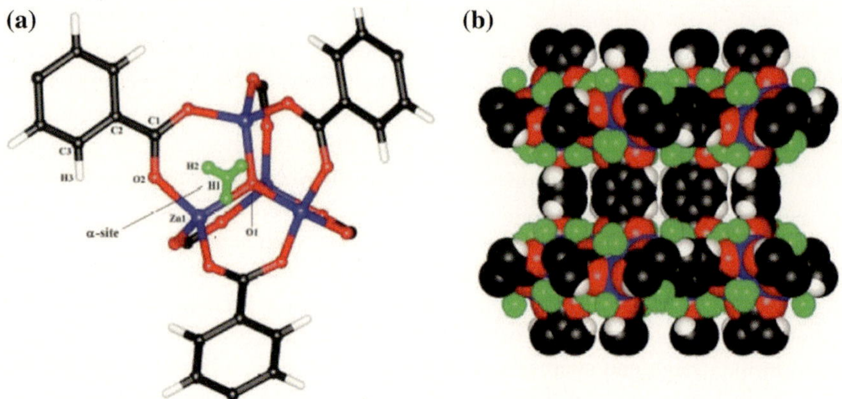

Fig. 8.7 H_2 binding in MOF-5 as determined using single-crystal neutron diffraction. From Ref. [96]

be reduced. Diffraction patterns can be collected at different loadings of D_2 and interpreted under the assumption that sorption sites become occupied roughly in order of their respective binding energies. The molecule itself cannot, however, be refined as two separate atoms, so that only its centre-of-mass is found. In addition to the locations of the D_2 molecules the degree that these sites are occupied is refined in the analysis, along with isotropic atomic displacement parameters all within the symmetry of the parent structure.

$Cu_3(btc)_2$ (where btc = 1,3,5-benzenetricarboxylate) and also known as HKUST-1, was one of the earliest MOFs reported and has been extensively characterized since then, including several neutron diffraction and inelastic neutron-scattering studies, as well as computational studies. Perhaps the most interesting structural feature of this material is the well-known dicopper acetate "paddlewheel" unit, which has subsequently been used in many MOFs. The apical ligand on the Cu usually is a weakly bound water molecule, which can easily be removed, and thereby opens up a metal site for presumably stronger adsorption of hydrogen. The most favourable binding site in HKUST-1 was indeed found to be, using neutron powder diffraction, adjacent to the open Cu site ($D_2(1)$ in Fig. 8.8) [98, 99]. This would be expected on the basis of the relatively-strong interaction of the hydrogen molecule with the metal, so that these sites are largely responsible for the improved heats of adsorption found in this material. The hydrogen molecule was located [98] at a distance of 2.39(1) Å from the Cu site, which is, however, far too great to allow for a direct electronic interaction with the metal. Additional binding sites at various points in the host structure were identified in this study, some of which are shown in Fig. 8.8. Site 2, for example, is adjacent to the benzene rings of the btc unit, while D_2 in site 3 is close to six btc oxygen atoms in the small window leading to the 5 Å pores.

The enhanced interaction of H_2 with such open-metal sites has been demonstrated in a number of other MOFs, most notably MOF-74 [100, 101], also known as CPO-27 [102], for which a number of isostructural variants with different metal centres

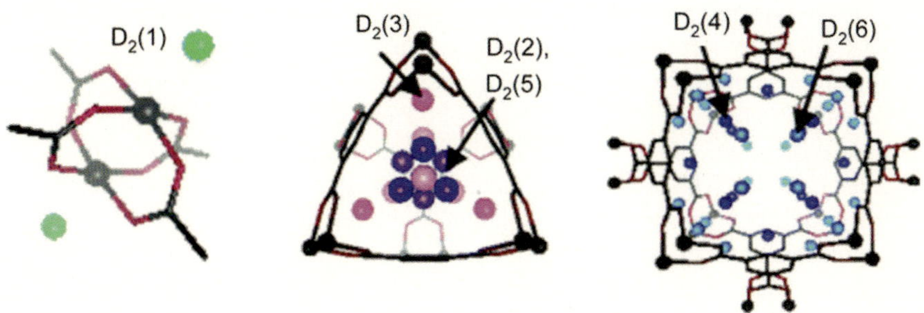

Fig. 8.8 Binding of D_2 in HKUST-1, with the sites labelled sequentially in order of binding strength. H atoms are omitted from the framework for clarity. From Ref. (V.K. Peterson, Y. Liu, C. M. Brown, C.J. Kepert, J. Am. Chem. Soc. **128**, 15578 (2006)) [98]. Copyright (2006) American Chemical Society

(Zn, Mg, Mn, Co, Ni) can be synthesized [102–104]. The materials consist of a linear chain of inorganic metal clusters connected with organic linkers to form hexagonal channels, M_2(dhtp), (dhtp = dihydroxyterephthalate). H_2 binding energies indeed depend strongly on the type of metal centre, and do reach about 13 kJ/mol for Ni, but only until all these sites are filled. A powder neutron diffraction study [137] on the Zn analog finds that the D_2 molecule is close enough (d(Zn-D_2) $\sim$ 2.6 Å) to be affected by polarization from the metal, but again too far for a direct electronic interaction. A somewhat closer metal-H_2 distance of 2.27 Å was found [105] by neutron powder diffraction in the Mn-based MOF formed from Mn_4Cl clusters and 1,3,5-benzenetristetrazolate bridging ligands. Desolvation of the parent material opens up one of the coordination sites on Mn^{2+}, which in turn interacts rather strongly with H_2 (10.1 kJ/mol near zero-loading), but again, not by direct electronic interaction, which requires H_2-mteal distances in the order of 1.8 Å.

One of the more important aspects of experimental efforts to locate the adsorbed hydrogen molecules in these systems is that this sort of information is critical for benchmarking the extensive theoretical studies of these systems that have been undertaken. A neutron powder diffraction study on hydrogen adsorbed in the zeolitic imidazolate framework ZIF-8 [106] was the first to rationalize the experimental findings with parallel DFT calculations of binding energies at various sites. The preferred adsorption site in this material in fact is not near the metal cluster but on top of the metal imidazolate linker. Most computational characterization of MOFs for hydrogen adsorption is accomplished by grand-canonical Monte Carlo simulations, or by molecular dynamics, both of which readily yield pair-distribution functions (PDFs) for the adsorbate molecules. These can be directly compared with diffraction experiments without any model dependent structure refinement, if the diffraction data are converted to direct space, i.e. to PDF form, as has been reported for D_2 in IRMOF-1 [107]. This approach would be especially important for those MOFs, where the adsorption sites are poorly defined, MOFs which have defect structures, or are in fact amorphous.

A much greater number of neutron scattering studies on sorbent hydrogen storage materials has been carried out by inelastic scattering from the bound hydrogen molecules, perhaps because of the relative ease with which these can be carried out, and because of the sensitivity of the method. A detailed interpretation of such data is, however, very challenging, and ultimately requires extensive computational studies. The rotational dynamics of the adsorbed hydrogen molecule are a highly sensitive measure of its interactions with the host material and are readily accessible by the inelastic scattering of neutrons. The reason for this is that the rotational energy levels of the quantum rotor H_2, E, are strongly perturbed from those of the free quantum rotor (E = BJ(J + 1), where B is the rotational constant of the molecule and J is the quantum number for total rotational angular momentum) by the barrier to rotation resulting from the influence of the adsorbent.

Transitions between these hindered-rotor levels can be directly observed by incoherent INS and give an indication of the height and shape of the barrier to rotation. Such a quantitative analysis of these measurements must however employ a phenomenological model for this barrier, whose parameters may be determined from the pattern of observed transitions.

The simplest such model is one of a double-minimum potential with the two angular degrees of freedom for the H_2 molecule,

$$V(\theta, \Phi) = 1/2 V_2 (1 - \cos 2\theta) \tag{8.6}$$

The energy levels for this potential can then be obtained by solving the Schrödinger equation:

$$-\frac{\hbar^2}{2I} \left(\frac{1}{\sin\theta} \frac{\partial}{\partial\theta} \sin\theta \frac{\partial\psi}{\partial\theta} + \frac{1}{\sin^2\theta} \frac{\partial^2\psi}{\partial\phi^2} \right) + \left(\frac{1}{2} V_2 \right) (1 - \cos 2\theta)\psi = E\psi \tag{8.7}$$

for various values of V_2, with the result that the observed transition(s) yield a value for the height of the barrier. It is important to note that the energy-level diagram (Fig. 8.9) shows that the spacing between the two lowest levels *decreases* with *increasing* barrier height, which is indicative of rotational tunnelling, where the tunnelling probability has an approximately-exponential inverse dependence on the barrier height. Therein lies the great sensitivity of this type of measurement.

A number of more elaborate phenomenological models for the H_2 potential-energy surface (PES) have also been used to interpret the INS data, and these include separate barriers to in plane and out of plane reorientation (or some ratio between the two) plus a separable centre-of-mass vibration perpendicular to the surface [109]. Such models, however, require the observation of several transitions in order to determine the parameters in the form of the potential. While the use of such a phenomenological potential is essential for the assignment of the observed rotational transitions to hydrogen molecules absorbed at different sites, it does not, however provide direct insight into the atomic level origins of the rotational barriers. What we would like to learn from these experiments is the origin and nature of the interatomic interactions at each binding site that give rise to the PES, which

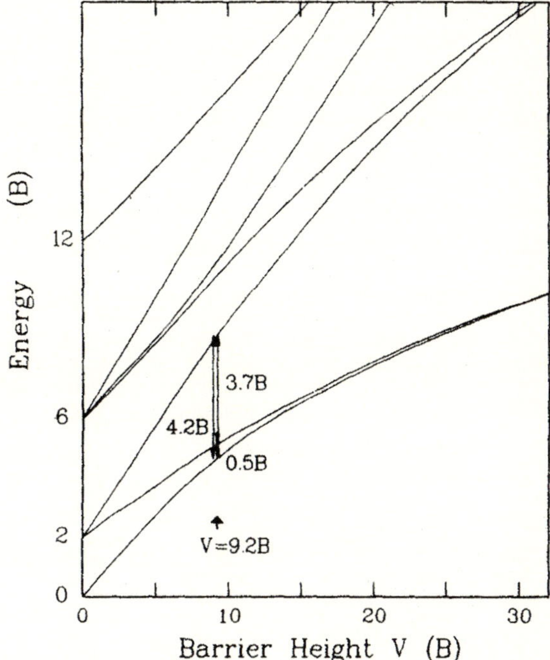

Fig. 8.9 Rotational energy-level diagram for a dumbbell molecule in a double-minimum potential. B is the rotational constant. Arrows indicate observed transitions for H_2 in zeolite CoNaA, and indicate the manner in which the barrier to rotation is derived. Figure taken from [108]

in turn gives rise to the observed rotational energy levels. This can, of course, be done by ab initio computational studies as long as the structural details of the binding sites are known, preferably from a neutron diffraction experiment. The dynamics of the hydrogen molecule within such a theoretical PES then has to be treated exactly, i.e. with quantum dynamics for both the rotational and (centre of mass) translational excitations. This has been accomplished for H_2 in clathrates, various kinds of fullerenes, and just one MOF, namely MOF-5, which will be described below.

The seminal work of this type [110] on H_2 adsorbed in MOF-5 helped to rationalize the large storage capacity of this material in terms of the availability of several different types of binding site. The INS spectra are collected at 10 K (or less) as a function of loading (carried out at 77 K) in a direct relationship to the numbers of equivalent sites of particular types available to facilitate a tentative identification of specific spectral features with binding sites (Fig. 8.10) by making use of the phenomenological PES described above. Three additional sites could be inferred from the INS spectra, which may now be matched with information available from the diffraction experiments as well as computational studies, i.e. the β, γ, and δ sites.

Similar experiments have been carried out on hydrogen on various forms of carbon, some of which were at one time considered to be promising hydrogen storage materials. The INS spectra, however, generally show just one broad peak centred near the free rotor value of 14.7 meV, which is representative of a broad

Fig. 8.10 INS spectra of H_2 in MOF-5 at two loadings, and their difference. Inset shows the MOF-5 structure. Taken from [110]

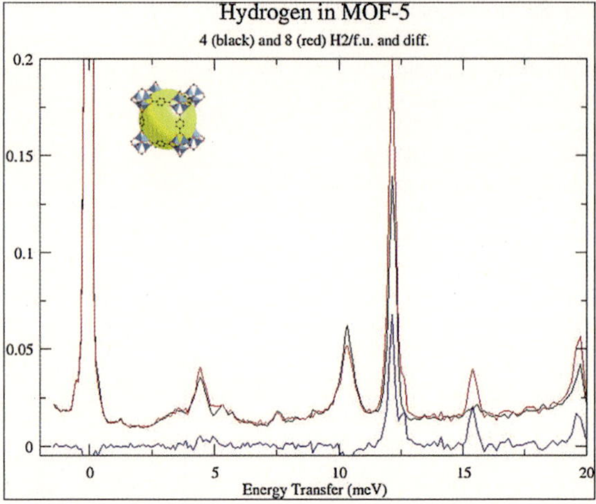

Fig. 8.11 INS spectra of H_2 adsorbed on (*top to bottom*) two forms of activated carbon, nanotubes, and nanofibers. Reprinted from (H.G. Schimmel, G. J. Kearley, M.G. Nijkamp, C.T. Visser, K.P. Dejong, F.K. Mulder, Chem. Eur. J. **9**, 4764 (2003)) [111] with permission

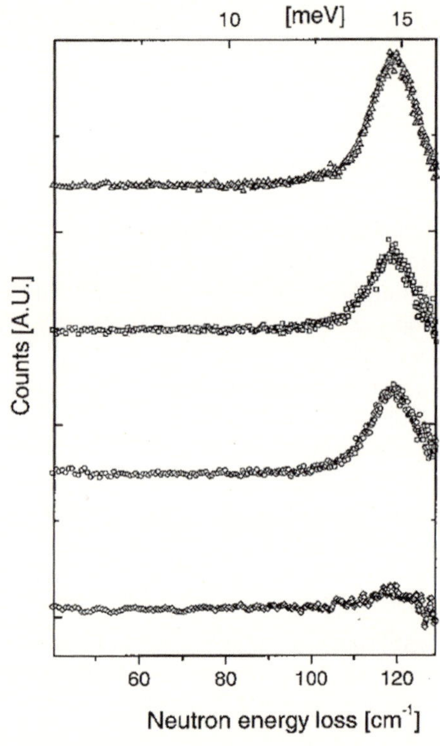

distribution of binding sites with very low barriers to rotation, or weak interaction with the host. These spectra clearly show that there is very little difference in the interaction of H_2 and various forms of carbons, be it activated carbon, or nanotubes [111, 112] (Fig. 8.11).

These results may be contrasted with those for MOF-5, where the INS spectrum exhibits a number of sharp transitions indicative of different, well-defined binding sites, some of which have higher barriers to rotation of H_2 than do carbons.

A considerable number of metal-organic materials (MOMs) have been investigated using rotational-tunnelling INS spectroscopy of the adsorbed H_2 in an effort to determine which structural features may improve the observed heats of adsorption. Here we utilize the fact that a smaller rotational tunnel splitting implies a higher barrier to rotation, which in turn should, in general, signal a stronger interaction between the adsorbed H_2 and the host and a particular binding site. This type of information can only be extracted if particular spectral features can be associated with H_2 a known sites, which can be quite difficult in the presence of simultaneous multiple-site occupancy, as will be described below.

Previous studies of H_2 adsorption in zeolites [113] had shown the strong effect of a charged framework along with the presence of charge-balancing extra framework cations on the binding energies of hydrogen molecules. The vast majority of MOMs, however, possess neutral frameworks, but some MOFs with anionic or cationic frameworks have indeed been characterized. Zeolitic MOFs, ZMOFs, are such a class of anionic MOFs built from metal centres with imidazole dicarboxylate ligands to form zeolitic structures (e.g. *rho*-ZMOF, Fig. 8.12) with much larger pores than their SiO_2 analogs. This structural similarity arises from the fact that the Si–O–Si bond angle in the zeolite is nearly the same as the angle between the two N atoms in the imidazole dicarboxylate. The negatively-charged framework is compensated by exchangeable extra framework cations, as in zeolites. In the case of ZMOFs, however, these cations are hydrated and bound to the framework by H-bonds from the water ligands to carboxylate oxygens. INS studies on *rho*-ZMOFs with different

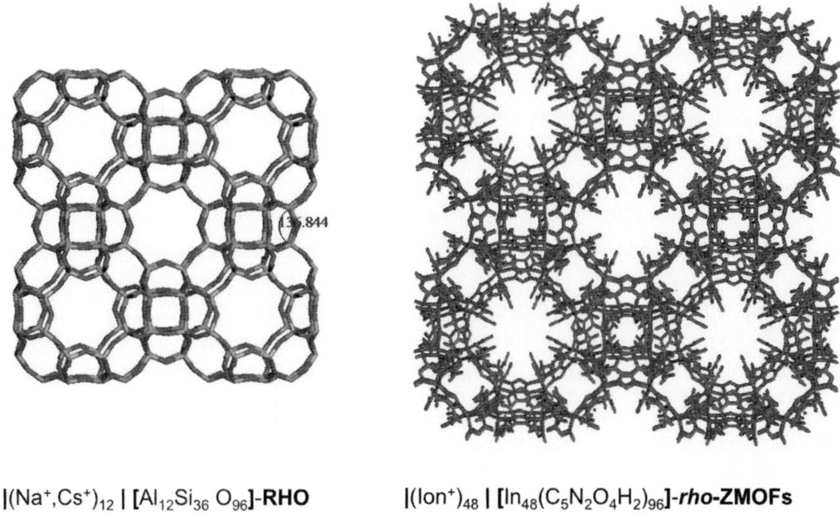

Fig. 8.12 *Left* Zeolite *rho*. *Right* A *rho*-ZMOF. Taken from [114]

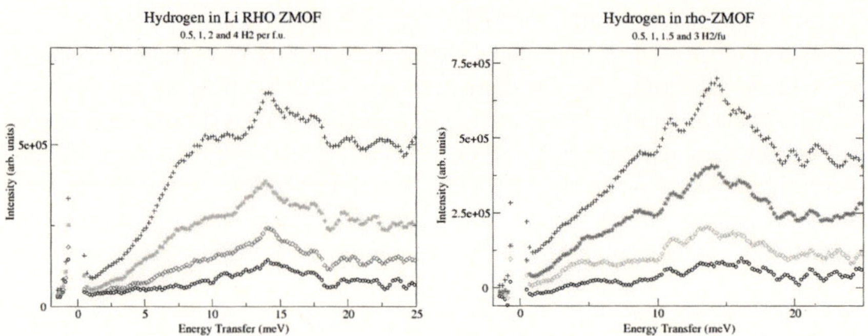

Fig. 8.13 INS spectra of H_2 in *rho*-ZMOF with Li (*left*), as synthesized with DMA (*right*) [108]

cations revealed rather broad spectra with few well-defined peaks, which do not appear to depend very much on the nature of the cation. Much of the intensity in these spectra at lower loadings may be found well below 12 meV, the main peak in the spectrum for H_2 in MOF-5 and therefore indicate stronger interaction (higher barriers to rotation) with the charged framework (Fig. 8.13). A wide range of binding sites seem to be occupied by hydrogen at all loadings, which appears to be characteristic of this environment of large, hydrated cations in an anionic framework. The isosteric heats of adsorption are indeed some 50 % greater than in the neutral framework MOF-5 at low loadings, but virtually independent of the nature of the cation.

Many of the original MOFs were noted for their very large pore sizes and hence extremely-low densities. This property is not, however, optimal for storing large amounts of hydrogen under ambient conditions because much of the interior of the pores may only bind H_2 to other H_2 molecules. If, however, the pore size of a MOM were to be reduced to values not much bigger than that the size of a hydrogen molecule, it could interact with many framework atoms all around rather than just those on a surface. We can therefore readily conclude that a MOM with uniformly small pores would be highly beneficial to hydrogen binding. This was convincingly demonstrated in recent work by Pham et al. [115] on a MOM, which occurs in two forms, one with interpenetrating identical frameworks, the other in a non-inter-penetrating form. SiF_6 units and bipyridyl acetylene (bpa) linkers connect the Cu centres in this structure (Fig. 8.14). We note that the channel dimensions of the interpenetrating structure of 3.5 × 3.5 Å are only slighty larger than the size of the hydrogen molecule of 2.9 Å, which would suggest that H_2 will have appreciable interaction with many of the surrounding framework atoms.

The INS spectrum in fact is rather unique for H_2 in pores in that it only exhibits one strong peak at all loadings with a shoulder on the low energy side (Fig. 8.15), both of which simply increase intensity until the pores are filled. Moreover, this peak occurs at a lower energy (6.3 meV) than that for hydrogen near open-metal sites and hence indicates an appreciably-stronger overall interaction with the framework. The two similar binding sites found in a computational study [115] are

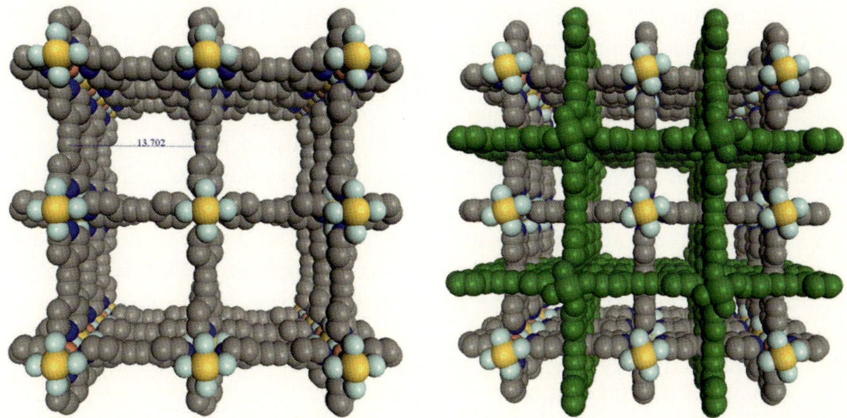

Fig. 8.14 Structures of [Cu(bpa)$_2$(SiF$_6$)]$_n$, non-interpenetrating (*left*) and interpenetrating (*right*). Channel dimensions are 10.3 × 10.3 and 3.5 × 3.5 Å, respectively. Surface areas are 3247 and 735 m^{2}g^{-1}, respectively. Reprinted with permission from (T. Pham, K.A. Forrest, K. McLaughlin, B. Tudor, P. Nugent, A. Hogan, A. Mullen, C.R. Cioce, M.J. Zaworotko, B. Space, J. Phys. Chem. C **117**, 9970 (2013)) [115]. Copyright (2013) American Chemical Society

Fig. 8.15 INS spectra for H$_2$ in [Cu(bpa)$_2$(SiF$_6$)]$_n$, the interpenetrating form. From [116]

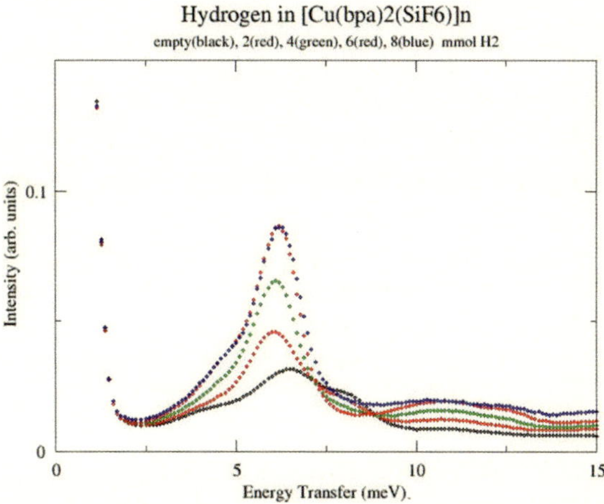

adjacent to each other in the narrow channels, and hence differ little in their interactions with framework atoms.

The analogous binding sites in the non-interpenetrating form are open towards the large pore, so that binding energies and barriers to rotation are much lower, and hence the rotational-tunnel transitions in the usual range of MOFs without open-metal sites, i.e. above 10 meV. The isosteric heat of adsorption [115] in the interpenetrating structure (8.3 kJmol^{-1}) is nearly twice that of the non-interpenetrating structure and constant over most of the range of gas loading. The latter is an important requirement for an adsorption-based hydrogen storage system, and one that few such materials to date can satisfy.

A direct electronic interaction of H_2 with metal centres is, of course, known from inorganic and organometallic chemistry to be very strong and does in fact lead to dissociative binding of H_2 in most cases. The ability to create coordinatively-unsaturated metal sites in a MOM by removal of a weakly-bound ligand might therefore have been expected to significantly enhance hydrogen binding-energies. INS studies of H_2 adsorbed at such sites in a variety of MOFs have indeed revealed stronger interaction by the observation of larger barriers to rotation with smaller rotational-tunnelling splittings. The effect is not, however, nearly as large as anticipated, since the H_2 molecule does not approach the metal site very closely. This was shown directly by powder neutron diffraction experiments on HKUST-1 and on MOF-74 [98, 99, 104, 137].

INS spectra of H_2 in HKUST-1 [111] do show evidence for the stronger inter-action with this open-metal site, with a peak at about 9 meV (Fig. 8.16), which can then be associated with sorption near the open Cu site. This value is appreciably lower (and hence the barrier to rotation higher) than in MOF-5, which in turn reflects the respective isosteric heats of adsorption of 6.8 and 4.8 $kJmol^{-1}$, respectively, at the lowest loadings.

Additional features in the INS spectrum have been attributed to secondary binding-sites in accordance with the structural information, although this process is not unambiguous. For example, the peak at 12.5 meV may be attributed to hydrogen at the so-called small pore site, while at higher loadings a peak near the free-rotor transition may be indicative of several binding sites farther inside the pores. The precise attribution of these spectral features has, however, been subject to different interpretations, which can be difficult to resolve.

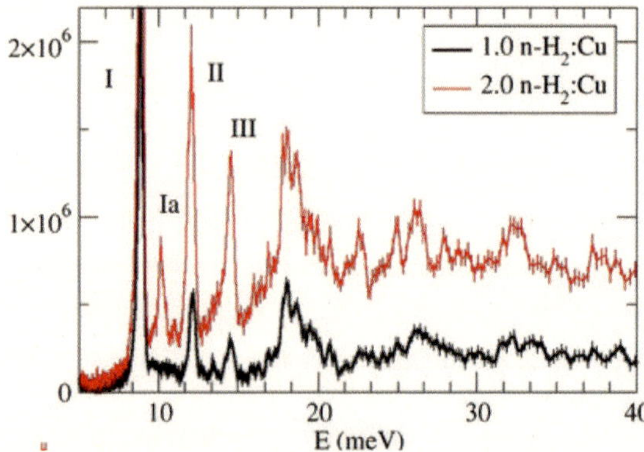

Fig. 8.16 INS spectra for two loadings of H_2 in HKUST-1. Reprinted with permission from (C.M. Brown, Y. Liu, T. Yildirim, V.K. Peterson, C.J. Kepert, Hydrogen adsorption inHKUST-1: a combined inelastic neutron scattering and first-principles study,Nanotechnology **20**, 204025 (2009)) [118]

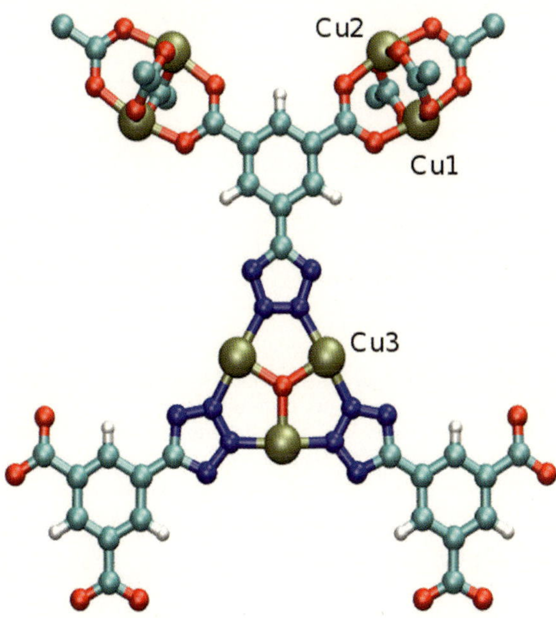

Fig. 8.17 5-tetrazolylisophthalic acid ligands, copper paddlewheel and Cu₃O trimer units in *rht*-MOF-1 [120]

The Cu paddlewheel structural building unit (Fig. 8.8, left) can be found in many MOFs, some of which have been investigated by INS on adsorbed H_2. The rotational transitions for H_2 at this open Cu site vary somewhat from that in HKUST-1 depending on structural details of the environment, as, for example, the close proximity of another paddlewheel unit in $Cu_6(mdip)_3$ (mdip = 5,5'-methylene diisophthalate), also known as PCN-12 [119], where these are observed 7.7 and 8.6 meV.

A remarkable new class of 'designer MOFs' have structures with so-called *rht* net, which are synthesized from metal ions coordinated to a C_3 symmetric ligand with three coplanar isophthalate moieties. Each isophthalate group is linked to metal ions to form a square paddlewheel cluster. In *rht*-MOF-1 a trigonal Cu_3O trimer is liked to three 5-tetrazolylisophthalic acid ligands (Fig. 8.17). This compound therefore contains at least two different open Cu sites, which were identified from the rotational-tunnelling spectra and computational studies [120]. In this case the transition associated with the Cu paddlewheel was found to be at 9.7 mev, compared with about 7 meV for binding sites near the Cu_3O trimer.

The increased interaction strength of hydrogen with these open-metal sites arises primarily from polarization, but falls far short of what one might expect if the molecule were to coordinate to the metal as in organometallic complexes. To date this has definitively been achieved only in the post-synthesis modified zeolite $Na_nAl_nSi_{96}-nO_{192}$ where $0 < n < 27$, also known as ZSM-5. Highly undercoordinated Cu sites can be created by deposition of Cu in some form from the vapour or

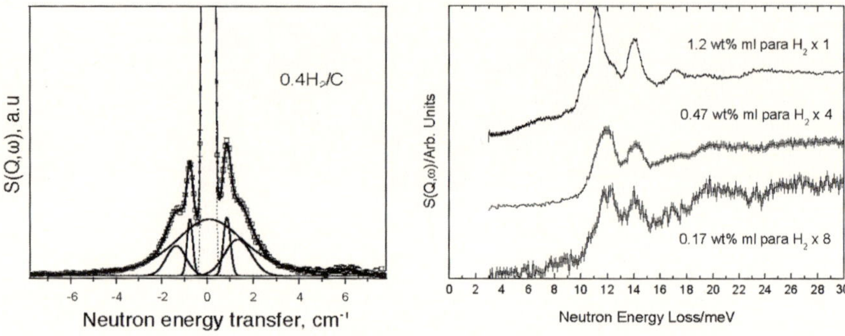

Fig. 8.18 H_2 in Cu-ZSM-5 **a** coordinated the Cu(I) and **b** physisorbed on the walls of chabazite. The rotational-tunnelling transitions in the two cases differ by a factor of nearly 100. Figure taken from [123]. Reprinted with permission from (A.J. Ramirez-Cuesta, P.C.H. Mitchell, Cat. Today **120**, 368 (2007)) [124]. Copyright (2007) Elsevier

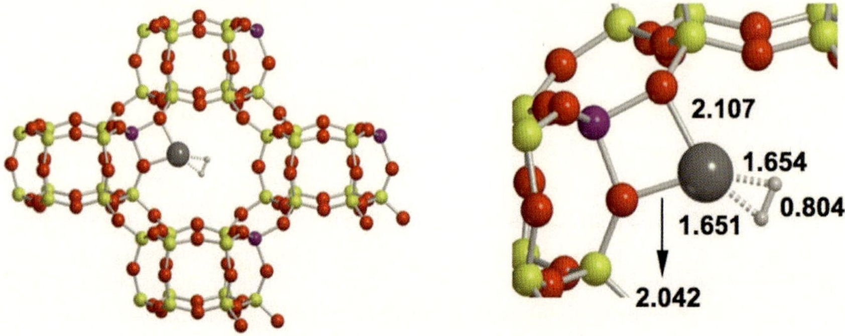

Fig. 8.19 Ab initio computational model for H_2 in the zeolite chabazite. For details see [122]

solution in the pores of the material, whereby Cu-ZSM-5 is created. Cu (II) ions evidently coordinate to framework O atoms and can be reduced to Cu(I) by thermal activation. H_2 can coordinate to these open Cu ions and form a dihydrogen complex similar to some organometallic compounds [121], which tend to show very small rotational tunnelling splittings (Fig. 8.18a) because of the chemical bond formed with the metal centre. This also requires that the H_2 be able to approach the metal to within at least 1.7–1.8 Å as is indeed shown in the computational model developed for the open Cu site using the zeolite chabazite (Fig. 8.19) [122].

The coordinated H_2 is located at 1.653 Å from the Cu centre, and the H–H bond has been activated as shown by an elongated H–H distance of 0.804 Å. Heats of adsorption for binding to Cu(I) in the zeolite have been observed and calculated [122, 123], to be as high as 70 kJmol^{-1}, more than ten times the typical H_2 physisorption energies. The rotational-tunnelling splitting of about ~ 1 cm^{-1} for coordinated H_2 is about 1/100 of that of hydrogen physisorbed on the walls

($\sim$11.5 meV) in the same zeolite [124]. However, only about 40 % of the Cu sites in the zeolite appear to be able to molecularly chemisorb H_2, so that the overall isosteric heat of adsorption decreases rapidly when the H_2 loading increases much beyond that. Nonetheless, this result may be viewed as a proof of principle in that it demonstrates that this type of molecular chemisorption offers the greatest hope to boost binding energies in sorption-based systems to the point where they can store adequate amounts of hydrogen for ambient temperature operation.

For MOMs to enable this type of binding the metal sites must be more open than what is provided by just having one vacant binding site. This may be more easily achieved in the presence of large numbers of defects, or in an amorphous material, and efforts to this end are well underway [125].

A hybrid type of hydrogen storage medium, which combines sorption of H_2 and binding of H in the solid utilizes the so-called hydrogen spillover effect [126]. While neutron scattering methods would again seem to offer the best approach for the study of this phenomenon, experiments carried out to date have encountered difficulties because of the complexity of such media, e.g. metal particles on carbon supports, and the spillover process itself. The approach taken has been to monitor changes in the rotational transition band for H_2 on the carbon support, and attempt to detect the dissociated H by INS vibrational spectroscopy [127].

One of the great strengths of neutron scattering methods is that the experimental results can be connected in a very direct manner to computational studies on account of the simple form of the interaction of neutrons with the atomic nuclei. This aspect does not play a role when it comes to determining to location of the adsorbate molecules, as the experimental probe does not enter the picture, but it does so in spectroscopy. In this case one may compare actual intensities for the various transitions or entire spectral profiles with experiment, in addition to their frequencies.

A number of approaches have been taken in the present case of adsorbed H_2, the simplest of which may be that of rotational barriers. Energies are calculated for stepwise rotation of the hydrogen molecule with fixed centre-of-mass to obtain the rotational barrier height. This value is compared with a barrier height given by a assuming a phenomenological potential which is adjusted to account for the experimentally observed transitions. For H_2 in Cu-ZSM-5 the barriers are 15 and 9 kJmol^{-1}, respectively.

It is, of course, preferable to avoid having to choose a phenomenological potential for interpreting the experimental transition-frequencies by calculating those directly from a theoretical potential and making the comparison on the basis of frequencies, instead of derived barrier-heights. This requires not only an extremely accurate PES for the adsorbed H_2, but also a full quantum dynamical treatment of all of its five degrees of freedom (two rotational and three translational). A simplification of this rather challenging computation can be achieved if one makes the assumption that the rotational degrees of freedom are separable from the centre-of-mass translational motions. This approach has been taken a number of times, including for HKUST-1 and MOF-74. Translational motion is often treated separately using a harmonic oscillator model as an approximation. For H_2 in MOF-74, for example, the PES was derived from a periodic DFT calculation using a van der Waals density functional method [128], from which the calculated rotational transitions for the main binding sites near the open metal, as well as two other sites, were obtained in reasonable agreement with experiment. Translational excitations were calculated separately in a harmonic model with anharmonic corrections.

The computational methodology for the full five-dimensional, coupled translational-rotational quantum dynamics for a bound H_2 was developed by Bačić and collaborators [129] and originally applied to cases where the H_2 was essentially trapped in a cage, and where the interaction potentials are well known, such as H_2 in C_{60} [130], and in ice clathrates [131]. In the more general case for molecules with six degrees of freedom, such as methane, the Schrödinger equation with the following Hamiltonian is solved numerically:

$$H = -\frac{\hbar}{2m}\left(\frac{\partial^2}{\partial x^2} + \frac{\partial^2}{\partial y^2} + \frac{\partial^2}{\partial z^2}\right) + Bj^2 + V(x, y, z, \theta, \phi, \chi) \tag{8.8}$$

where the potential V has to be determined either from empirical forces, or high-level ab initio calculations, or some combination which includes all relevant interactions between the adsorbed molecule and the host. To do this for all five or six degrees of freedom is, however, a lengthy undertaking for many reasons, not the least of which is the fact the H_2 interacts very weakly with its surroundings.

The first application of this methodology to hydrogen adsorbed in a MOF used PESs for MOF-5 derived in two different ways [132]. While both potentials reproduced the structural aspects (binding geometries) fairly well, they were found that a low barrier could facilitate translational motion from the α to the neighbouring γ sites. This is, however, in disagreement both with diffraction experiment, and INS, which show single site occupancy at low temperature and low loadings. The full five-dimensional PES was therefore calculated with the γ sites blocked by other H_2 molecules, and translational-rotational transitions calculated within the PES surface shown (Fig. 8.20). The transition energies obtained agree well with the experimental. spectrum, and do confirm the conclusion reached from infrared experiments [133] that the lowest energy peak at 10 meV in the INS spectrum is in fact a translational excitation. Translational and rotational excitations were found to be coupled only for transitions to higher levels, as has generally been assumed in the analysis of the INS spectra described above at low frequencies.

While the comparison based on observed and calculated transition frequencies described above is indeed rather powerful, the ideal connection between experiment and theory is one where the computation is extended to produce what is actually observed in an experiment, namely in this case the INS spectrum itself. This is quite commonly done for INS vibrational spectroscopy, but has only recently been accomplished [135] for the full five-dimensional translational-rotational quantum dynamics of H_2, in this case when trapped in the small cage of clathrate hydrate (Fig. 8.21).

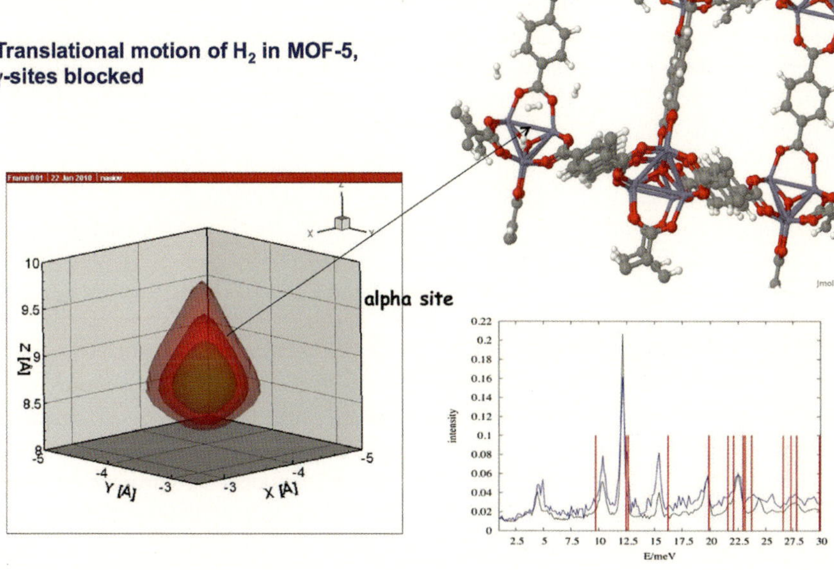

Fig. 8.20 Translational PES (*left*) for H_2 in the α-site of MOF-5 (*upper right*) when the γ-sites are occupied. Calculated five-dimensional translational-rotational spectrum (stick diagram) for this case, superimposed on the experimental data [132]

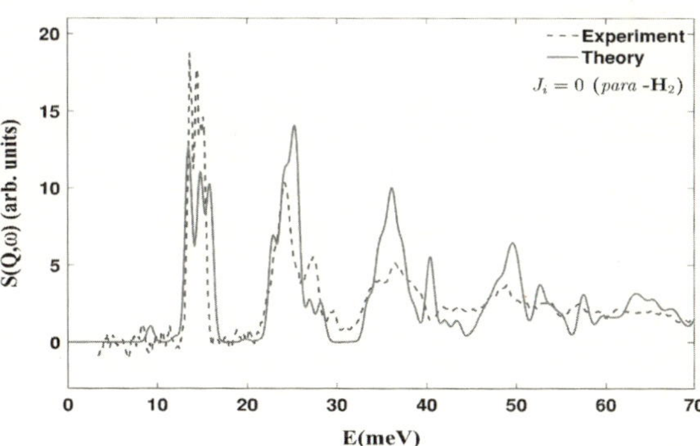

Fig. 8.21 Comparison of the calculated [135] and measured [134] INS spectrum of H_2 in the small cage of clathrate hydrate

This is a relatively more favourable case for application of this methodology that for H_2 on interior surface sites in MOF-5, as the molecule is trapped in a well-defined site, and its interaction with the host, i.e. ice, is rather better known than with those of MOFs. Nonetheless, this approach holds significant promise for extracting the maximum amount of information on the interaction of H_2 with the host material from the INS spectra. The significant computational effort involved in obtaining accurate, multi-dimensional PESs for each binding site, along with solving the five-dimensional quantum dynamics problem for the energy levels and displacements does, however, limit its application to select cases at the present time.

While quasielastic neutron-scattering (QENS) studies have been crucial in the extensive efforts to reach an understanding of the function of complex hydrides such experiments have rarely been carried out on sorption based systems. Translational diffusion of hydrogen in MOFs can readily be observed by QENS [136], but this may not be of sufficient interest for these system, as it is not in any way a rate-limiting factor in the operation of a typical sorption-based storage medium. Investigations of materials with very small pores, or those including hydrogen spillover, may well require detailed QENS studies in the future for their understanding.

8.4 Concluding Remarks

The use of neutron-scattering methods in atomic, and molecular level characterization of virtually all types of potential hydrogen storage materials has provided remarkably fine detail of their structures and the associated hydrogen dynamics which would be impossible to obtain by any other single experimental technique. Hydrogen storage materials are, of course, ideal subjects for neutron scattering studies on account of the neutron's sensitivity to hydrogen, the fact that both diffraction and spectroscopic investigations can readily performed, and that this can be done under a wide range of experimental conditions (pressure, temperature, hydrogen content) with relative ease.

In the case of complex and chemical hydrides it is the knowledge of hydrogen positions in the various crystal structures, identification of the nature of the H-species, in part by INS studies together with computation, nature of diffusive processes, either self-diffusion or aided by reorientations, which have helped to develop some tentative picture of possible discharge-processes in a number of important systems, and in some cases regeneration paths as well. A large number of important questions remain to be answered in detail, most notably perhaps, the role of the catalyst where it is necessary for the function of the material.

Neutron scattering studies of sorption-based storage systems have provided a wealth of detail of the mainly weak interactions of hydrogen molecules with the various host systems. Perhaps the most noteworthy aspect of these investigations is the remarkable sensitivity of the rotational tunnelling transition of the adsorbed H_2 molecules, that can readily be observed by INS, and are found to range over more than two orders of magnitude. While we typically observe a higher barrier to rotation for a more strongly-bound molecule, we note that these barriers are not directly related to the centre-of-mass binding energies of the hydrogen molecule, but to angle-dependent interactions with the host. Nonetheless, these can be coupled by computational studies, which include the development of multi-dimensional potential energy surfaces. The effort to obtain such PESs and to understand hydrogen sorption in MOFs and other porous media by computational analysis has therefore spawned an extensive theoretical work to improve treatment of non-bonded, or mainly dispersive, interactions in DFT methods, which have long been overdue in the study of many other weakly-interacting systems.

Much of the detailed interpretation and computational analysis of some of the results described above depend on the fact the vast majority of the materials under investigation are crystalline so that the positions of the atoms can be known. These must be known, for example, as a starting point for most theoretical modelling studies. Difficulties do arise from the presence of disorder or various types, and the very large zero-point and thermal motion of the weakly bound hydrogen molecules in sorption-based systems. This problem becomes less and less tractable for storage systems that are amorphous, or for the liquid carriers, including those, which have chemical hydrides in suspension. Structural studies of amorphous and liquid systems by the well-established pair distribution function (PDF) approach are therefore likely to become more common, as these approaches to hydrogen storage gain in importance. Structural results from these techniques (preferably a combination of neutron PDF for H and other light atoms, and X-ray PDF for heavier atoms) are probabilistic in nature rather than giving accurate positional parameters, but do bear a direct relationship to molecular dynamics simulations by way of comparisons of observed and calculated pair-correlation function. Spectroscopic studies by INS will suffer from inhomogeneous broadening of the lines from the disordered structures, and the fact that it is usually best done at low temperatures. Quasielastic neutron scattering may become more important in these cases as well, but may depend even more critically on molecular-dynamics simulations.

We can certainly expect that neutron scattering methods in conjunction with computational analyses will continue to play the pivotal role in the ultimate development and design of a practical, materials-based hydrogen storage system for mobile applications, which at this time is viewed as a long-range solution to this problem. In the near term, however, highly compressed hydrogen gas at room temperature is the storage medium of choice for most of the experimental, and commercial (starting in 2015) hydrogen powered fuel-cell cars, including battery hybrids, on the road.

References

1. B. Bogdanovic, M. Schwickardi, J. Alloys Compd. **254**, 1 (1997)
2. A. Zuttel, A. Borgschulte, S.I. Orimo, Scripta Mater. **56**, 823 (2007)
3. S.I. Orimo, Y. Nakamori, J.R. Eliseo, A. Zuttel, C.M. Jensen, Chem. Rev. **107**, 4111 (2007)
4. D.L. Anton, J. Alloys Compd. **356**, 400 (2003)
5. M. Fichtner, O. Fuhr, O. Kircher, J. Rothe, Nanotechnology **14**, 778 (2003)
6. J. Graetz, J.J. Reilly, J. Johnson, A.Y. Ignatov, T.A. Tyson, Appl. Phys. Lett. **85**, 500 (2004)
7. T. Ichikawa, N. Hanada, S. Isobe, H.Y. Leng, Mater. Trans. **46**, 1 (2005)
8. B.C. Hauback, H.W. Brinks, C.M. Jensen, K. Murphy, A.J. Maeland, J. Alloys Compd. **358**, 142 (2003)
9. E. Rönnebro, D. Noreus, K. Kadir, A. Reiser, B. Bogdanović, J. Alloys Compd. **299**, 101 (2000)
10. P. Claudy, B. Bonnetot, G. Chahine, J.M. Letoffe, Thermochim. Acta **38**, 75 (1980)
11. T. Vegge, Phys. Chem. Chem. Phys. **8**, 4853 (2006)
12. G. Krishna, P. Dathar, D.S. Mainardi, J. Phys. Chem. C **114**, 8026 (2010)

13. J.M. Bellosta von Colbe, W. Schmidt, M. Felderhoff, B. Bogdanović, F. Schüth, Angew. Chemie Int. Ed. **45**, 3663 (2006)
14. K. Sakaki, Y. Nakamura, E. Akiba, M.T. Kuba, C.M. Jensen, J. Phys. Chem. C **114**, 6869 (2010)
15. K. Bai, P. Wu, Appl. Phys. Lett. **89**, 201904 (2006)
16. S. Singh, S.W.H. Eijt, J. Huot, W. Kockelmann, M. Wagemaker, F.M. Mulder, Acta Mater. **55**, 5549 (2007)
17. H.W. Brinks, C.M. Jensen, S.S. Srinivasan, B.C. Hauback, D. Blanchard, K. Murphy, J. Alloy. Compd. **376** 215 (2004)
18. H.W. Brinks, M. Sulic, C.M. Jensen, B.C. Hauback, J. Phys. Chem. C **110**, 2740 (2006)
19. P. Canton, M. Fichtner, C. Frommen, A. Leon, J. Phys. Ch **110**, 3051 (2006)
20. J. Íñiguez, T. Yildirim, T. Udovic, M. Sulic, C. Jensen, Phys. Rev. B **70**, 060101 (2004)
21. O. Løvvik, S. Opalka, Phys. Rev. B **71**, 054103 (2005)
22. K.J. Michel, V. Ozolin, J. Phys. Chem. C **115**, 21454 (2011)
23. A.J. Du, S. Smith, G. Lu, Phys. Rev. B **74**, 193405 (2006)
24. A. Marashdeh, R.A. Olsen, O.M. Løvvik, G. Kroes, J. Phys. Chem. C **111**, 8206 (2007)
25. Z. Łodziana, Z. Andreas, P. Zielinski **4**, 465210 (2008)
26. Q. Shi, J. Voss, H.S. Jacobsen, K. Lefmann, M. Zamponi, T. Vegge, J. Alloys Compd. **447**, 469 (2007)
27. J. Voss, Q. Shi, H.S. Jacobsen, M. Zamponi, K. Lefmann, T. Vegge, J. Phys. Chem. B **111**, 3886 (2007)
28. Q.J. Fu, A. Ramirez-Cuesta, S.C. Tsang, J. Phys. Chem. B **110**, 711 (2006)
29. A. Albinati, D. Colognesi, P.A. Georgiev, C.M. Jensen, A.J. Ramirez-Cuesta, J Alloy. Compd. **523**, 108 (2012)
30. Q.J. Fu, A. Ramirez-Cuesta, S.C. Tsang, J. Phys. Chem. B **110**, 711 (2006)
31. O. Friedrichs, F. Buchter, A. Borgschulte, A. Remhof, C.N. Zwicky, P. Mauron, M. Bielmann, A. Zuttel, Acta Mater. **56**, 949 (2008)
32. E. Ronnebro, E.H. Majzoub, J. Phys. Chem. B **111**, 12045 (2007)
33. G. Severa, E. Ronnebro, C.M. Jensen, Chem. Commun. **46**, 421 (2010)
34. T. Matsunaga et al., J. Alloy. Compd. **459**, 583 (2008)
35. Y. Filinchuk, D. Chernyshov, R. Cerny, J. Phys. Chem. C **112**, 10579 (2008)
36. J.-P. Soulie, G. Renaudin, R. Cerný, K. Yvon, J. Alloy. Compd. **346**, 200 (2002)
37. F. Buchter, Z. Łodziana, P. Mauron, A. Remhof, O. Friedrichs, A. Borgschulte, A. Züttel, D. Sheptyakov, T. Strässle, A. Ramirez-Cuesta, Phys. Rev. B **78**, 1 (2008)
38. A. Remhof, Z. Łodziana, P. Martelli, O. Friedrichs, A. Züttel, A. Skripov, J. Embs, T. Strässle, Phys. Rev. B **81**, 1 (2010)
39. K. Jimura, S. Hayashi, J. Phys. Chem. C **116**, 4483 (2012)
40. A. Remhof, A. Zuttel, T. (A.J.) Ramirez-Cuesta, V. Garcia-Sakai, B. Frick, Chem. Phys. **427**, 18 (2013)
41. N. Verdal, T.J. Udovic, J.J. Rush, J. Phys. Chem. C **116**, 1614 (2012)
42. P. Martelli, A. Remhof, A. Borgschulte, P. Mauron, D. Wallacher, E. Kemner, M. Russina, F. Pendolino, A. Zuttel, J. Phys. Chem. A **114**, 10117 (2010)
43. M. Matsuo, S. Orimo, Adv. Energy Mater. **1**, 161 (2011)
44. N. Verdal, T.J. Udovic, J.J. Rush, H. Wu, A.V Skripov, J. Phys. Chem. C **117**, 12010 (2013)
45. P. Fischer, A. Züttel, Mater. Sci. Forum **443–444**, 287 (2004)
46. G. Renaudin, S. Gomes, H. Hagemann, L. Keller, K. Yvon, J. Alloy. Compd. **375**, 98 (2004)
47. N. Verdal, M.R. Hartman, T. Jenkins, D.J. Devries, J.J. Rush, T.J. Udovic, J. Phys. Chem. C **114**, 10027 (2010)
48. C.C. Stephenson, D.W. Rice, W.H. Stockmayer, J. Chem. Phys. **23**, 1960 (1955)
49. K. Chłopek, C. Frommen, A. Léon, O. Zabara, M. Fichtner, J Mater. Chem. **17**, 3496 (2007)
50. R. Cerný, Y. Filinchuk, H. Hagemann, K. Yvon, Angew. Chem. **119**, 5867 (2007)
51. J.H. Her, P.W. Stephens, Y. Gao, G.L. Soloveichik, J. Rijssenbeek, M. Andrus, J.C. Zhao, Acta Crystallogr. Sect. B Struct. Sci. **63**, 561 (2007)

52. Y. Filinchuk, B. Richter, T.R. Jensen, V. Dmitriev, D. Chernyshov, H. Hagemann, Angew. Chem. Int.'l Ed. **50**, 11162 (2011)
53. M. Paskevicius, M.P. Pitt, C.J. Webb, D.A. Sheppard, U. Filsø, E.M. Gray, C.E. Buckley, J. Phys. Chem. C **116**, 15231 (2012)
54. B. Dai, D.S. Sholl, J.K. Johnson, J. Phys. Chem. C **112**, 4391 (2008)
55. Y. Zhang, E. Majzoub, V. Ozoliņš, C. Wolverton, Phys. Rev. B **82**, 174107 (2010)
56. P. Martelli, D. Blanchard, J.B. Maronsson, M.D. Riktor, J. Kheres, D. Sveinbj, E.G. Bardají, J. Phys.Chem. C **116**, 2013 (2013)
57. A. Giannasi, D. Colognesi, L. Ulivi, M. Zoppi, A.J. Raminez-Cuesta, E.G. Bardají, E. Roehm, M. Fichtner, J. Phys. Chem. A **114**, 2788 (2010)
58. F. Buchter, Z. Łodziana, A. Remhof, O. Friedrichs, A. Borgschulte, P. Mauron, A. Zu, D. Sheptyakov, G. Barkhordarian, R. Bormann, K. Chłopek, M. Fichtner, M. Sørby, M. Riktor, B. Hauback, S. Orimo, J. Phys. Chem. B **112**, 8042 (2008)
59. T. Noritake, M. Aoki, M. Matsumoto, K. Miwa, S. Towata, H.-W. Li, S. Orimo, J. Alloy. Compd. **491**, 57 (2010)
60. Y. Filinchuk, E. Ronnebro, D. Chandra, Acta Mater. **57**, 732 (2009)
61. T.J. Frankcombe, J. Phys. Chem. C **114**, 9503 (2010)
62. A. Borgschulte, R. Gremaud, P. Martelli, A. Remhof, A. Ramirez-Cuesta, K. Refson, E. Bardaij, W. Lohstroh, M. Fichtner, H. Hagemann, M. Ernst, Phys. Rev. B **83**, 024102 (2011)
63. I. Llamas-Jansa, O. Friedrichs, M. Fichtner, E.G. Bardaji, A. Zu, B.C. Hauback, J. Phys. Chem. C **116**, 13472 (2012)
64. D. Blanchard, M.D. Riktor, J.B. Maronsson, H.S. Jacobsen, J. Kehres, D. Sveinbjornsson, E.G. Bardajı, A. Leon, F. Juranyi, J. Wuttke, B.C. Hauback, M. Fichtner, T. Vegge, J. Phys. Chem. C **114**, 20249 (2010)
65. E.G. Bardají, Z. Zhao-Karger, N. Boucharat, A. Nale, M.J. van Setten, W. Lohstroh, M. Fichtner, J. Phys. Chem. C **115**, 6095 (2011)
66. J.Y. Lee, D. Ravnsbæk, Y.S. Lee, Y. Kim, Y. Cerenius, J.H. Shim, Y.W. Cho J. Phys. Chem C **113**, 15080 (2009)
67. H. Hagemann, M. Longhini, J.W. Kaminski, T.A. Wesolowski, R. Černý, N. Penin, C.M. Jensen, J. Phys. Chem. A, **112**, 7551 (2008)
68. H.W. Li, S. Orimo, Y. Nakamori, K. Miwa, N. Ohba, S. Towata, A. Züttel, J. Alloy. Compd. **446**, 315 (2007)
69. Y. Nakamori, S.I. Orimo, J. Alloy. Compd. **370**, 271 (2004)
70. P. Chen, Z. Xiong, J. Luo, J. Lin, K.L. Tan, Interaction of hydrogen with metal nitrides and imides. Nature **420**, 302 (2002)
71. W. Luo, E. Rönnebro, Towards a viable hydrogen storage system for transportation application. J. Alloy. Compd. **404–406**, 392 (2005)
72. W. Luo, S. Sickafoose, J. Alloy. Compd. **407**, 274 (2006)
73. W. Luo, J. Alloy. Compd. **381**, 284 (2004)
74. W. Lohstroh, M. Fichtner, J. Alloy. Compd. **446–447**, 332 (2007)
75. R. Juza, K. Opp, Z. Anog, Allg. Chem. **266**, 313 (1951)
76. M.H. Sørby, Y. Nakamura, H.W. Brinks, T. Ichikawa, S. Hino, H. Fujii, B.C. Hauback, J. Alloy. Compd. **428**, 297 (2007)
77. T. Noritake, H. Nozaki, M. Aoki, S. Towata, G. Kitahara, Y. Nakamori, S. Orimo, J. Alloy. Compd. **393**, 264 (2005)
78. K. Ohoyama, Y. Nakamori, S. Orimo, K. Yamada, J. Phys. Soc. Jpn **74**, 483 (2005)
79. J. Rijssenbeek, Y. Gao, J. Hanson, Q. Huang, C. Jones, B. Toby, J. Alloy. Compd. **454**, 233 (2008)
80. F. Dolci, E. Napolitano, E. Weidner, S. Enzo, P. Moretto, M. Brunelli, T. Hansen, M. Fichtner, W. Lohstroh, In Situ **50**, 1116 (2011)
81. Y. Wang, M. Chou, Phys. Rev. B **76**, 014116 (2007)
82. E. Weidner, F. Dolci, J. Hu, W. Lohstroh, T. Hansen, D.J. Bull, M. Fichtner, J. Phys. Chem. C **113**, 15772 (2009)

83. F. Dolci, E. Weidner, M. Hoelzel, T. Hansen, P. Moretto, C. Pistidda, M. Brunelli, M. Fichtner, W. Lohstroh, Int. J. Hydrogen Energy **35**, 5448 (2010)
84. M. Aoki, T. Noritake, G. Kitahara, Y. Nakamori, S. Towata, S. Orimo, J. Alloy. Compd. **428**, 307 (2007)
85. E. Weidner, D.J. Bull, I.L. Shabalin, S.G. Keens, M.T.F. Telling, D.K. Ross, Chem. Phys. Lett. **444**, 76 (2007)
86. W.I.F. David, M.O. Jones, D.H. Gregory, C.M. Jewell, S.R. Johnson, A. Walton, P.P. Edwards, J. Am. Chem. Soc. **129**, 1594 (2007)
87. F. Zhang, Y. Wang, M.Y. Chou, Phys. Rev. B **82**, 094112 (2010)
88. K.J. Michel, A.R. Akbarzadeh, V. Ozolins, J. Phys. Chem. C **113**, 14551 (2009)
89. J. Hu, E. Weidner, M. Hoelzel, M. Fichtner, Dalton Trans. **39**, 9100 (2010)
90. Y.E. Filinchuk, K. Yvon, G.P. Meisner, F.E. Pinkerton, M.P. Balogh, Inorg. Chem. **45**, 1433 (2006)
91. T. Noritake, M. Aoki, S. Towata, A. Ninomiya, Y. Nakamori, S. Orimo, Appl. Phys. A **83**, 277 (2006)
92. G.P. Meisner, M.L. Scullin, M.P. Balogh, F.E. Pinkerton, M.S. Meyer, J. Phys. Chem. B **110**, 4186 (2006)
93. M. Aoki, K. Miwa, T. Noritake, G. Kitahara, Y. Nakamori, S. Orimo, S. Towata, Appl. Phys. A **80**, 1409 (2005)
94. H. Freimuth, H. Wiechert, H.P. Schildberg, H.J. Lauter, Phys. Rev. B **42**, 587 (1990)
95. J.L.C. Rowsell, E.C. Spencer, J. Eckert, J.A.K. Howard, O. Yaghi, Science **309**, 1350 (2005)
96. E. Spencer, J.A.K. Howard, G.J. McIntyre, J.L.C. Rowsell, O.M. Yaghi, Chem. Comm. 278 (2006)
97. T. Yildirim, M.R. Hartmann, Phys. Rev. Lett. **95**, 215504 (2006)
98. V.K. Peterson, Y. Liu, C.M. Brown, C.J. Kepert, J. Am. Chem. Soc. **128**, 15578 (2006)
99. V.K. Peterson, Y. Liu, C.M. Brown, C.J. Kepert, J. Phys. Chem. C **115**, 8851 (2011)
100. N.L. Rosi, J. Kim, M. Eddaoudi, B. Chen, M. O'Keefe, O.M. Yaghi, J. Am. Chem. Soc. **127**, 1504 (2005)
101. P.D.C. Dietzel, R.E. Johnson, R. Blom, H. Fjellvåg, Chem. Eur. J. **14**, 2389 (2008)
102. P.D.C. Dietzel, P.A. Georgiev, J. Eckert, R. Blom, T. Strässle, T. Unruh, Chem. Comm. **46**, 1 (2010)
103. W. Zhou, H. Wu, T. Yildirim, J. Am. Chem. Soc. **130**, 15268 (2008)
104. Y. Liu, H. Kabbour, C.M. Brown, D.A. Neumann, C.C. Ahn, Langmuir **24**, 4772 (2008)
105. M. Dinca, A. Dailly, Y. Liu, C.M. Brown, D.A. Neumann, J.R. Long, J. Am. Chem. Soc. **128**, 16876 (2006)
106. H. Wu, W. Zhou, T. Yildirim, J. Am. Chem. Soc. **129**, 5314 (2007)
107. I. Kanoya, T. Furuta, R. Sakamoto, M. Hosoe, M. Ichikawa, K. Itoh, T. Fukunaga, J. Appl. Phys. **108**, 074310 (2010)
108. J.M. Nicol, J. Eckert, J. Howard, J. Phys. Chem. **92**, 7717 (1988)
109. A.J. Ramierz-Cuesta, P.C.H. Mitchell, S.F. Parker, J. Mol. Catal. A: Chemical **167**, 217 (2001)
110. N.L. Rosi, J. Eckert, M. Eddaoudi, D.T. Vodak, J. Kim, M. O'Keefe, O.M. Yaghi, Science **300**, 1127 (2003)
111. H.G. Schimmel, G.J. Kearley, M.G. Nijkamp, C.T. Visser, K.P. Dejong, F.K. Mulder, Chem. Eur. J. **9**, 4764 (2003)
112. P.A. Georgiev, D.K. Ross, A. De Monte, U. Montaretto-Marullo, R.A.H. Edwards, A.J. Ramirez-Cuesta, D. Colognesi, J. Phys.: Cond. Matter **16**, L73 (2004)
113. J. Eckert, F.R. Trouw, B. Mojet, P. Forster, R. Lobo, J. Nanosci. Nanotechnol. **9**, 1 (2009)
114. F. Nouar, J. Eckert, J.F. Eubank, P. Forster, M. Eddaoudi, J. Am. Chem. Soc. **131**, 2864 (2009)
115. T. Pham, K.A. Forrest, K. McLaughlin, B. Tudor, P. Nugent, A. Hogan, A. Mullen, C.R. Cioce, M.J. Zaworotko, B. Space, J. Phys. Chem. C **117**, 9970 (2013)
116. P. Nugent, T. Pham, K. McLaughlin, P.A. Georgiev, W. Lohstroh, J.P. Embs, M.J. Zaworotko, B. Space, J. Eckert, J. Mater. Chem. A **2**, 13884 (2014)

117. P. Nugent, Y. Belmabkhout, S.D. Burd, A.J. Cairns, R. Luebke, K.A. Forrest, T. Pham, S. Ma, B. Space, L. Wojtas, M. Eddaoudi, M.J. Zaworotko, Nature **495**, 80 (2013)
118. C.M. Brown, Y. Liu, T. Yildirim, V.K. Peterson, C.J. Kepert, Hydrogen adsorption in HKUST-1: a combined inelastic neutron scattering and first-principles study, Nanotechnology **20**, 204025 (2009)
119. X.-S. Wang, S. Ma, P.M. Forster, D. Yuan, J. Eckert, J.J. Lopez, B.J. Murphy, J.B. Parise, H.-C. Zhou, Angew. Chem. Int.'l Ed. **47**, 1 (2008)
120. T. Pham, K.A. Forrest, A. Hogan, K. McLaughlin, J.L. Belof, J. Eckert, B. Space, J. Mater. Chem. A **2**, 2088 (2014)
121. J. Eckert, G.J. Kubas, J. Phys. Chem. **97**, 2378 (1993)
122. M. Sodupe, X. Solans-Montfort, J. Eckert, J. Phys. Chem. C **114**, 13926 (2010)
123. P.A. Georgiev, A. Albinati, B.L. Mojet, J. Olivier, J. Eckert, J. Am. Chem. Soc. **129**, 8086 (2007)
124. A.J. Ramirez-Cuesta, P.C.H. Mitchell, Cat. Today **120**, 368 (2007)
125. T.K.A. Hoang, L. Morris, J. Sun, M.L. Trudeau, D.M. Antonelli, J. Mat. Chem. A **1**, 1947 (2013)
126. R. Prins, Chem. Rev. **112**, 2714 (2012)
127. Y. Liu, C.M. Brown, D.A. Neumann, D.B. Geohegan, A.A. Puretzky, C.M. Rouleau, H. Hu, D. Styers-Barret, P.O. Krasnov, B.I. Yakobsen, Carbon **50**, 4953 (2012)
128. L. Kong, G. Roman-Perez, J.M. Soler, D.C. Langreth, Phys. Rev. Lett. **103**, 096103 (2009)
129. S. Liu, Z. Bacic, J.W. Moskowitz, K.E. Schmidt, J. Chem. Phys. **103**, 1829 (2005)
130. M. Xu, F. Sebastianelli, Z. Bacic, R. Lawler, N.J. Turro, J. Chem. Phys. **128**, 011101 (2008)
131. M. Xu, F. Sebastianelli, Z. Bacic, J. Chem. Phys. **128**, 244715 (2008)
132. I. Matanovic, J. Belof, B. Space, K. Sillar, J. Sauer, J. Eckert, Z. Bacic, J. Chem. Phys. **137**, 014701 (2012)
133. S.A. Fitzgerald, K. Allen, P. Landermann, J. Hopkins, J. Matters, J.L.C. Rowsell, Phys. Rev. B **77**, 224301 (2008)
134. K.T. Tait, F. Trouw, Y. Zhao, C.M. Brown, R.T. Downs, J. Chem. Phys. **127**, 134505 (2007)
135. M. Xu, L. Ulivi, M. Celli, D. Colognesi, Z. Bačić, Phys. Rev. B **83**, 241403 (2011)
136. H. Jobic, Catalysis Chapter, this volume

Part III
Energy Use

Extracting the electrical energy from the simple reaction between hydrogen and oxygen to produce water is an extremely attractive proposition, which is exactly what a fuel cell does:

$2H_2 + O_2 \longrightarrow 2H_2O + \text{electricity} + \text{heat}$

A basic fuel cell consists of an anode and a cathode separated by an electrolyte as shown in Fig. 9.1 of the following chapter. At the anode, hydrogen is separated into protons and electrons, and because the electrolyte only conducts protons, electrons are forced through an external circuit, providing the current to do work. At the cathode, oxygen reacts with protons and electrons to produce water and heat. It is possible to capture the heat for cogeneration, leading to overall efficiencies as high as 90%. Fuel cells are fundamentally more efficient than combustion systems, and efficiencies of 40–50% are gained in today's applications. We note that the ubiquitous internal combustion engine used in transport has an efficiency below 20%. The main challenge facing fuel cell use is probably system cost, with estimates for automotive applications being $\sim\$50/kW$, compared with the $\sim\$30/kW$ that is required to make fuel-cells favourable over internal combustion. Internal combustion engines are highly developed and matching their durability, reliability, weight and infrastructure with fuel cells is a considerable hurdle to the commercial uptake of fuel cells, but one which can be overcome by understanding the fundamental limitations of fuel cells.

Fuel cells are typically distinguished by the type of electrolyte used in charge transport. The major classes of fuel cells include: alkaline fuel cells (AFC), solid oxide fuel cells (SOFC), phosphoric acid fuel cells (PAFC), molten carbonate fuel cells (MCFC) and proton-exchange (or polyelectrolyte) membrane fuel cells (PEMFCs). In Chap. 9 we concentrate on solid oxide fuel cells (SOFC) which use ceramic materials as the electrolyte, enabling their operation at high temperatures using a variety of fuels. Higher temperatures are required for adequate diffusion rates of protons and other charged species, which are measurable via neutron scattering, but impose materials problems and significant start-up delays. Chapter 10 presents neutron scattering studies of the operation of PEM fuel cells, where the aqueous environment provides rapid diffusion of the charged species at much lower temperatures, but poses other challenges. These two fuel cell chapters make use of a wide range of neutron techniques because they are concerned with structure and dynamics over a variety of length scales, from atomic through to macroscopic. The reader is referred to Chap. 1 for an outline of these techniques for a more thorough description. Chapter 9, in particular, is an excellent demonstration of how combining information from a variety of neutron techniques of analysis leads to a more complete understanding of structure/dynamics-function relations.

Chapter 9
Neutron Scattering of Proton-Conducting Ceramics

Maths Karlsson

Abstract This chapter aims to demonstrate the important role that neutron scattering now plays in advancing the current understanding of the basic properties of proton-conducting ceramic separator-materials for future intermediate-temperature fuel cells. In particular, the breadth of contemporary neutron scattering work on proton-conducting perovskite-type oxides, hydrated alkali thio-hydroxogermanates, solid acids, and gallium-based oxides, is highlighted to illustrate the range of information that can be obtained. Crucial materials properties that are examined include crystal structure, proton sites, hydrogen bonding interactions, proton dynamics, proton concentrations, and nanoionics. Furthermore, the prospectives for future neutron studies within this field, particularly in view of the latest developments of neutron methods and the advent of new sources and their combination with other techniques, are discussed.

9.1 High-Temperature Fuel Cells and the Strive Towards Intermediate Temperatures

A particularly promising, yet challenging, clean-energy technology is the solid oxide fuel-cell (SOFC) [1–9]. At the heart of this device is an oxide-ion or proton-conducting oxide electrolyte, which is sandwiched between two porous electrodes (the anode and cathode). The working principle of the SOFC is based on the chemical reaction of hydrogen (at the anode) and oxygen (at the cathode) to produce electricity and water, working temperatures being in the range $\sim 800\text{–}1,000\,^{\circ}\mathrm{C}$. The electrolyte does not conduct electrons, but is permeable to the diffusion of either oxide ions or protons. Schematics of SOFCs based on oxide-ion and proton-conducting electrolytes are shown in Fig. 9.1.

M. Karlsson (✉)
Chalmers University of Technology, 412 96 Göteborg, Sweden
e-mail: maths.karlsson@chalmers.se

© Springer International Publishing Switzerland 2015
G.J. Kearley and V.K. Peterson (eds.), *Neutron Applications in Materials
for Energy*, Neutron Scattering Applications and Techniques,
DOI 10.1007/978-3-319-06656-1_9

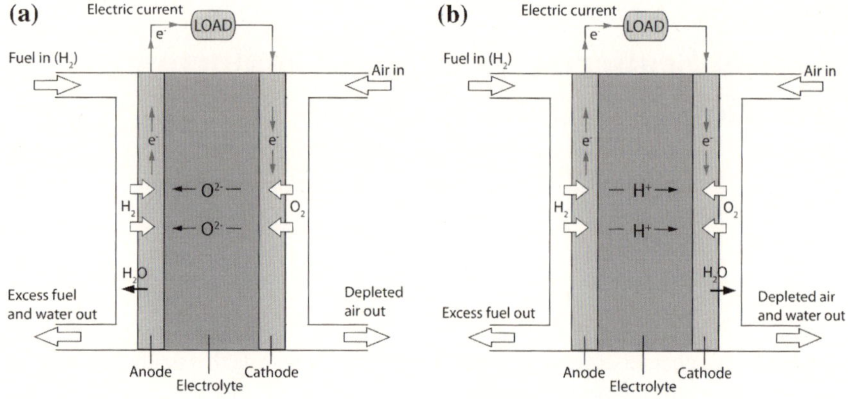

Fig. 9.1 Schematics of the operation of a SOFC utilizing **a** oxide-ion and **b** proton-conducting electrolyte

In the 1990s the Siemens Westinghouse SOFC, consisting of an oxide-ion conducting yttria-stabilized zirconia (YSZ) electrolyte, a lanthanum strontium manganite cathode, and a nickel-YSZ cermet anode, in a unique tubular design, was developed and is now commercially available [1, 10, 11]. Thanks to its high operation temperature, which offers the advantages of fuel flexibility (existing fossil fuels can be used) and high energy-conversion efficiency, the SOFC is particularly attractive for use in combined heat and power applications or efficiently coupled with gas turbines. However, the high operation-temperature also has disadvantages, such as a long startup time, durability issues, and the need for relatively expensive component materials. Therefore, in more recent years, much research has focused on trying to reduce the operation temperature of the SOFC to the so-called *intermediate temperature range* between approximately 200 and 500 °C. Such a reduction in operating temperature would have a beneficial impact on the total cost and the durability of the fuel cell, as problems associated with thermal cycling and performance degradation would be reduced and it would be possible to use cheaper materials in interconnects and heat exchangers [12]. Lowering the temperature would also shorten the startup time of the fuel cell, which is of particular importance for mobile applications. Further, in comparison to low temperature (<100 °C) polymer-electrolyte membrane fuel cells (PEMFCs) [13], a temperature of 200 °C is high enough in order to allow for the use of smaller amounts of expensive platinum to catalyze the electrode reactions [12]. In fact, intermediate-temperature fuel-cell technology has the unique potential to be used both for stationary combined heat and power systems and for hybrid and plug-in hybrid vehicles and are indeed expected to produce energy densities per volume and specific energy per weight significantly larger than state-of-the-art Li ion and Ni metal hydride batteries [14].

However, it is discouraging that any decrease of the operation temperature of SOFCs results in a decreased power density of the device, mainly due to a lowering of the ionic conductivity of the electrolyte. Targeted conductivities exceed

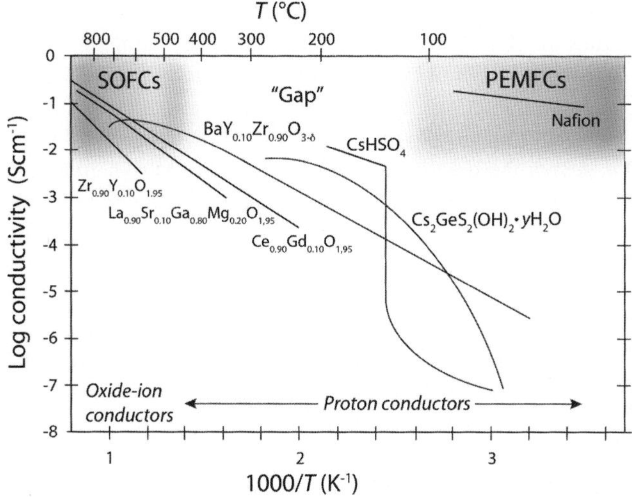

Fig. 9.2 Conductivity of state-of-the-art electrolytes over a wide temperature range. Note the gap of highly-conducting materials in the temperature range $\sim$ 100–500 °C. The figure is redrawn and modified from Ref. [15], copyright Wiley, 2002. The conductivity of $Cs_2GeS_2(OH)_2{\cdot}yH_2O$ is taken from Ref. [16]

0.01 Scm^{-1},[1] and in this respect proton-conducting ceramics emerge as the main candidate electrolytes for SOFCs. However, despite intense research, the conductivities of even the best proton-conducting ceramics are still one to two orders of magnitude below the target. This can be appreciated from Fig. 9.2, which displays the conductivities for the state-of-the-art ionic conductors over a wide temperature range, including electrolytes for low temperature PEMFCs and high-temperature SOFCs, i.e. the more mature fuel-cell technologies. A significant advancement in relation to the conductivity of present day proton-conducting ceramics is therefore critical to future breakthroughs in the development of next-generation fuel-cell technology, operating in the intermediate temperature range. Such an enhancement depends on increasing the understanding of the fundamental science of key materials aspects such as crystal structure and proton-conduction mechanisms, in the most promising classes of materials, and the exploration of novel and completely new systems. For this purpose, neutron scattering is a powerful tool that has been applied successfully to studies of both oxide-ion and proton-conducting materials, and its role in this area of research is vast.

---

[1] To ensure that the total internal resistance (electrolyte + electrodes) of a fuel cell is sufficiently small, the target value for the areal specific-resistivity of the electrolyte is set at 0.15 Ωcm^2. Oxide films can be reliably produced using conventional ceramic fabrication routes at thicknesses down to $\sim$ 15 μm. It follows that the specific conductivity of the electrolyte must exceed 0.01 Scm^{-1} [1].

The aim of this chapter is not to give an exhaustive account for all the neutron scattering studies in this field, nor is it aimed at providing a review of SOFC technology, as excellent reviews can be found elsewhere [1–9]. Rather, this chapter centres on proton-conducting ceramics as separator materials for intermediate-temperature fuel cells and aims to give a flavour of how a variety of neutron methods can be used to reveal the details of structures and proton dynamics in this class of energy-relevant materials. In particular, the scope of contemporary structural and dynamical studies of *perovskite-type oxides*, known as the most promising class of proton-conducting ceramics for intermediate temperature fuel-cell applications, will be reviewed. This is followed by a concise review of selected examples of studies of other classes of promising candidate materials, such as hydrated alkali thio-hydroxogermanates, solid acids, and lanthanum gallates, to illustrate the breadth of information that can be obtained. The chapter finishes with my personal thoughts on prospectives for future work in this timely area of research. The chapter may be viewed as an extended version of my recent publication in Dalton Trans. [17].

9.2 Proton-Conducting Perovskites

The first report on proton conduction in perovskite-structured oxides dates back to 1981, when Iwahara and co-workers [18] showed that the perovskites $SrCeO_3$ and $BaCeO_3$ exhibit high proton-conductivities in hydrogen-containing atmospheres and may be applicable as electrolytic membrane in electrochemical cells such as fuel cells, steam electrolyzers, and gas sensors. Although $BaCeO_3$ has remained a benchmark for high proton-conductivity in perovskites, many other related materials have been developed and found to be proton conducting. Examples include more complex perovskites such as $Ba_3Ca_{1.18}Nb_{1.82}O_{8.73}$ [19–21], rare earth oxides such as Er_2O_3 [22], and pyrochlore-type oxides such as $La_2Zr_2O_7$ [23, 24].

9.2.1 The Perovskite Structure

The basic chemical formula of perovskite-type oxides is ABO_3, where A represents a relatively large cation with oxidation state +1, +2, or +3, and B is a cation with oxidation state between +1 and +7. For proton-conducting perovskites, A is usually +2 (e.g. Sr^{2+} or Ba^{2+}) and B is usually +4 (e.g. Ce^{4+} or Zr^{4+}). The "ideal" perovskite structure is defined by a cubic lattice of corner-sharing BO_6 octahedra and 12-fold coordinated A site ions, see Fig. 9.3a. A cubic structure is, however, typically observed only if the sizes of the cations are compatible with the sizes of their respective interstices, and if not, the perovskite adopts a structure of lower symmetry. The deviation from cubic symmetry may be quantified with the

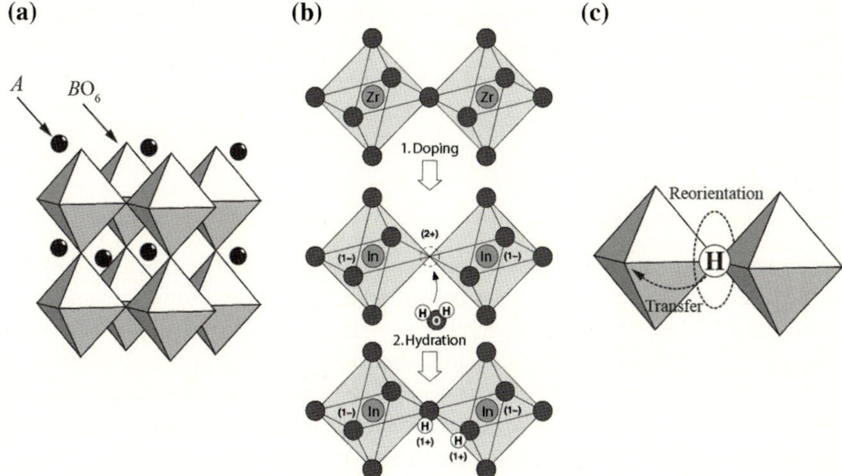

Fig. 9.3 **a** Schematic of the cubic ABO_3-type perovskite structure showing the octahedrally coordinated B^{4+} ions within BO_6 octahedra and the 12-fold coordinated A^{2+} ions. **b** Schematic of doping with In^{3+} at the Zr^{4+} site in cubic-structured $BaZrO_3$, followed by the incorporation of protons through hydration in a humid atmosphere. **c** Schematic of the two principal steps of the proton-conduction mechanism in hydrated perovskites

Goldschmidt factor, t_G, which has a value of one or close to one for a cubic perovskite [25].[2]

9.2.2 The Incorporation of Protons

Loading protons into the perovskite structure depends on acceptor doping at the B site [26]. Acceptor doping, such as when In is substituted for Zr in $BaZrO_3$, creates an oxygen-deficient structure, which under elevated temperature and humidity absorbs water dissociatively so that oxygen vacancies fill with hydroxyl groups and the remaining protons bind to other oxygens in the structure [26]. In the Kröger-Vink notation [27] this reaction is written

$$H_2O + V_O^{\cdot\cdot} + O_O^{\times} \Leftrightarrow 2(OH_O^{\cdot}),\qquad(9.1)$$

where $V_O^{\cdot\cdot}$ denotes an oxygen vacancy, $O_O^{\times}$ denotes a lattice oxygen, and $OH_O^{\cdot}$ denotes a proton bound to a lattice oxygen (the superscripts $\cdot$ and $\times$ denote positive and neutral charges, respectively). The protonation process is shown schematically

---

[2] $t_G = (R_A + R_O)/\sqrt{2}(R_B + R_O)$, where R_A is the ionic radius of the A ion, R_B is the ionic radius of the B ion, and R_O is the ionic radius of oxygen [25].

in Fig. 9.3b. At elevated temperature, the protons are not stuck to any particular oxygen but move rather freely from one oxygen to another, as described further below, resulting in a high conductivity of the material.

9.2.3 Proton Mobility

The dynamical picture of the proton-conduction mechanism in hydrated perovskites started to emerge in the mid-nineties and is mainly based on results obtained from molecular-dynamics simulations [28–31] and quasielastic neutron scattering [32, 33]. On a local scale, the protons jump between neighbouring oxygens, with an intermediate reorientational motion of the −OH group in between jumps (a schematic is shown in Fig. 9.3c), whereas at the longer length-scale the protons diffuse as a series of such jumps and reorientations. However, the effect of dopant atoms on the local chemistry and structure, as well as the resultant symmetry reduction and proton-defect interactions, complicate the description of the proton conductivity and such effects are not completely understood for even the simplest perovskite systems.

A key parameter for long-range proton diffusion is the hydrogen bonding experienced between a proton and a neighbouring oxygen, since the transfer between neighbouring oxygens is a hydrogen-bond mediated process, whereas the reorientational motion of the −OH group requires the breaking of such bonds. The proton diffusivity is further affected by the vibrational dynamics of the proton. More specifically, the proton performs localized O–H stretch and O–H wag vibrations, which may be seen as precursors to the transfer and reorientational step, respectively. There is also a coupling of the proton motions to the phonons of the perovskite host lattice [30, 34]. Hence, the hydrogen bonding, vibrational dynamics, and long-range proton diffusion are, most likely, strongly correlated.

9.2.4 State-of-the-Art of Proton-Conducting Perovskites

The highest proton-conductivities in polycrystalline samples of perovskite-structured oxides are generally found in barium cerates ($BaCeO_3$-based compounds), however, these materials react with CO_2 and/or H_2O at intermediate temperatures to form $BaCO_3$ (or $Ba(OH)_2$) and CeO_2 and therefore degrade with time. Hence, they are poorly suited for use in fuel cells [26]. Strategies to increase the chemical stability of such materials include substitution (doping) of different types of atoms, but improvements are generally small. In comparison, barium zirconates ($BaZrO_3$-based compounds) show excellent chemical stability in CO_2 and H_2O-containing atmospheres and are in this respect more suitable for use in fuel cells from an application point of view. The conductivity of one specific barium zirconate, namely 10 % Y-doped $BaZrO_3$ (included in Fig. 9.2), indeed exhibits the highest bulk proton-conductivity reported for any oxide material [26]. However, barium zirconates are difficult to sinter, which implies that barium zirconate samples contain a relatively-large volume of grain boundaries, which decrease the total conductivity [26, 35–38]. The sinterability of barium zirconates may be enhanced with the use of sintering aids [36, 39] or by the introduction of a second dopant at the B site of the perovskite [40, 41], however, the bulk conductivity is then typically lowered.

9.3 Neutron Scattering of Proton-Conducting Perovskites

The development of new perovskite structures with improved conductivity, thermodynamical stability, and sinterability, depends on the exploration of new classes of compounds as well as an increased understanding of the basic science of those materials already known. Such investigations should elucidate key material detail such as crystal structure, proton sites, proton concentrations, hydrogen-bonding interactions, and the mechanics of proton dynamics on different time- and length-scales, as well as to clarify how these details correlate with each other. For this purpose, neutron scattering offers the unique potential to access simultaneously information in both space and time, through the momentum ($\hbar Q$) and energy ($\hbar E$) transferred in the scattering event, respectively. This combination makes neutron scattering a powerful tool for investigating structures (using neutron diffraction (ND)), vibrational dynamics and hydrogen-bonding interactions (using inelastic neutron scattering (INS)), and diffusional dynamics (using quasielastic neutron scattering (QENS)). Neutrons can also be used to obtain details concerning the proton concentration in the sample through prompt-gamma activation analysis (PGAA). Therefore, neutrons offer good opportunities to advance the understanding of state-of-the-art proton-conducting perovskites. In this context, the remainder of this chapter aims to give a flavour of the important role that neutron methods play in providing a deep insight into the functionality of these materials. Emphasis is put on barium zirconates, due to their great promise for fuel-cell applications.

Examples of recent neutron work on proton-conducting perovskites, which are here briefly reviewed, include studies of proton sites and local structures using ND (Sect. 9.3.1), studies of vibrational proton dynamics and hydrogen-bonding interactions using INS (Sect. 9.3.2), studies of proton diffusion using QENS (Sect. 9.3.3), and studies of proton concentrations using PGAA (Sect. 9.3.4). No attempt has been made to be complete in this work, with the aim rather to highlight the different types of information that can be obtained using neutrons.

9.3.1 Neutron Diffraction

ND has been widely applied to study proton-conducting perovskites. More frequently, Rietveld refinement using *conventional* neutron powder diffraction data is applied, and in some cases, is used to understand how the structure may change with temperature and/or surrounding atmosphere. ND is of paramount importance for studies of proton-conducting perovskites due to its ability to determine proton sites, with both Rietveld and pair-distribution function (PDF) analysis of neutron total-scattering data being applied.

9.3.1.1 Determination of Proton Sites

Knowledge about proton sites is essential to understand the properties of proton-conducting oxides, whether it be the local proton-dynamics, hydrogen-bonding interactions, or macroscopic proton-conductivity. Some early examples of ND measurements on proton-conducting perovskites in this area include by Knight [42] who reported on $BaCe_{0.9}Y_{0.1}O_{2.95}$, Sata et al. [43] who reported on Sc-doped $SrTiO_3$, Sosnowska et al. [44] who reported on $Ba_3Ca_{1.18}Nb_{1.82}O_{9-\delta}$, and Kendrick et al. [45] who reported on $La_{0.6}Ba_{0.4}ScO_{2.8}$. More recently, Ahmed et al. [46] reported on $BaZr_{0.50}In_{0.50}O_{3-\delta}$, whilst Azad et al. [47] reported on $BaCe_{0.4}Zr_{0.4}Sc_{0.2}O_{2.9}$. One may note that most of these studies were performed on samples which were deuterated rather than hydrated, in order to reduce the incoherent scattering and transform the scattering into a useful signal, which increases the chance of locating the positions of protons even in systems with low proton-concentration.

To give a representative example of the previous work we turn to the neutron powder diffraction measurements by Ahmed et al. [46] on $BaZr_{0.50}In_{0.50}O_{3-\delta}$. As pointed out by the authors, the Rietveld analysis of the ND patterns (Fig. 9.4a) indicated no departure from cubic $Pm\bar{3}m$ symmetry. However, further analysis showed that the material is phase-separated into a deuterium-rich and non-deuterated phase with phase fractions of 85 and 15 %, respectively.

To determine the deuterium positions in the deuterium-rich phase, the authors calculated the Fourier-difference maps by taking the difference between simulated (with no deuterium in the structural model) and experimental data in reciprocal space; the Fourier-difference map taken at $z = 0$ is shown in Fig. 9.4b. Here, the positive peak at approximately (0.5, 0.2, 0) in Fig. 9.4b is consistent with the positive scattering-length of deuterium, suggesting that this is the deuterium site. However, from a closer analysis of the neutron data, it was found that the missing scattering density is distributed anisotropically within the *ab* plane, which suggests instead delocalization of the deuterium atom at the $24k$ site. It follows that there are eight equivalent deuterium-sites around each oxygen, which are tilted towards a neighbouring oxygen [46]. Such tilting increases the tendency for the formation of a strong hydrogen-bond between the deuterium and the oxygen towards which it is

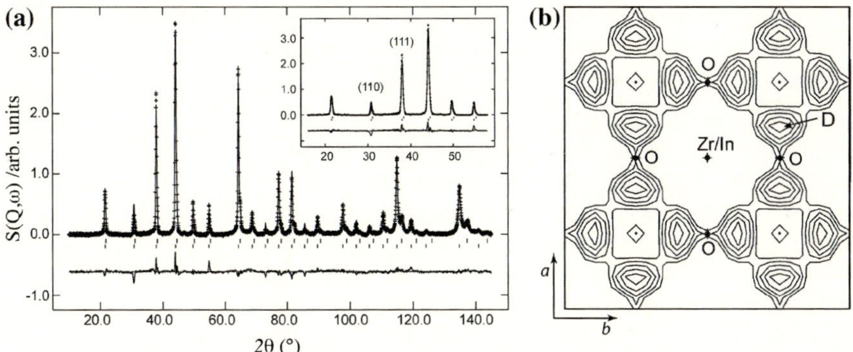

Fig. 9.4 **a** ND pattern for deuterated $BaZr_{0.50}In_{50}O_{3-\delta}$, obtained on the D2B instrument at the Institut Laue-Langevin in Grenoble, France, as reported by Ahmed et al. [46]. Crosses are experimental data, lines are calculated and difference plots (through the data and at the bottom, respectively). The lower set of tick marks indicates the position of calculated reflections for the deuterated phase with no deuterium position defined in the structural model, whereas the upper set of tick marks are those of the non-deuterated phase. The inset shows a close-up of the significant discrepancies between calculated and observed intensities reflecting the absence of D in the structural model. **b** Fourier-difference maps taken at $z = 0$, showing missing positive scattering-density attributed to D. The positive peak, marked as D, is located at approximately (0.5, 0.2, 0). Map size from centre of edge is set to the unit-cell parameter, ~ 4.2 Å. Contour lines are drawn at 0.0132, 0.0264, 0.0396, 0.0528, and 0.0660 fm Å^{-1}. Reprinted with permission from (I. Ahmed, C.S. Knee, M. Karlsson, S.G. Eriksson, P.F. Henry, A. Matic, D. Engberg,L. Börjesson, J. Alloy Compd. **450**, 103 (2008)) Ref. [46], copyright Elsevier

tilted (*c.f.* Fig. 9.5a) and is consistent with results obtained both from first-principles calculations and infrared spectroscopy [48].

Further structural refinements based on the diffraction patterns revealed highly-anisotropic atomic displacement parameters (ADPs) of the oxygen atoms, as illustrated in Fig. 9.5b, as well as the deuterium site occupancy. The large ADPs

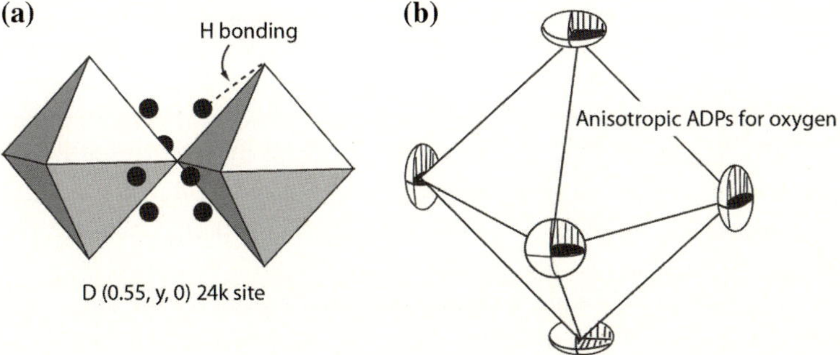

Fig. 9.5 **a** Representation of the refined 24k structural site for the deuteron in the deuterium-rich phase of the deuterated sample of $BaZr_{0.50}In_{0.50}O_{3-\delta}$ at 5 K. **b** Schematic picture of the ADPs of oxygen ions in deuterated $BaZr_{0.50}In_{0.50}O_{3-\delta}$. Reprinted with permission from (I. Ahmed, C.S. Knee, M. Karlsson, S.G. Eriksson, P.F. Henry, A. Matic, D. Engberg,L. Börjesson, J. Alloy Compd. **450**, 103 (2008)) Ref. [46], copyright Elsevier

reflect static, localized, displacements of the oxygen ions around their ideal site in a cubic structure and most likely result from the structural disorder as induced by the difference in size between Zr^{4+} and In^{3+} [46]. Similarly, large ADPs have been obtained elsewhere for the composition $BaZr_{0.33}In_{0.67}O_{3-\delta}$ and those results were also attributed to static-disorder effects [49]. To conclude, the authors not only succeeded in determining the location and concentration of protons (deuterons) in the perovskite structure, but also revealed the presence of pronounced short-range structural distortions of the average cubic perovskite-structure, which are likely to affect considerably the proton transport in the material. More detailed information about the local structure, however, needs the use of more local probes, such as PDF analysis of neutron total-scattering data, which is highlighted below.

9.3.1.2 Local Structural Studies with Neutron Total Scattering

The "extra" scattering between Bragg peaks adds information about the structure on a local scale and is therefore of high importance for structural studies of proton-conducting perovskites when the material is not fully periodic, such as when disorder is present. In particular, the pair-correlation function $G(r)$ is sensitive to key structural details, such as bond distances and angles, the symmetry of structural distortions, and oxygen and/or cation ordering, for example. The initial neutron total-scattering experiments coupled to PDF analysis of proton-conducting perovskites were done by Malavasi et al. [50, 51] on undoped and Y-doped $BaCeO_3$. For the undoped material, the neutron total-scattering data suggests no difference between the long-range orthorhombic *Pnma* structure as determined by Rietveld refinement and the short-range structure as determined from PDF analysis, regardless if the sample is hydrated or not [51].

The good agreement between the long-range and short-range structures may be appreciated from Fig. 9.6, which, for the dry undoped material, shows the good fit of the *Pnma* model obtained from Rietveld refinement to $G(r)$. For the dry Y-doped material, however, the fit of the *Pnma* model is less satisfactory, *c.f.* Fig. 9.7a. For this material, the authors instead found an excellent agreement between the $G(r)$ and a structural model based on the lower-symmetry space group $P2_1$ (see Fig. 9.7b), indicating local regions around the Y dopant of such symmetry [51]. Moreover, it was found that the Y-doped material returns to the orthorhombic *Pnma* structure upon hydration, suggesting that the source of local structural distortion is mainly due to Y-induced oxygen vacancies and not linked directly to the substitution of cations [51]. Since the local structure around the proton can be expected to correlate strongly with the mechanistic detail of proton dynamics, such local structural information is of high interest and indeed crucial for the tailoring of new materials with higher proton-conductivities. The importance of understanding the details of local structure may be exemplified by the 10 % Y- and Sc-doped $BaZrO_3$ materials, which both exhibit an *average* cubic structure, but for which the proton conductivity differs by several orders of magnitude [52].

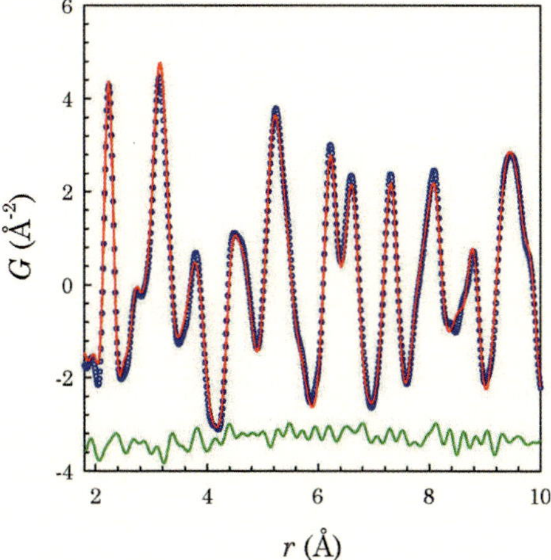

Fig. 9.6 Fit of the pair-correlation function of dry $BaCeO_3$, using the *Pnma* model obtained from Rietveld refinement. *Blue circles* represent the experimental data, the *red line* the calculated data, and the *green line* the difference between the two. Reprinted with permission from (L. Malavasi, H.J. Kim, T. Proffen, J. Appl. Phys. **105**, 123519 (2009)) Ref. [51], American Institute of Physics

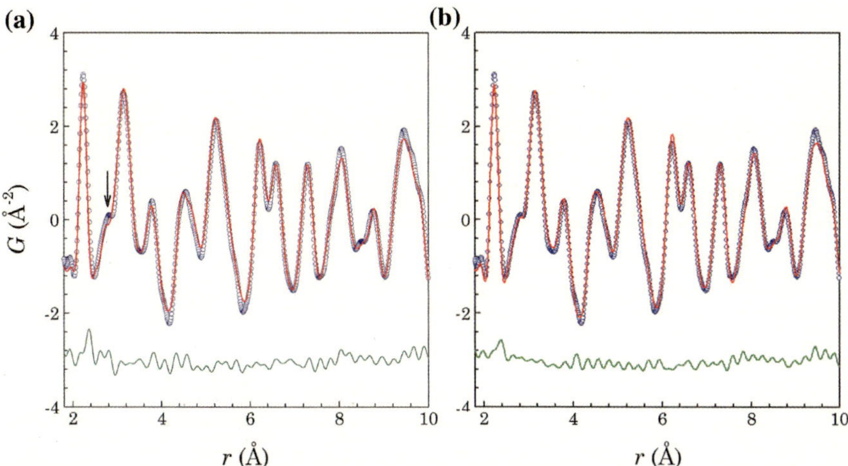

Fig. 9.7 Fit to $G(r)$ of dry Y-doped $BaCeO_3$, using a *Pnma* model **a** and $P2_1$ model **b**. Regions with a marked difference between experimental data (*blue line*) and calculated quantity (*red line*) are highlighted with *arrows*. Note the better agreement when using the $P2_1$ model. The *green line* is the difference between the model and the calculation. Reprinted with permission from (L. Malavasi, H.J. Kim, T. Proffen, J. Appl. Phys. **105**, 123519 (2009)) Ref. [51], American Institute of Physics

9.3.2 Inelastic Neutron Scattering Studies of Vibrational Proton-Dynamics

The first INS study on proton-conducting perovskites was done by Karmonik et al. [53] who investigated vibrational proton-dynamics in $SrCe_{0.95}M_{0.05}O_{3-\delta}$ (M = Sc, Ho, Nd). The INS spectra of the materials are reprinted in Fig. 9.8a and reveal O–H wag modes at around 115 (Sc) and 105 (Ho) meV. For the Nd-doped equivalent, the O–H wag band overlaps with a band at ~ 80 meV. It appears that the frequency of the O–H wag mode shifts to higher wavenumbers with decreasing size of the dopant cation (Nd $\rightarrow$ Ho $\rightarrow$ Sc), indicating that this band is related to protons in the vicinity of such atoms [53]. This behaviour was later validated by Yildrim et al. [54], who performed lattice-dynamics calculations on a $\sqrt{2} \times \sqrt{2} \times 1$ supercell of $SrCeO_3$, replacing one Ce by Sc + H to give a supercell of $Sr_8Ce_7ScHO_{24}$, whose composition is close to that of the real material. In particular, the authors calculated the vibrational spectrum for the hydrogen at the undoped (U) and doped (D) site and by comparing the experimental and calculated spectra (Fig. 9.8b) it could be confirmed that the O–H wag mode at ~ 120 meV indeed is associated with protons close to dopant (Sc) atoms, whereas the O–H wag mode at lower frequency, ~ 80 meV, is associated with protons in the vicinity of host-lattice Ce atoms [54].

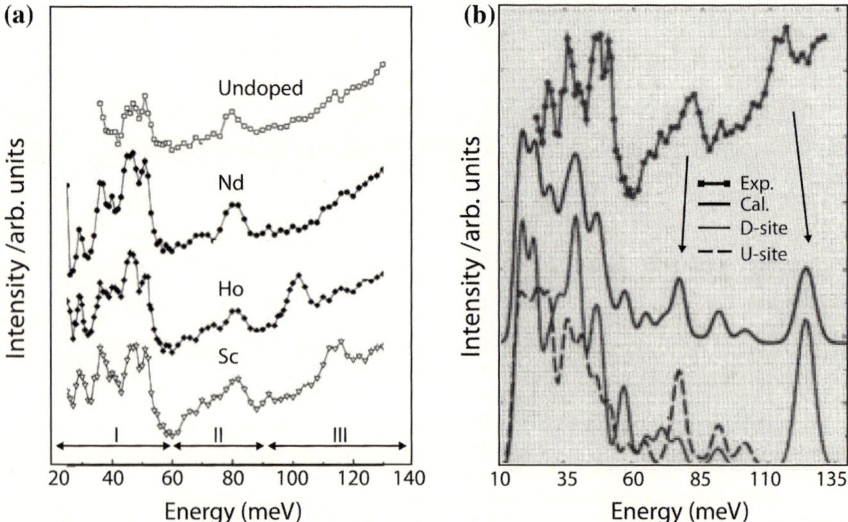

Fig. 9.8 a INS spectra of hydrated samples of $SrCeO_3$ and $SrCe_{0.95}M_{0.05}O_{3-\delta}$ (M = Sc, Ho, Nd), revealing three well-defined vibrational bands in the energy ranges 20–60, 60–90, and 100–140 meV. **b** Comparison of the INS spectrum of hydrated $SrCe_{0.95}M_{0.05}O_{3-\delta}$ (*top*) and the calculated spectrum of $Sr_8Ce_7ScHO_{24}$. *Lines* shown at the *bottom* are the contributions from the $H–MO_6$ clusters, where M = Ce (U site) and M = Sc (D site). The figure is modified and reprinted with permission from (T. Yildrim, B. Reisner, T.J. Udovic, D.A. Neumann, Solid State Ionics **145**, 429 (2001)) Ref. [54], copyright Elsevier

More recently, Karlsson et al. [55] addressed the question of how the O–H wag frequency depends on the dopant concentration. Specifically, the authors performed a systematic INS study of the $BaIn_xZr_{1-x}O_{3-x/2}$ ($x = 0.20$, 0.50, and 0.75) system, which exhibits an average cubic $Pm\bar{3}m$ symmetry independent of the In concentration. The INS spectra are shown in Fig. 9.9a. It can be seen that the O–H wag vibrations show up as a strong, broad, band between approximately 600 and 1,300 cm^{-1}, whilst the peak-fit analysis presented in Fig. 9.9b shows that this band can be decomposed into three Gaussian components. Figure 9.9c shows the In concentration dependence of the relative intensities of the three peak fitted Gaussians. A significant redistribution of intensity amongst the three Gaussians as the In concentration is varied can be observed (the total integrated-intensity of the O–H wag band increases linearly with increasing In concentration [55]). Most interestingly, there is an increased contribution from the two high-frequency components to the overall spectrum, reflected by a band broadening towards higher frequencies, whereas the width and position of each individual Gaussian are found to be essentially independent of the In concentration [55]. The increase in total intensity of the O–H wag band results from the increasing concentration of protons in the sample, whereas the increased contribution of the high-frequency modes is due to an increased fraction of protons in more or less strongly hydrogen-bonding configurations [55]. The formation of strong hydrogen-bonds is believed to be the result of dopant atoms and/or oxygen vacancies in the vicinity of the protons, which act as charged defects, pushing the proton towards a neighbouring oxygen and increasing the tendency for hydrogen-bond formation [55]. However, the presence of such strongly hydrogen-bonding configurations may equally well be the result of tilts and/or rotations of oxygen octahedra induced by doping at the acceptor-atom site, which is of purely static origin [56]. Whatever the case, the formation and breaking of hydrogen bonds are crucially important for long-range proton transport, since proton transfer is a hydrogen-bond mediated process. Thus, information about the nature of hydrogen bonds, which can be derived from the O–H stretch and O–H wag frequencies, and how they link to the structural and dynamical details of the material, is highly valuable. Moreover, O–H stretch and O–H wag mode frequencies are useful in computer simulations, where they are utilized as prefactors in transition-state models to estimate the rates of proton transfer and −OH reorientational motion, respectively [57].

Further information about the behaviour of protons in the perovskite lattice may be derived from the temperature dependence of the INS spectra. In this regard, Karlsson et al. [55] performed a variable-temperature study on $BaZr_{1-x}In_xO_{3-x/2}$ ($x = 0.20$). The INS spectra measured at $T = 30$, 100, 200, and 300 K are shown in Fig. 9.9d. As can be seen the spectra measured at the four different temperatures look essentially the same, which suggests that there is only a small change of the Debye-Waller factor as the temperature is raised from 30 to 300 K, i.e. the total root mean-square displacement, U_T, increases only slightly within this temperature range [55]. The weak temperature-dependence of U_T indicates that there is no particular difference between the proton dynamics in this material at 30 and 300 K.

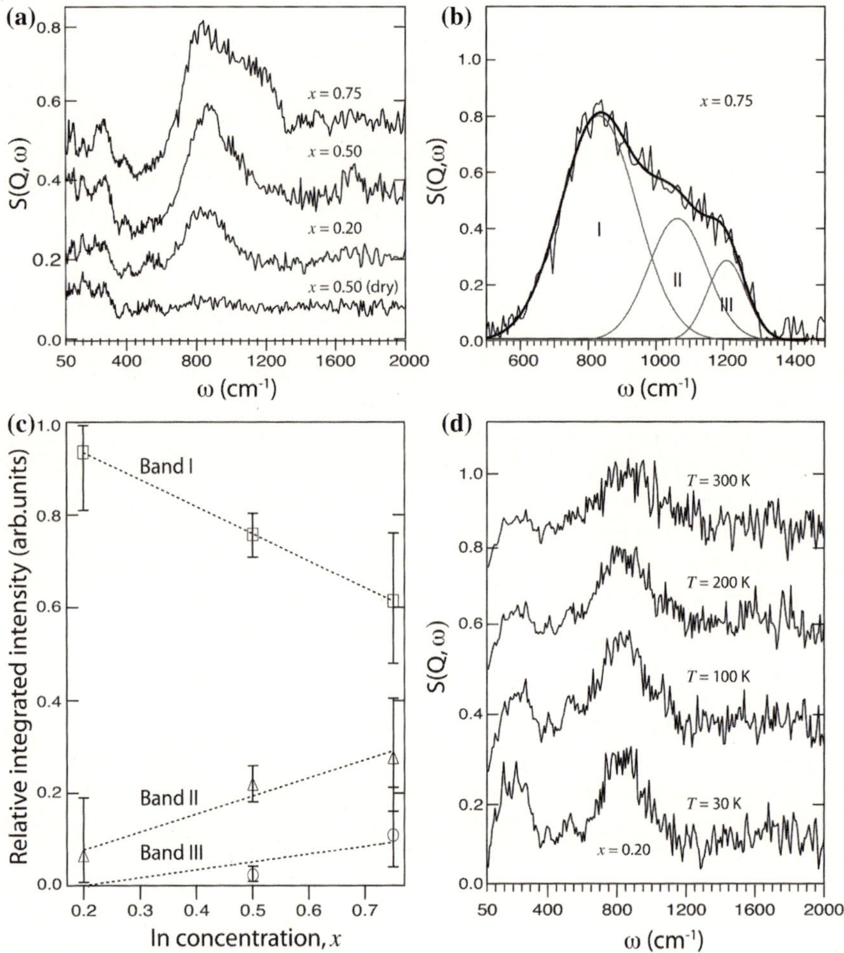

Fig. 9.9 **a** INS spectra for hydrated and dry samples of the $BaIn_xZr_{1-x}O_{3-x/2}$ ($x = 0.20$, 0.50, and 0.75) perovskite system, measured at $T = 30$ K. **b** Peak fit of the baseline-corrected spectrum of $BaIn_xZr_{1-x}O_{3-x/2}$ ($x = 0.75$). **c** Relative integrated-intensities of the three Gaussian components as a function of x. Lines are linear fits and serve as guides for the eye. **d** The INS spectrum of $BaIn_xZr_{1-x}O_{3-x/2}$ ($x = 0.20$) shown for $T = 30$, 100, 200, and 300 K. The spectra have been separated vertically. Reprinted with permission from (M. Karlsson, A. Matic, S.F. Parker, I. Ahmed, L. Börjesson, S.G. Eriksson, Phys. Rev. B **77**, 104302 (2008)) [55], copyright American Physical Society

Since it is unlikely that the protons undergo long-range diffusion at 30 K, it follows that this is also the case at 300 K. This is in agreement with the generally low proton-conductivity for barium zirconates at these temperatures [26, 52].

9.3.3 *Quasielastic Neutron Scattering*

Quasielastic neutron scattering (QENS) has played a central role in understanding the nature of proton dynamics in proton-conducting perovskites. The usefulness of the technique comes from the fact that it gives access to the relevant time- and length-scales on which the atomic-scale dynamics of protons typically occur. In addition, the very large neutron-scattering cross section of protons provides a good contrast in experiments and enables studies of systems containing only small amounts of protons.

9.3.3.1 Studies of Local-Diffusional Proton-Dynamics

The majority of QENS studies on proton-conducting perovskites have been performed with the use of either time-of-flight or backscattering methods [58]. These methods give access to the picosecond timescale, extended to ~ 1 nanosecond in some cases, in which local diffusional-dynamics have been observed, although data have also been interpreted in terms of translational diffusion [59, 60]. The first work was done by Hempelmann et al. [32, 33] for $SrYb_{0.05}Ce_{0.95}O_{2.975}$, where rotational-diffusional motion of the $-OH$ group was observed. These results gave support for molecular-dynamics simulations, which suggested that the proton-conduction mechanism in hydrated perovskites involves proton jumps between neighbouring oxygens and rotational diffusion of the $-OH$ group between proton transfers [28–31]. Later, Groß et al. [61] reported on localized diffusional proton-dynamics in $BaZr_{0.85}M_{0.15}O_{2.925}$ ($M = $ Y, In, and Ga), Pionke et al. [59] reported the proton self-diffusion constant for protons in $Ba[Ca_{0.39}Nb_{0.61}]O_{2.91}$, and similarly, Wilmer et al. [62] presented results for $BaY_{0.10}Zr_{0.90}O_{2.95}$. Braun et al. [60], reported two different activation energies for proton diffusion in $BaY_{0.10}Zr_{0.90}O_{2.95}$ at different temperature ranges, Colomban et al. [63] reported a change in local proton-dynamics across a structural phase transition of $(Ba/Sr)Zr_{1-x}Ln_xO_{3-\delta}$, whilst Karlsson et al. [64] reported a relatively-small difference in the activation energy for local proton-dynamics depending on the choice of dopant atom in $BaM_{0.10}Zr_{0.90}O_{2.95}$ ($M = $ Y and Sc). This collection of examples illustrates the success of time-of-flight and backscattering methods to study the local diffusional proton-dynamics in proton-conducting oxides. However, to reach the long time-scale of several nanoseconds needed to study the long-range translational proton-diffusion on an atomic length-scale (~ 1–30 Å), another QENS method, namely neutron spin-echo (NSE) spectroscopy, is required.

9.3.3.2 Studies of Long-Range Diffusional Proton Dynamics

NSE offers a unique opportunity to obtain information about dynamical processes on different timescales, e.g. from the elementary processes of the proton-conduction mechanism occurring on the picosecond timescale to the long-range translational diffusion of protons occurring on the nanosecond timescale, simultaneously. Despite these advantages, the first application of NSE to investigate proton-conducting ceramics was relatively recent, in 2010 when Karlsson et al. [65] reported proton dynamics in hydrated $BaY_{0.10}Zr_{0.90}O_{2.95}$. Figure 9.10a shows the intermediate-scattering function, $I(Q,t)$, at different temperatures (521–650 K) at the Q-value of 0.3 Å^{-1}, within the time-range of 0.2–50 ns. The $I(Q,t)$ is characterized by a decay with time, which is related to the proton motions in the material. In particular, from the shape of this decay and how it depends on temperature and momentum transfer, $\hbar Q$, information about the mechanistic detail of proton motions, such as the timescale, activation energy, and spatial geometry, can be obtained.

From Fig. 9.10a, it can be seen that the $I(Q,t)$ is described well by a single exponential function (solid lines) with a relaxation time τ and a relaxation rate τ^{-1} that exhibits a Q^2-dependence (inset), which indicates that the relaxational decay is related to long-range translational diffusion. To further justify a result obtained using a single exponential function, the authors modelled the scattering function, $I_{calc.}(Q,t)$, using a kinetic model based on first-principles calculations. Figure 9.10b shows these results for momentum transfers Q = 0.3, 0.5, 2.0 Å^{-1}, as well as for the long-range diffusion limit $Q \to 0$, at a temperature $T = 563$ K. In the latter case, the scattering function is given by a single exponential with a characteristic relaxation rate $\tau^{-1}(Q) = DQ^2$, where D is the diffusion constant. Since the

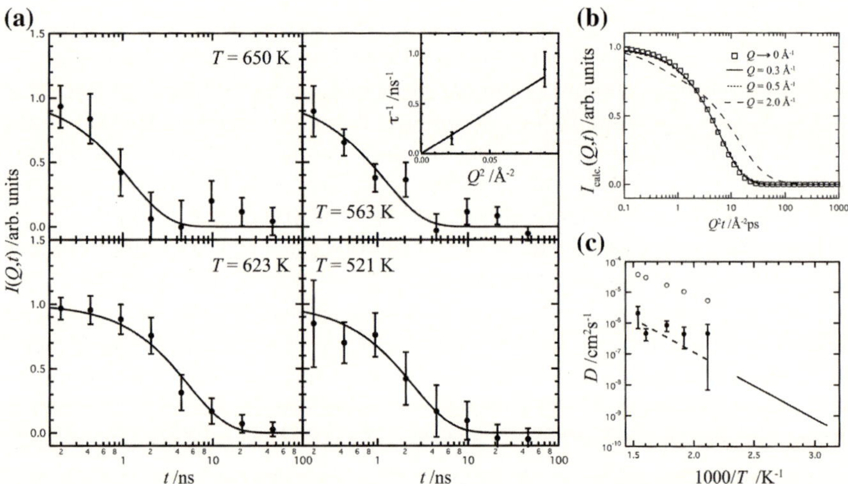

Fig. 9.10 **a** Exponential fits to $I(Q,t)$ at Q = 0.3 Å^{-1} for $T = 521$–650 K. *Upper right panel* shows the Q^2-dependence of the relaxation rate at $T = 563$ K where the solid line represents a fit to $\tau^{-1} = DQ^2$, with $D = 8.53 \cdot 10^{-7}$ cm^{2}s^{-1}. **b** Calculated intermediate scattering-functions at $T = 563$ K. **c** Plot of the diffusion constants obtained from NSE spectroscopy (*bullet*), first-principles calculations (*white bullet*), and conductivity measurements via the Nernst-Einstein relationship (*line*) [52]; the *dotted line* is an extrapolation of the data obtained from conductivity measurements at lower temperatures. Reprinted with permission from (M. Karlsson, D. Engberg, M.E. Björketun, A. Matic, G. Wahnström, P.G. Sundell, P. Berastegui, I. Ahmed, P. Falus, B. Farago, L. Börjesson, S. Eriksson, Chem. Mater. **22**, 740 (2010)) [65], copyright American Chemical Society

$I_{calc.}(Q, t)$s are plotted against $Q^2 t$ they collapse onto a single curve in the long-range regime and this can be seen in Fig. 9.10b up to at least $Q = 0.5$ Å^{-1}. These results suggest that the long-range translational diffusion of protons occurs and on this basis, a diffusion coefficient assuming a $\tau^{-1}(Q) = DQ^2$ dependence was extracted. Temperature-dependent results are shown in the Arrhenius plot in Fig. 9.10c, from which it is evident that the diffusion constant is consistent between different temperatures, showing that the analysis is physically reasonable. Included in this figure is also the diffusion constant extracted from conductivity experiments and derived from first-principles calculations. It is evident that the diffusion constant obtained from NSE and conductivity experiments are comparable, which implies that already on a length-scale as short as ~ 20 Å the effect of potential local traps or other "imperfections" in the structure that can be expected to affect the proton dynamics, has averaged out. That is, there are no new features revealed on a larger length-scale that have not been experienced by the proton on the shorter length-scale probed by NSE. However, by extending the Q range to higher Q values it should be possible to observe the crossover from single-exponential behaviour at low Q values, typical for long-range proton diffusion, to a more complex behaviour at larger Q values, suggesting that several processes are taking place. Further work along these lines is likely to give answers to questions like how the type and concentration of dopant atoms correlate with the macroscopic proton-conductivity, which as discussed above is a topic of some controversy.

Several researchers claim that the dopant atoms act as localized trapping-centres where the proton spends an extended time before it diffuses further throughout the material. This view was first introduced by Hempelmann et al. [32, 33] on the basis of their QENS data for $SrYb_{0.05}Ce_{0.95}O_{2.975}$, which could be described by a so-called "two-state" model, suggesting that the proton migration takes place through a sequence of trapping and release events (Fig. 9.11 (left)). This view later found support from muon spin relaxation experiments [66] and computational studies of proton dynamics in perovskite-type oxides [34, 67–70], and most recently from a combined thermogravimetric and a.c. impedance spectroscopy study [71], as well as from luminescence spectroscopy measurements [72]. Converse to this picture, Kreuer et al. [73, 74] proposed that the dopant atoms may affect the proton transport in a more nonlocal fashion (Fig. 9.11 (right)). This suggestion is based on conductivity data for Y-doped $BaCeO_3$ [73, 74], which shows that the proton conductivity increases with dopant level, but not as a result of a decrease of the pre-exponential factor, D_0, in the expression for the diffusivity $D = D_0 \exp(-E_a/k_B T)$ as anticipated by the two-state model, but rather as a result of an increased activation energy, E_a [73, 74].[3] Moreover, Mulliken population analyses of the electron densities at the oxygens showed that the additional negative charge introduced by

---

[3] Assuming that the two-state model is true and the proton spends an average time, t_1, in the defect-free region and a time in a trap, $t_1 + t_0$, then the diffusivity is scaled by a factor, $t_1/(t_1 + t_0)$ [32]. Increasing the concentration of traps then leads to a decrease of t_1. It implies that in the two-state model the diffusivity depends on the concentration of traps but not on the activation energy.

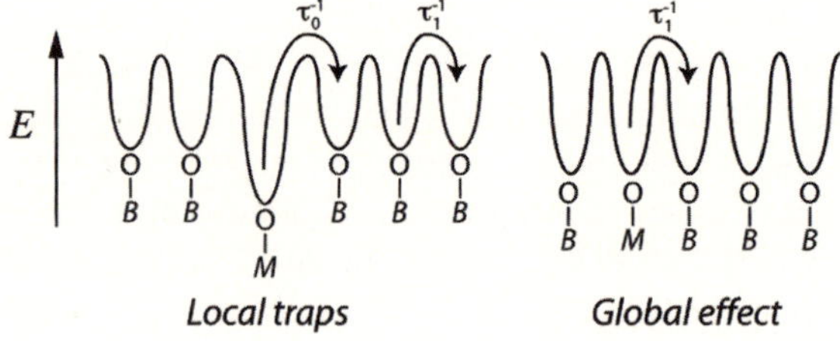

Fig. 9.11 Schematic of two different views of the potential-energy landscape experienced by the proton in hydrated $AB_{1-x}M_xO_3$ type perovskites. τ_0^{-1} and τ_1^{-1} are the escape rates from a trap and a regular oxygen site, respectively, i.e. $\tau_0^{-1} < \tau_1^{-1}$

the dopants is distributed rather homogeneously over the oxygen lattice, which results in stronger bonding of the protons with increased proton-transfer barriers in general [73].

Irrespective of whether the dopant atoms influence the proton diffusion in a spatially restricted way or more non-locally, further QENS investigations using time-of-flight, backscattering, and spin-echo methods, is likely the only way to experimentally elucidate the mechanistic detail of proton dynamics in proton-conducting perovskites. Such information is crucial for the development of strategies for the strategic design of new materials with conductivities beyond the current state-of-the-art materials and hence are critical for future breakthroughs in the development of intermediate-temperature fuel-cell technology.

9.3.4 Neutron Prompt-Gamma Activation Analysis

In order to make correct statements about the role of protons and oxygen vacancies on the structure and dynamics of proton-conducting oxides, an accurate measure of the proton concentration in the material is necessary. Determinations of proton concentrations are done routinely using thermogravimetric methods by measuring the weight change of the sample during dehydration on heating [26]. However, the use of thermogravimetric methods is not suitable for all types of materials. For example, hydrated perovskites containing elements that may change oxidation state upon heating may decrease in mass because of oxygen loss in addition to the evaporation of water molecules. An alternative technique for the analysis of proton concentration in such systems is neutron prompt-gamma activation analysis (PGAA), which can indeed be used for determining the presence and amount of elements in materials, irrespective of oxidation state.

An example of a PGAA study of a proton-conducting perovskite is the work by Jones et al. [75] on undoped and Y-doped BaPrO$_3$. The PGAA spectra of dry and hydrated (saturated) $BaY_{0.1}Pr_{0.9}O_{3-\delta}$ are shown in Fig. 9.12a. The sensitivity to hydrogen and the effect of hydration are clearly visible. The proton concentration,

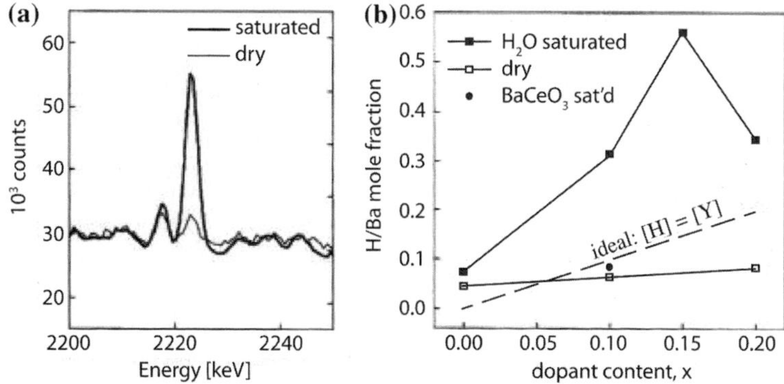

Fig. 9.12 **a** Neutron PGAA spectra of dry and hydrated (saturated) $BaY_{0.1}Pr_{0.9}O_{3-\delta}$. **b** Proton concentration in dry (*white squares*) and hydrated (*black squares*) $BaY_{0.1}Pr_{0.9}O_{3-\delta}$, determined from the PGAA spectra. A comparison with hydrated $BaY_{0.1}Ce_{0.9}O_{3-\delta}$ (*bullets*) is included. The figure is modified and reprinted with permission from (C.Y. Jones, J. Wu, L. Li, S.M. Haile, J. Appl. Phys. **97**, 114908 (2005)) [75], copyright American Institute of Physics

as derived from the PGAA spectra, is shown in Fig. 9.12b for both samples. A key result of this study is that the proton concentration in the hydrated Y-doped $BaPrO_3$ sample is as much as three times larger than the dopant concentration. The unexpectedly-high proton concentration in the hydrated sample is thought to occur as a result of the change of the Pr oxidation state from +4 to +3. This implies that the Pr ions act as self-dopants, which form intrinsic oxygen-vacancies that add to the oxygen vacancies formed by the replacement of Pr for Y. Hence, the material can accommodate more −OH groups and therefore take up more protons than is expected from the dopant concentration alone. This information is naturally of paramount importance for making the correct analysis and conclusions from data obtained in both structural and dynamical studies.

9.4 Case Studies of Other Classes of Proton-Conducting Ceramics

It is clear that research on proton-conducting ceramics continues to be focused on archetypical ABO_3-type perovskites. However, a variety of other classes of proton-conducting ceramics which show appreciable proton-conductivities at intermediate to high temperatures have been developed, and are receiving increased attention. Examples of these include more complex perovskites of the form $Ba_3Ca_{1.18}Nb_{1.82}O_{8.72}$ and $Sr_3CaZr_{0.5}Ta_{1.5}O_{8.75}$ [19–21], which possess cation ordering leading to a doubling of the unit-cell, and perovskite-related phases such as brownmillerite-structured oxides (e.g. $Ba_2In_2O_5$ [49]). Other examples include gallium-based oxides (e.g. $LaBaGaO_4$ [76]), pyrochlores (e.g. $La_2Zr_2O_7$ [23, 24]), phosphates (e.g. $LaPO_4$ [77, 78]), niobates and tantalates (e.g. $LaNbO_4$ [79]), tungstates (e.g. La_6WO_{12} [80]), solid acids (e.g. $CsHSO_4$ [81]), hydrated alkali thio-hydroxogermanates (e.g. $Cs_2GeS_2(OH)_2 \cdot yH_2O$ [16, 82]), tungsten-bronze titanate/niobate systems (e.g. $Ba_{0.6}Mg_{0.067}Nb_{0.933}O_3$ [83]), and cupsidine systems (e.g. $La_4(Ga_{2-x}Ti_xO_{7+x/2})O_2$, $x = 0$–2 [84]).

In this section, we follow some examples of recent neutron-scattering studies of hydrated alkali thio-hydoxogermanates, solid acids, and gallium-based oxides, to further highlight the breath of information that can be obtained with neutron scattering.

9.4.1 ND Study of Hydrated Alkali Thio-Hydroxogermanates

Hydrated alkali thio-hydroxogermanates, $M_xGeS_x(OH)_{4-x} \cdot yH_2O$, where M = Na, K, Rb or Cs; $1 < x < 4$; $y \sim 1$, represent a novel class of amorphous proton-conducting materials, which were first synthesized by Poling et al. [16, 82] at Iowa State University. The conductivities of these materials typically reach a maximum of the order of 10^{-2} Scm^{-1} in the intermediate-temperature range of 100–300 °C [16], which competes with even the best perovskite-type oxides. Karlsson et al. [85] reported a structural investigation of these materials using a combination of neutron diffraction and first-principles calculations. A key result from the experiment was that the neutron structure-factors of the hydrated and dehydrated materials (Fig. 9.13) are overall similar to each other, indicating that there are no dramatic structural changes such as phase transitions or structure degradation as the materials are dehydrated. In order to gain understanding for what such a structure may look like, the authors proceeded their analysis by generating a candidate three-dimensional structure of the Cs-based compound, by taking the orthorhombic crystal structure of $Na_2GeS_2(OH)_2 \cdot 5H_2O$ as the starting point in the calculations, replacing the Na ions with Cs ions and reducing the number of water molecules from five to one in order to agree with the real composition. Snapshots of the generated structures of dehydrated and hydrated materials are shown in Fig. 9.14, whereas Fig. 9.15 shows a close up of the local configuration of the hydrated material.

In the hydrated state (Figs. 9.14a and 9.15), the calculations suggest a structure built of thio-hydroxogermanate anion dimers connected to each other via hydrogen bonds to water molecules located between the dimers. In the dehydrated state (Fig. 9.14b), the calculations suggest instead that the thio-hydroxogermanate anions form an extended network through the creation of inter-dimer hydrogen bonds, whereas the alkali ions are found to act as "space fillers" in "voids" formed by the thio-hydroxogermanate anion dimers in both the hydrated and dehydrated state. These generated structures are justified by comparing the experimental and calculated pair-correlation functions, which are shown in Fig. 9.16. It can be appreciated that the experimental and calculated pair-correlation functions are overall similar

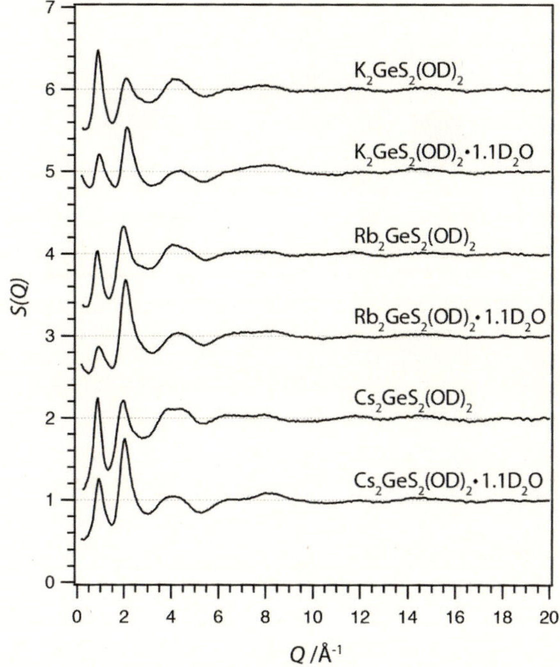

Fig. 9.13 Experimental neutron static structure-factors of hydrated and dehydrated Cs_2GeS_2 $(OD)_2 \cdot yD_2O$, $Rb_2GeS_2(OD)_2 \cdot yD_2O$, and $K_2GeS_2(OD)_2 \cdot yD_2O$. For clarity, the diffractograms have been vertically shifted by unity. The figure is reprinted with permission from (M. Karlsson, A. Matic, I. Panas, D.T. Bowron, S.W. Martin, C.R. Nelson, C.A. Martindale, A. Hall, L. Börjesson, Chem. Mater. **20**, 6014 (2008)) [85], copyright American Chemical Society

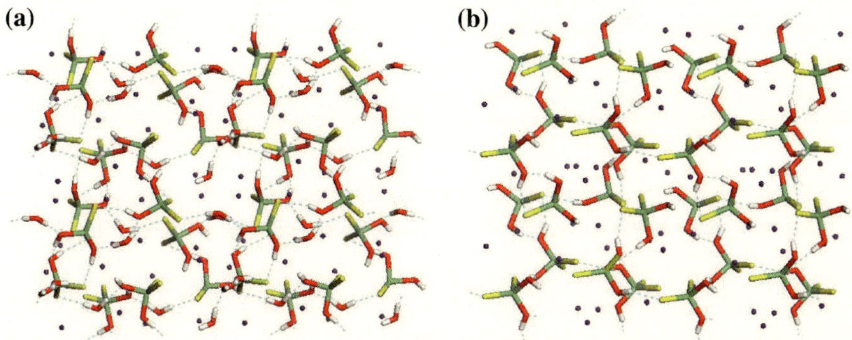

Fig. 9.14 Snapshots of the modelled structure of **a** Hydrated and **b** Dehydrated $Cs_2GeS_2(OH)_2 \cdot y$ H_2O [85]. Oxygen is shown in *red*, sulfur in *yellow*, hydrogen in *white*, and cesium in *violet*. *Dashed lines* are hydrogen bonds. The figure is reprinted with permission from (M. Karlsson, A. Matic, I. Panas, D.T. Bowron, S.W. Martin, C.R. Nelson, C.A. Martindale, A. Hall, L. Börjesson, Chem. Mater. **20**, 6014 (2008)) [85], copyright American Chemical Society

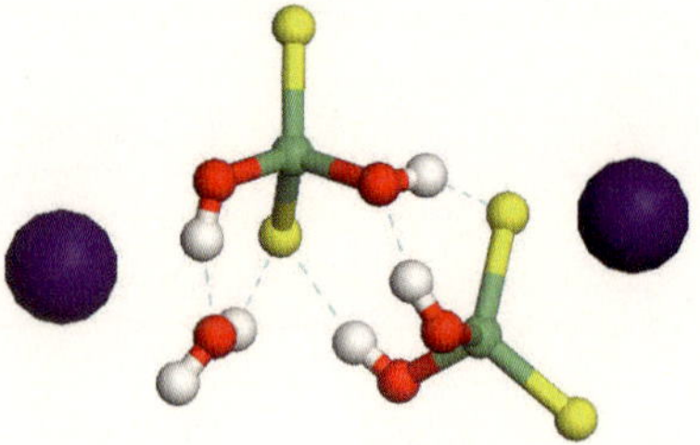

Fig. 9.15 Part of the $Cs_2GeS_2(OH)_2 \cdot yH_2O$ structure obtained from modelling [85]. Oxygen is *red*, sulfur is *yellow*, hydrogen is *white*, and cesium and germanium ions are *violet* and *green*, respectively. *Dashed lines* are hydrogen bonds. The figure is reprinted with permission from (M. Karlsson, A. Matic, I. Panas, D.T. Bowron, S.W. Martin, C.R. Nelson, C.A. Martindale,A. Hall, L. Börjesson, Chem. Mater. **20**, 6014 (2008)) [85], copyright American Chemical Society

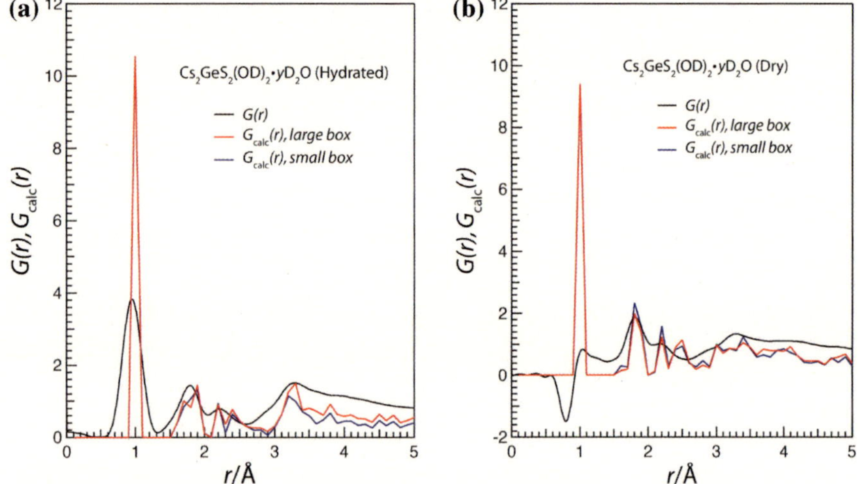

Fig. 9.16 Comparison of pair-correlation functions for **a** Hydrated and **b** Dehydrated Cs_2GeS_2 $(OD)_2 \cdot yD_2O$, obtained from ND and first-principles calculations. *Red* depicts the larger and *blue* the smaller, computational boxes. The figure is reprinted with permission from (M. Karlsson, A. Matic, I. Panas, D.T. Bowron, S.W. Martin, C.R. Nelson, C.A. Martindale,A. Hall, L. Börjesson, Chem. Mater. **20**, 6014 (2008)) [85], copyright American Chemical Society

which suggests that the real structures are at least reasonably-well described by the structural models for both the hydrated and dehydrated materials. However, to obtain a complete picture of the structure and elucidate how the structure depends on the type and concentration of alkali ion, further investigations, such as reverse Monte-Carlo simulations of diffraction data, are necessary.

9.4.2 QENS Study of Nanoionic Proton-Mobility in Solid Acids

Proton-conducting solid acids are compounds, such as $KHSO_4$ and $CsHSO_4$, that feature spectacular phase transitions during heating for which the proton conductivity increases by several orders of magnitude [81, 86, 87]. $CsHSO_4$, for example, has a phase-transition temperature of 414 K [81]. Below this temperature, $CsHSO_4$ has a monoclinic structure in which the number of protons is equal to the number of proton sites. Consequently, the hydrogen atoms are localized within rigid hydrogen bonds between SO_4 tetrahedra and hence their mobility is low. In the high-temperature phase, the SO_4 tetrahedra can rotate rather freely between crystallographically identical positions, creating six times as many possible proton sites as there are protons available. As a result, an almost isotropic and dynamic hydrogen-bonding network between the different sulfate groups is created, where all oxygens are involved in hydrogen bonding. In this hydrogen-bonded network, proton diffusion is a fast process which occurs through proton jumps between neighbouring SO_4 groups, as assisted by rotational motion of these groups.

Chan et al. [88] addressed the question of how the addition of nanoparticles, such as SiO_2 and TiO_2, impacts on the $CsHSO_4$ phase-transition temperature and proton conductivity. Using QENS the authors showed that nanostructuring has the twin effect of lowering the superprotonic phase-transition temperature and increasing the local diffusional-dynamics in the superprotonic phase. The results are summarized in Fig. 9.17, which shows (a) the QENS spectra of bulk $CsHSO_4$ and nanocomposite $CsHSO_4$ with SiO_2 (7 nm), (b) the quasielastic width, and (c) the fraction of mobile protons as derived from the quasielastic intensity as a function of temperature. As can be seen in Fig. 9.17b, the phase-transition temperature for the nanostructured

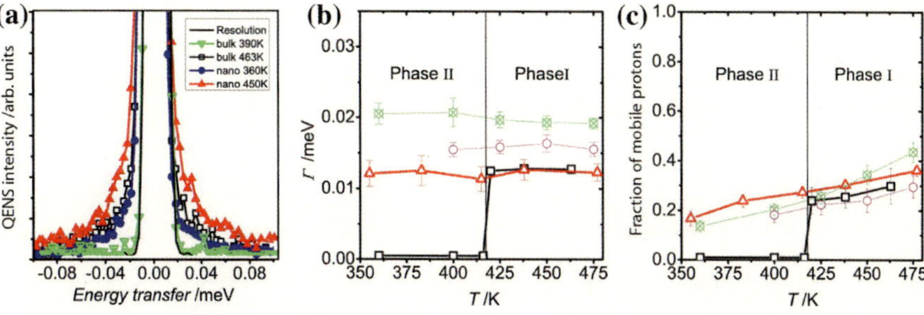

Fig. 9.17 **a** QENS spectra of bulk $CsHSO_4$ and SiO_2 (7 nm) nanocomposite samples with molar ratio 1:2 at $Q = 0.61$ Å^{-1}. **b** Width of the quasielastic signal, Γ, as related to proton mobility at $Q = 1$ Å^{-1} for four different samples. **c** Fraction of mobile protons derived from the integrated intensity of the quasielastic scattering. *Black squares* bulk $CsHSO_4$, *red triangles* nanocomposite $CsHSO_4$ with 24 nm TiO_2 particles, *purple circles* nanocomposite $CsHSO_4$ with 40 nm SiO_2 particles, *green hexagons* nanocomposite $CsHSO_4$ with 7 nm SiO_2 particles. The figure is reprinted with permission from (W.K. Chan, L.A. Haverkate, W.J.H. Borghols, M. Wagemaker, S. J. Picken, E.R.H. van Eck, A.P.M. Kentgens, M.R. Johnson, G.J. Kearley, F.M. Mulder, Adv. Funct. Mater. **21**, 1364 (2011)) [88], copyright Wiley

samples is reduced to a least 360 K. As suggested by Chan et al. [88], this behaviour may be linked to the creation of space-charge layers between the conducting phases and the nanoparticles. The creation of such space-charge layers would lead to an increase of the number of vacant sites for the protons to move to, hence allowing a larger fraction of the protons to become as mobile as they are in the superprotonic phase. Indeed, Fig. 9.17c shows that up to 25 % of the protons are mobile in the nanocomposite sample below the phase-transition temperature of 414 K, whereas no sign of proton mobility can be seen in the bulk sample at those temperatures. The results suggest that nanostructuring may be a rewarding research direction in the pursuit of optimized proton-conductivity.

9.4.3 ND Studies of Structure and Proton Sites in Lanthanum Gallates

Lanthanum barium gallates, i.e. $LaBaGaO_4$-based compounds, represent a novel class of proton-conducting oxides [89]. The structure of these materials consists of discrete GaO_4 tetrahedra which are charge balanced by Ba/La ions, as shown in Fig. 9.18a. Increasing the Ba:La ratio results in the formation of oxygen vacancies, and similarly to perovskite-type oxides, such vacancies can be filled with $-OH$ groups in a humid atmosphere [89]. In this regard, Kendrick et al. [89] raised the question of how the oxygen vacancies are accommodated in the structure, with the knowledge that the inclusion of oxygen vacancies may result in energetically unfavourable three-coordinated Ga atoms, which would limit the oxygen vacancy concentration and hence the degree of hydration. Using computer-modelling methods, the authors found that the oxygen vacancies are accommodated via considerable relaxation of neighbouring GaO_4 tetrahedra, resulting in the formation of

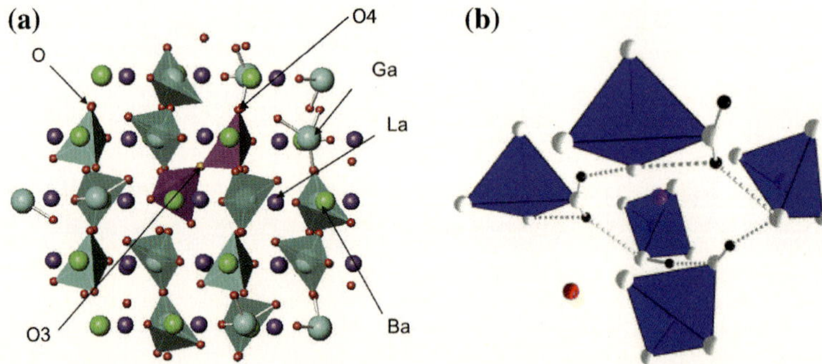

Fig. 9.18 a Crystal structure of $LaBaGaO_4$, illustrating tetrahedrally coordinated Ga and the presence of Ga_2O_7 units (*purple*). The figure is taken from Ref. [90] **b** Hydrogen-bonding interactions in $La_{0.8}Ba_{1.2}GaO_{3.96}H_{0.12}$. *Tetrahedra* GaO_4, *pink sphere* La, *red sphere* Ba, *white spheres* O, *black spheres* H. The figure is reprinted with permission from Ref. (E. Kendrick, K.S. Knight, M.S. Islam, P.R. Slater, J. Mater. Chem. **20**, 10412 (2010)) [91], copyright The Royal Society of Chemistry

Ga_2O_7 groups so that the Ga retains tetrahedral coordination (see purple unit in Fig. 9.18a) [89]. These computational results were supported by ND results which showed a splitting of the oxygen site, consistent with the presence of both GaO_4 and Ga_2O_7 [89]. Hydration of the material can lead to the breaking of such units [89].

On the basis of these results, later investigations focussed on proton positions within the hydrated form of the lanthanum barium gallate structure, using ND coupled with difference Fourier-density analysis of protonated and deuterated samples. Notably, three stable proton positions, one located adjacent to the O3 site and two located adjacent to the O4 site, were found (Fig. 9.18a). The presence of several proton sites is in agreement with modelling work suggesting little difference in energy between proton sites for different oxygens [89]. Further, determination of Ga–O–H bond angles indicated that protons point almost perpendicularly to the Ga–O bond direction and that the protons experience a mixture of intra- and inter-tetrahedral hydrogen-bonding, as shown in Fig. 9.18b [91]. The highest degree of hydrogen bonding is found for O3–H⋯O3 linkages, in agreement with previous modelling work [91]. Modelling also suggests facile inter-tetrahedron proton-conducting pathways due to hydrogen bonding, whereas intra-tetrahedron pathways are found to be less favourable and hence rate-limiting for proton conduction [89].

9.5 Prospectives of Future Proton-Conducting Ceramics Research

Independent of the direction of future research, a major leap in the development of next-generation fuel cells depends on the exploration of new classes of materials and a better understanding of those already known. This chapter has given a flavour of past and current research in the area of proton-conducting ceramics, targeted as electrolytes for future intermediate-temperature fuel-cell technology, and demonstrated the important role neutron scattering plays in elucidating the fundamental science of these materials. In particular, a collection of contemporary neutron studies on proton-conducting perovskite type oxides, hydrated alkali thio-hydroxogermanates, solid acids, and gallium-based oxides, using a range of different neutron methods, has been reviewed in order to illustrate the breadth of information that can be obtained.

In the future, it is clear that there exists great scope for further neutron studies to explore and understand the basic science of structural and dynamical aspects of such classes of proton-conducting oxides. In particular, I foresee an increasing use of PDF analysis and reverse Monte-Carlo modelling [92, 93] of neutron total-scattering data for the investigation (and re-investigation) of local-structural details, such as bond distances and angles, proton sites, and oxygen vacancy and/or cation ordering, for example, of both traditional and new materials. The influence of interactions between oxygen vacancies and dopant atoms on the conductivity of oxide-ion conducting yttria-doped zirconia has been observed [94, 95] and their extension to the broader class of proton-conducting perovskite-structured analogues is an interesting direction of research. In parallel, QENS will play an increasingly important role in elucidating the detail of the proton-conduction mechanism and how it depends on the local-structural details as explored with diffraction methods. For this purpose, I foresee an increasing use of the neutron spin-echo technique in particular, which offers the twin advantages of reaching the long time-scales needed to observe the translational proton-diffusion on an atomic length-scale whilst covering a very large time-range, so that it may be possible to observe and analyse different types of proton motions in a single measurement.

The importance of exploring nanoionic and thin film phenomena is also noted, as nanostructuring and thin film properties may be, and often are, different from the properties of the bulk. In this context, the use of neutron reflectivity, a technique which, to the best of my knowledge, thus far has been neglected in studies of proton-conducting oxides, offers unique possibilities to obtain information about surface and near-surface states and may yield information such as the properties of interfaces and distribution of protons across a single electrolytic-membrane or membrane-electrode assembly. Such information would certainly help in understanding the role of interfaces and, in particular, the reason for the reduced proton conductivity across grain boundaries (GBs). Two explanations for low GB conductivity have been put forward, the first being a structural misalignment in the GB region, and the second being the appearance of a space-charge layer around the GB core, leading to Schottky barriers and the depletion of mobile protons. Presently, the latter explanation predominates research [96, 97], however, details of the GB core are neither well understood nor sufficiently explored.

From a more technical point of view, the recent development of in situ conductivity and humidification cells for ND now allow relatively small features in conductivity to be related to concurrent changes in structure and/or level of hydration [98]. In the near future, the development of in situ cells may also enable investigations of materials under conditions that mimic those under operating fuel-cell conditions and therefore also present the potential to bridge the gap between fundamental scientific problems and applied research. In the longer term, this research can expect also to benefit from the development of completely-new instrumental concepts. An example of this is the recent demonstration of using dynamic nuclear-polarization techniques [99] coupled with ND, where the Bragg peaks can be enhanced or diminished significantly and the incoherent background is reduced [100]. This method offers unique possibilities to tune continuously the contrast of the Bragg reflections and thereby represents a new tool for increasing substantially the signal-to-noise ratio in ND patterns of hydrogenous matter, including proton-conducting oxides.

9.6 Concluding Remarks

This chapter has aimed to demonstrate the valuable role that neutron scattering now plays in providing deeper insight into the fundamental aspects of structure and dynamics of proton-conducting ceramics, with a view to their implementation as electrolytes in next-generation intermediate-temperature fuel cells. Future research in this area is likely to include new challenges with respect to key fundamental properties such as the mechanistic detail of average crystal-structures, local-structural disorder, hydrogen-bonding interactions, proton dynamics, proton incorporation, and nanoionic phenonmena, in archetypical proton-conducting perovskite type oxides as well as in more modern classes of proton conductors.

Acknowledgments   Financial support from the Swedish Research Council (Grant no. 2011-4887) is gratefully acknowledged. P. Slater is thanked for the provision of Fig. 9.18a.

References

1. B.C.H. Steele, A. Heinzel, Nature **414**, 345 (2001)
2. D.J.L. Brett, A. Atkinson, N.P. Brandon, S.J. Skinner, Chem. Soc. Rev. **37**, 1568 (2008)
3. A. Orera, P.R. Slater, Chem. Mater. **22**, 675 (2010)
4. A.J. Jacobson, Chem. Mater. **22**, 660 (2010)
5. J.A. Kilner, Faraday Discuss. **134**, 9 (2007)
6. V.V. Kharton, F.M.B. Marques, A. Atkinson, Solid State Ionics **174**, 135 (2004)
7. H.L. Tuller, Phys. Chem. Chem. Phys. **11**, 3023 (2009)
8. N.Q. Minh, Solid State Ionics **174**, 271 (2004)
9. J.B. Goodenough, Annu. Rev. Mater. Res. **33**, 91 (2003)
10. L. Carrette, K.A. Friedrich, U. Stimming, ChemPhysChem **1**, 162 (2000)
11. J. Larminie, A. Dicks, *Fuel Cell Systems Explained*, 2nd edn. (Wiley, new York, 2003)
12. A. Weber, E. Ivers-Tiffée, J. Power Sources **127**, 273 (2004)
13. K.D. Kreuer, J. Membr. Sci. **185**, 29 (2001)
14. E. Traversa, Electrochem. Soc. Interface **18**, 49 (2009)
15. K.D. Kreuer, ChemPhysChem **3**, 771 (2002)
16. S. Poling, C. Nelson, S.W. Martin, Chem. Mater. **17**, 1728 (2005)
17. M. Karlsson, Dalton Trans. **42**(2), 317 (2013)
18. H. Iwahara, T. Esaka, H. Uchida, N. Maeda, Solid State Ionics **3–4**, 359 (1981)
19. K.C. Liang, A.S. Nowick, Solid State Ionics **61**, 77 (1993)
20. K.C. Liang, Y. Du, A.S. Nowick, Solid State Ionics **69**, 117 (1994)
21. H.G. Bohn, T. Schober, T. Mono, W. Shilling, Solid State Ionics **117**, 219 (1999)
22. Y. Larring, T. Norby, Solid State Ionics **77**, 147 (1995)
23. T. Shimura, M. Kumori, H. Iwahara, Solid State Ionics **86–88**, 685 (1996)
24. T. Omata, K. Okuda, S. Tsugimoto, S. Otsuka-Yao-Matsuo, Solid State Ionics **104**, 249 (1997)
25. V.M. Goldschmidt, Naturwissenschaften **14**, 477 (1926)
26. K.D. Kreuer, Ann. Rev. Mater. Res. **33**, 333 (2003)
27. F.A. Kröger, H.J. Vink, *Solid State Physics: Advances in Research and Applications* (Academic Press, New York, 1956)
28. W. Münch, G. Seifert, K.D. Kreuer, J. Maier, Solid State Ionics **97**, 39 (1997)
29. K.D. Kreuer, W. Munch, U. Traub, J. Maier, Ber. Bunsenges. Phys. Chem. **102**, 552 (1998)

30. W. Münch, G. Seifert, K.D. Kreuer, J. Maier, Solid State Ionics **86–88**, 647 (1996)
31. F. Shimojo, K. Hoshino, H. Okazaki, J. Phys. Soc. Jpn. **66**, 8 (1997)
32. R. Hempelmann, C. Karmonik, T. Matzke, M. Cappadonia, U. Stimming, T. Springer, M.A. Adams, Solid State Ionics **77**, 152 (1995)
33. T. Matzke, U. Stimming, C. Karmonik, M. Soetramo, R. Hempelmann, F. Güthoff, Solid State Ionics **86–88**, 621 (1996)
34. M.S. Islam, R.A. Davies, J.D. Gale, Chem. Mater. **13**, 2049 (2001)
35. H.G. Bohn, T. Schober, J. Am. Ceram. Soc. **83**, 768 (2000)
36. P. Babilo, S.M. Haile, J. Am. Ceram. Soc. **88**, 2362 (2005)
37. F.M.M. Snijkers, A. Buekenhoudt, J. Cooymans, J.J. Luyten, Scripta Mater. **50**, 655 (2004)
38. P. Babilo, T. Uda, S.M. Haile, J. Mater. Res. **22**, 1322 (2007)
39. S. Tao, J.T.S. Irvine, Adv. Mat. **18**, 1581 (2006)
40. N. Ito, H. Matsumoto, Y. Kawasaki, S. Okada, T. Ishihara, Solid State Ionics **179**, 324 (2008)
41. S. Imashuku, T. Uda, Y. Nose, K. Kishida, S. Harada, H. Inui, Y. Awakura, J. Electrochem. Soc. **155**, B581 (2008)
42. K.S. Knight, Solid State Ionics **127**, 43 (2000)
43. N. Sata, K. Hiramotom, M. Ishigame, S. Hosoya, N. Niimura, S. Shin, Phys. Rev. B **54**, 15795 (1996)
44. I. Sosnowska, R. Prezenioslo, W. Schafer, W. Kockelmann, R. Hempelmann, K. Wysocki, J. Alloy Compd. **328**, 226 (2001)
45. E. Kendrick, K.S. Knight, M.S. Islam, P.R. Slater, Solid State Ionics **178**, 943 (2007)
46. I. Ahmed, C.S. Knee, M. Karlsson, S.G. Eriksson, P.F. Henry, A. Matic, D. Engberg, L. Börjesson, J. Alloy Compd. **450**, 103 (2008)
47. A.K. Azad, J.T.S. Irvine, Chem. Mater. **21**, 215 (2009)
48. M. Karlsson, M.E. Björketun, P.G. Sundell, A. Matic, G. Wahnström, D. Engberg, L. Börjesson, I. Ahmed, S.G. Eriksson, P. Berastegui, Phys. Rev. B **72**, 1 (2005)
49. P. Berastegui, S. Hull, F.J. García-García, S.-G. Eriksson, J. Solid State Chem. **164**, 119 (2002)
50. L. Malavasi, H.J. Kim, T. Proffen, J. Appl. Phys. **9**, 2309 (2008)
51. L. Malavasi, H.J. Kim, T. Proffen, J. Appl. Phys. **105**, 123519 (2009)
52. K.D. Kreuer, S. Adams, W. Münch, A. Fuchs, U. Klock, J. Maier, Solid State Ionics **145**, 295 (2001)
53. C. Karmonik, T.J. Udovic, R.L. Paul, J.J. Rush, K. Lind, R. Hempelmann, Solid State Ionics **109**, 207 (1998)
54. T. Yildrim, B. Reisner, T.J. Udovic, D.A. Neumann, Solid State Ionics **145**, 429 (2001)
55. M. Karlsson, A. Matic, S.F. Parker, I. Ahmed, L. Börjesson, S.G. Eriksson, Phys. Rev. B **77**, 104302 (2008)
56. M. Karlsson, A. Matic, C.S. Knee, I. Ahmed, L. Börjesson, S.G. Eriksson, Chem. Mater. **20**, 3480 (2008)
57. M.E. Björketun, P.G. Sundell, G. Wahnström, D. Engberg, Solid State Ionics **176**, 3035 (2005)
58. G.L. Squires, *Introduction to the Theory of Thermal Neutron Scattering* (Cambridge University Press, Cambridge, 1978)
59. M. Pionke, T. Mono, W. Schweika, T. Springer, H. Schober, Solid State Ionics **97**, 497 (1997)
60. A. Braun, S. Duval, P. Ried, J. Embs, F. Juranyi, T. Strässle, U. Stimming, R. Hempelmann, P. Holtappels, T. Graule, J. Appl. Electrochem. **39**, 262103 (2009)
61. B. Groß, C. Beck, F. Meyer, T. Krajewski, R. Hempelmann, H. Altgeld, Solid State Ionics **145**, 325 (2001)
62. D. Wilmer, T. Seydel, K.D. Kreuer, Mater. Res. Soc. Proc. **972**, 15 (2007)
63. P. Colomban, A. Slodszyk, D. Lamago, G. Andre, O. Zaafrani, O. Lacroix, S. Willemin, B. Sala, J. Phys. Soc. Jpn. **79**, 1 (2010)
64. M. Karlsson, A. Matic, D. Engberg, M.E. Björketun, M.M. Koza, I. Ahmed, G. Wahnström, P. Berastegui, L. Börjesson, S.G. Eriksson, Solid State Ionics **180**, 22 (2009)

65. M. Karlsson, D. Engberg, M.E. Björketun, A. Matic, G. Wahnström, P.G. Sundell, P. Berastegui, I. Ahmed, P. Falus, B. Farago, L. Börjesson, S. Eriksson, Chem. Mater. **22**, 740 (2010)
66. R. Hempelmann, M. Soetratmo, O. Hartmann, R. Wäppling, Solid State Ionics **107**, 269 (1998)
67. R.A. Davies, M.S. Islam, J.D. Gale, Solid State Ionics **126**, 323 (1999)
68. M.S. Islam, P.R. Slater, J.R. Tolchard, T. Dinges, Dalton Trans. 3061 (2004)
69. M.E. Björketun, P.G. Sundell, G. Wahnström, Faraday Discuss. **134**, 247 (2007)
70. M.E. Björketun, P.G. Sundell, G. Wahnström, Phys. Rev. B: Condens. Matter **76**, 054307 (2007)
71. Y. Yamazaki, F. Blanc, Y. Okuyama, L. Buannic, J.C. Lucio-Vega, C.P. Grey, S.M. Haile, Nature Mater. **12**, 647 (2013)
72. P. Haro-Gonzáles, M. Karlsson, S. Gaita, C.S. Knee, M. Bettinelli, Solid State Ionics **247–248**, 94 (2013)
73. K.D. Kreuer, W. Münch, M. Ise, T. He, A. Fuchs, U. Traub, J. Maier, Ber. Bunsenges. Phys. Chem. **101**, 1344 (1997)
74. K.D. Kreuer, Solid State Ionics **125**, 285 (1999)
75. C.Y. Jones, J. Wu, L. Li, S.M. Haile, J. Appl. Phys. **97**, 114908 (2005)
76. F. Schönberger, E. Kendrick, M.S. Islam, P.R. Slater, Solid State Ionics **176**, 2951 (2005)
77. K. Amezawa, H. Maekawa, Y. Tomii, N. Yamamoto, Solid State Ionics **145**, 233 (2001)
78. N. Kitamura, K. Amezawa, Y. Tomii, N. Yakamoto, Solid State Ionics **162–163**, 161 (2003)
79. R. Haugsrud, T. Norby, Nat. Mater. **5**, 193 (2006)
80. A. Magraso, C. Frontera, D. Marrero-López, P.Núñez, Dalton Trans. **10**, 283 (2009)
81. S.M. Haile, D.A. Boysen, C.R.I. Chisholm, R.B. Merle, Nature **410**, 910 (2001)
82. S.A. Poling, C.R. Nelson, S.W. Martin, Mater. Lett. **60**, 23 (2006)
83. E. Kendrick, M.S. Islam, P.R. Slater, Solid State Ionics **176**, 2975 (2005)
84. M.C. Martin-Sedeno, D. Marrero-Lopez, E.R. Losilla, L. Leon-Reina, S. Bruque, P. Núñez, M.A.G. Aranda, Chem. Mater. **17**, 5989 (2005)
85. M. Karlsson, A. Matic, I. Panas, D.T. Bowron, S.W. Martin, C.R. Nelson, C.A. Martindale, A. Hall, L. Börjesson, Chem. Mater. **20**, 6014 (2008)
86. T. Norby, Nature **410**, 877 (2001)
87. A.I. Baranov, L.A. Shuvalov, N.M. Schagina, JETP LETT. **36**, 459 (1982)
88. W.K. Chan, L.A. Haverkate, W.J.H. Borghols, M. Wagemaker, S.J. Picken, E.R.H. van Eck, A.P.M. Kentgens, M.R. Johnson, G.J. Kearley, F.M. Mulder, Adv. Funct. Mater. **21**, 1364 (2011)
89. E. Kendrick, J. Kendrick, K.S. Knight, M.S. Islam, P.R. Slater, Nat. Mater. **6**, 871 (2007)
90. E. Kendrick, P.R. Slater, University of Surrey Chemistry Electronic Publication Library, http://epubs.surrey.ac.uk/1685 (2010)
91. E. Kendrick, K.S. Knight, M.S. Islam, P.R. Slater, J. Mater. Chem. **20**, 10412 (2010)
92. R.L. McGreevy, L. Pusztai, Mol. Simul. **1**, 359 (1988)
93. R.L. McGreevy, J. Phys.: Condens. Mat. **13**, 4111 (2001)
94. S.T. Norberg, S. Hull, I. Ahmed, S.G. Eriksson, D. Marrocchelli, P.A. Madden, P. Li, J.T.S. Irvine, Chem. Mater. **23**, 1356 (2011)
95. D. Marrocchelli, P.A. Madden, S.T. Norberg, S. Hull, Chem. Mater. **23**, 1365 (2011)
96. C.T. Chen, C.E. Danel, S. Kim, J. Mater. Chem. **21**, 5435 (2011)
97. M. Shirpour, R. Merkle, C.T. Lin, J. Maier, Phys. Chem. Chem. Phys. **14**, 730 (2012)
98. F.G. Kinyanjui, S.T. Norberg, I. Ahmed, S.G. Eriksson, S. Hull, Solid State Ionics **225**, 312 (2012)
99. A. Abragam, M. Goldman, *Nuclear Magnetism, Order and Disorder* (Oxford University Press, Oxford, 1961)
100. F.M. Piegsa, M. Karlsson, B. van den Brandt, C.J. Carlile, E.M. Forgan, P. Hautle, J.A. Konter, G.J. McIntyre, O. Zimmer, J. Appl. Cryst. **46**, 30 (2013)

Chapter 10
Neutron Techniques as a Probe of Structure, Dynamics, and Transport in Polyelectrolyte Membranes

Kirt A. Page, Joseph A. Dura, Sangcheol Kim, Brandon W. Rowe and Antonio Faraone

Abstract Polyelectrolyte membranes (PEMs) have been employed as solid electrolytes in fuel-cell technologies as early as the 1950s, when they were used in NASA's Gemini program. However, PEM materials have only gained wide-spread attention in the last two decades due to advancements in membrane electrode-assembly (MEA) formation and the synthesis of new and interesting materials. Over the past several decades, various neutron techniques have played an instrumental role in measuring the structure and transport properties of PEMs in order to develop a deeper understanding of structure-property and performance relationships in PEM materials for fuel-cell applications.

10.1 Introduction

Proton-exchange (or polyelectrolyte) membrane fuel-cells (PEMFCs) have received increasingly more attention over the last two decades and is the principle subject of this chapter [1]. More specifically, this chapter presents an overview of how neutron techniques have been used to study polyelectrolyte membrane (PEM) materials. For an historical perspective on the use of polymers in fuel-cell technologies, the reader is encouraged to consulting the existing body of literature on the matter.

K.A. Page (✉) · S. Kim · B.W. Rowe
Materials Science and Engineering Division, National Institute of Standards and Technology, Gaithersburg, MD 20899, USA
e-mail: kirt.page@nist.gov

J.A. Dura · A. Faraone
Center for Neutron Research, National Institute of Standards and Technology, Gaithersburg, MD 20899, USA

© Springer International Publishing Switzerland 2015
G.J. Kearley and V.K. Peterson (eds.), *Neutron Applications in Materials for Energy*, Neutron Scattering Applications and Techniques,
DOI 10.1007/978-3-319-06656-1_10

It is generally understood that one of the key material properties influencing the conduction of protons through the PEM is the morphology of the material. The size-scale of the morphological features typically present in PEMs make small-angle neutron scattering (SANS) an ideal tool for probing the morphology. Additionally, researchers have recently shown an increasing interest in probing the structures that are present in PEM materials at interfaces, as these materials are used as binders in the catalyst layer and can be confined at interfaces with various types of material surfaces. For this, researchers have employed neutron reflectometry (NR). While the morphology of the material serves to provide a pathway for proton conduction, it is also understood that the polymer dynamics can play a role in charge transport. Moreover, the presence of water and water transport/dynamics is critical for optimal fuel-cell performance, and is therefore vital to elucidate the role and interdependent relationship that polymer and water dynamics have on charge transport in PEM fuel cells. Researchers have turned to neutron spectroscopic techniques such as quasi-elastic neutron scattering (QENS) and neutron spin-echo spectroscopy (NSE) to investigate the polymer and water dynamics in hydrated PEM materials.

The following chapter is an overview of the efforts to use neutron-scattering methods to study the structure and transport/dynamics in PEM materials and is divided into three sections. The first section gives a very brief overview of the characteristics of PEMs and highlights the material most studied using neutron techniques. The second section summarizes the structural studies on PEMs to date and demonstrates how techniques such as SANS and NR have aided in characterizing the bulk morphology and structures at interfaces, respectively. The third section focuses on the transport and dynamics in these materials, specifically describing how neutron spectroscopic techniques have been used to study the ion and water dynamics in hydrated PEMs. This chapter is not a comprehensive review of PEM materials and neutron techniques, but is intended to provide the reader with a demonstration of the many ways in which neutron measurements can aid in the understanding of structure-property and performance relationships in PEM materials.

10.2 Polyelectrolyte Membrane Materials

While it is beyond the scope of this chapter to give an extensive review of the field of PEMs, it is necessary that reader be aware of the types of materials that are being developed for fuel-cell applications. Regardless of the particular application (e.g., stationary, portable, or automotive power), these materials must exhibit a certain set of chemical and physical properties that are critical for optimal fuel-cell perfor-mance. Any material being used in a fuel cell must exhibit a list of properties including, but not limited to (1) high proton-conductivity (i.e., a good electrolyte), (2) negligible electrical-conductivity, (3) permeability to ions, but allow only one type of charge, (4) resistance to permeation of uncharged gases, (5) variable membrane-area and thickness, and (6) good mechanical strength. Furthermore, the membrane must be of reasonable cost and durability. Ultimately, it is the polymer chemistry and microstructure that give rise to the macroscopic performance properties that are desired. A range of synthetic approaches have yielded materials that include, but are not limited to, poly(perflourosulfonic acid)s (PFSA), ion-containing polystyrene derivatives, polyarylene ethers, polysulfones, polyimides, and ion-containing block copolymers. There have been extensive studies on each of these classes of materials, but to date neutron techniques have been primarily used to study poly(perflourosulfonic acid)s, namely Nafion[®] [2]. Therefore, it is necessary to give a brief background on this particular material. Throughout this chapter, where specific breakthroughs have been made, studies involving other PEM materials using neutron techniques will be highlighted. On the whole, however, the overlap of neutron measurements and PEM materials has been dominated by PFSAs.

The most widely studied PFSA, Nafion[®], is a product of the E. I. Dupont Chemical Company having the structure given below.

$$-\left(CF_2-CF_2\right)_n\left(CF_2-CF\right)_m$$
$$O-CF_2-CF-CF_3$$
$$O-CF_2-CF_2-SO_3H$$

The polar perfluoroether side-chains containing the ionic sulfonate-groups have been shown to organize into aggregates, thus leading to a nanophase-separated morphology where the ionic domains, termed clusters [3], are distributed throughout the non-polar polytetrafluoroethylene (PTFE) matrix. In addition, the runs of tetrafluroethylene, of sufficient length, are capable of organizing into crystalline domains having unit-cell dimensions virtually identical to that of pure PTFE [4, 5]. The degree of crystallinity in PFSIs is generally less than *ca.* 10 % as a mass fraction in 1,100 equivalent-weight Nafion[®] (EW, the grams of dry polymer per equivalent number of SO_3^- groups) and has been shown to vary with EW. The complex, phase-separated morphology, consisting of crystalline, amorphous, and ionic domains, of PFSIs has been the focus of several investigations [3, 5–16]. Over the last 50 years, a wide variety of studies involving Nafion[®] have aimed to relate the thermal, mechanical, and fuel-cell performance properties (i.e., transport, ionic conductivity, and dielectric behaviour) to specific morphological features. Many of these studies involve neutron-scattering techniques and will be discussed in detail below.

10.3 PEM Structure and Neutron Scattering

10.3.1 Nanoscale Membrane Structure

By and large, the neutron technique most used to study PEM materials has been SANS. It is understood that the nanostructure of PEMs plays a significant role in water uptake and transport, which are materials with properties vital to the performance of a working fuel-cell. However, in order to design a material with desired features, one must have a detailed understanding of the interplay between the nanostructure of the membrane and the overall performance properties. SANS is a versatile tool in elucidating the structure of a variety of membrane materials and can also be used to study transport, which will be discussed in the section on water transport.

As previously mentioned, Nafion® has been the most widely-studied PEM material to date and SANS has been employed extensively to study the nanoscale structure of this complex material [8, 10, 11, 14–23]. The earliest structural studies of Nafion® utilizing SANS and small-angle X-ray scattering (SAXS) revealed a broad peak at a Q value between 0.1 and 0.2 Å^{-1}, called the ionomer peak, which has been attributed to the correlation between the nanophase-separated ionic domains, termed clusters. The crystalline component contributes to the scattering at multiple length scales including peaks in the Q range 0.6–2.0 Å^{-1}, owing to the structure of the amorphous and local crystalline-lattice, in addition to a broad peak entered at lower values of Q ($\approx$0.05 Å^{-1}), which is related to the inter-crystalline scattering (known as the long period). Moreover, ultra-small-angle scattering reveals an upturn that can be associated with large-scale heterogeneities. The scattering for hydrated Nafion® over a wide range of length-scales can be seen in a review by Gebel and Diat [19, 24].

An example of the scattering using neutrons can be seen in Fig. 10.1, for Nafion® films cast from a dispersion, annealed between 80 and 180 °C, and equilibrated in liquid water. When annealed below the alpha relaxation temperature (T_α) of Nafion® ($\approx$100 °C), the lack of a peak at low Q values is evidence that there is no apparent long-range crystalline order in the film. Above T_α, however, one observes a peak due to long-range crystalline order and a shift in the crystalline peak to lower Q values with increasing annealing temperature, indicating that higher annealing-temperatures result in larger, more widely separated crystalline domains. An analysis of the scattering curves can be seen in the inset in Fig. 10.1. Clearly, the spacing of the crystallites increases with increasing annealing temperature. Moreover, the spacing between the ionic domains decreases with increasing annealing temperature. It is known that the crystalline structure plays an integral part in the mechanical stability and durability of fuel-cell membranes, but these data also reveal the relationship between the crystalline structure and water uptake. The decreased spacing between the ionic domains and the lower incoherent-background with increasing annealing temperature are evidence that these annealed films have a lower water-uptake. In these materials water retention must be balanced with annealing temperature in order to achieve desirable proton-conductivity and mechanical integrity. This is just one example of how SANS can be used to probe structure-processing-property relationships.

Over the decades, development and advancement of state-of-the-art scattering techniques were able to reveal the many scattering features, over multiple length-scales, of Nafion® and other perfluorosulfonic acid membranes. As a result, there has been a progression in the complexity and characteristics of the many morphological models that have been proposed to explain the observed scattering in effort to gain a deeper fundamental insight into how the structure is related to

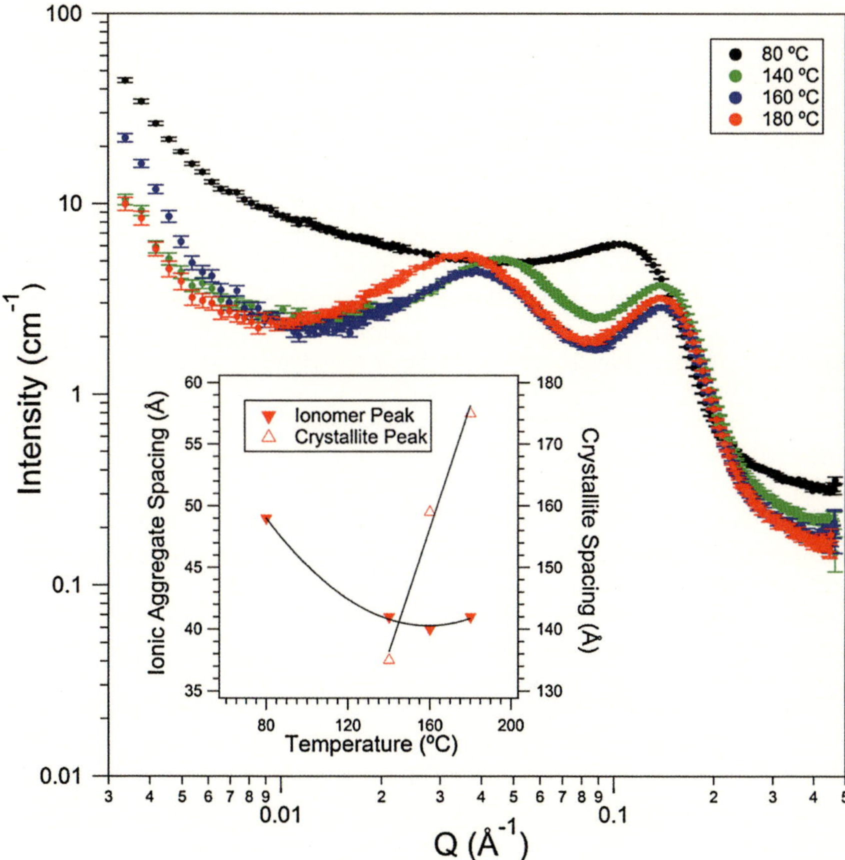

Fig. 10.1 Scattering intensity, $I(Q)$, and given as the macroscopic cross-section, measured by SANS for annealed, solution-cast Nafion® films equilibrated in liquid water. The inset shows how crystalline and ionic aggregate spacing is affected by annealing temperature. The lines serve as a guide to the eye

material performance. The proposed models generated have included a variety of structural units and range from the earliest spherical cluster-network model to other models including lamellar, sandwich-like, fringed micelle, rod-like, and ribbon-like, as well as a model which includes cylindrical water channels. Each of these models was able to account for the scattering to some degree reasonably well, making it difficult to discern which most accurately describes the morphology. Of course, the models able to capture the many scattering-features over a large Q range, with physically relevant fitting parameters are of the highest value. For the details of the structural models of Nafion® and their respective parameters, the reader is directed to the extensive literature concerning this topic.

In general, scattering techniques provide an excellent way to characterize the global structure of fuel-cell membranes [19, 24]. For polyelectrolytes such as Nafion® there are scattering features that are ubiquitous and considered to arise from favourable structures for fuel-cell membrane application, although that conventional wisdom is now being called into question by more recent studies. Typically, polyelectrolyte fuel-cell membranes contain ionic moieties that are able to conduct protons or hydroxide ions in alkaline fuel-cell membranes. These ion-containing, polar-groups phase separate from the more hydrophobic components of the polymer and can form ion-conducting channels which are responsible for ion transport. Quite often these ionic domains give rise to a scattering peak in SANS and, if other hierarchical structures are present, other scattering features are observed. This is especially true when block copolymers are used to provide a structural basis for the membrane. A recent review by Elabd and Hickner [25] has evaluated the state-of-the-art block-copolymer membranes by leveraging the self-assembled nanostructure of block copolymers as a template for creating well-defined transport pathways for use in fuel cells.

In addition to the work on Nafion®, there is a rich body of literature in which SANS has been used to probe the structure of a variety of PEM materials including sulfonated polyimides [26], sulfonated polyetherketones [24, 27], sulfonated trifluorostyrenes [28], poly(styrenesulfonic acid)-grafted cross-liked polytetrafluoroethylene [29], and a host of other materials [30–34]. Ultimately, one seeks to understand the role that molecular-level structure and chemistry play in the development of material nanostructure and how this nanostructure is correlated with performance properties such as water content and transport, as well as ion conductivity. For example, in the work by Iwase et al. SANS (in conjunction with SAXS) was used to investigate the hierarchical structure of graft-type PEMs synthesized from cross-linked PTFE [29]. The structure was studied over a large range of length scales (0.6 nm to 1.6 μm) as a function of the degree of grafting, X_g. It was determined that the structure of these materials consisted of conducting layers of polystyrene sulfonic acid (the grafted domains) arranged in lamellar stacks on the surface of the PTFE crystallites. Within the conducting layers, they observed scattering features consistent with correlations between sulfonic acid domains. With less than 15 % grafting the grafted domains were found to reside mainly in the amorphous domains between the PTFE crystalline lamellae. Within this regime, the lamellar spacing increased with increasing grafting content up to a value of X_g of about 5 % and remained constant until 15 %. Above 15 % the grafting domains appeared to phase separate from the hydrophobic matrix and become contiguous, thus forming a highly conductive domain around the crystallites.

While X-ray scattering is certainly more widely accessible for structural characterization of membrane materials, neutrons offer the unique benefit of contrast variation, or contrast matching, in the structural determination of systems with complex architectures. Owing to the large differences in scattering-length (SL) between deuterium and hydrogen, one can use isotopic replacement in the polymer, or the solvent, to highlight the scattering from various structural components or phases. One example of this can be found in the work by Gebel et al. [35] in which they used various mixtures of D_2O/H_2O to swell $N(CH_3)_4^+$-neutralized forms of Nafion® as a way of elucidating the nature of the scattering entities in these hydrated films. By varying the ratio of D_2O to H_2O and normalizing by the scattering of Nafion® in pure H_2O they were able to match out the structural component of the scattering due to Nafion® and to observe the counterion condensation at the interface between the hydrophobic components of the polymer and the hydrophilic water domains. This was the first measurement of condensation in a perfluorosulfonated ionomer. Using contrast variation to explore neutralized forms, different models could be applied to determine which structure accurately described the scattering curves. While this study was unable to determine the shape of the scattering particles (i.e., spherical or rod-like), it was determined that the features were aggregates of the polymer backbone surrounded by the electrolyte solution, as opposed to the scattering particles being cavities filled with the electrolyte solution.

Recently, a series of studies using in situ SANS, among other techniques, on block-copolymer electrolyte membranes consisting of a polymethylbutylene (PMB) block and polystyrenesulfonate (PSS) block have begun to call into question whether or not ionic aggregates are necessary for effective proton-transport, especially in the presence of structures established by block-copolymer morphology [33, 36, 37]. The composition of the block copolymer was varied in order to tune the size of the domains. Also varied was the degree of sulfonation of the polystyrene block, within a particular composition. This body of work represents an excellent example of the application of neutron scattering to elucidate the structure-property relationships of PEM materials in environments that are application relevant. These in situ measurements were achieved using a specially designed sample chamber at the National Institute of Standards and Technology (NIST) Center for Neutron Research wherein the humidity and temperature of the environment surrounding the sample could be controlled. Moreover, the water reservoir within the sample chamber could be filled with various mixtures of H_2O and D_2O, allowing contrast-variation experiments to be performed. The scattering was measured over a range of relative humidity and temperature values using D_2O. For one particular block copolymer composition, the scattering indicated the presence of a hexagonal phase over the entire range of relative humidity and temperature values studies (Fig. 10.2). At low temperatures (25 °C) and humidity (≈25 %), the scattering arises from the block-copolymer morphology. However, at 95 % relative humidity and 40 °C, a shoulder at approximately $Q = 1.8$ nm^{-1} was observed, which became more pronounced and intense upon further heating and humidification. It was acknowledged that this peak was similar to that of the ionomer peak observed in Nafion®, but was referred to as the 'water peak' as it was only visible upon hydration. The water domain-spacing was taken to be $2\pi/Q_{max}$ of the peak. For a given block-copolymer system under the same environmental conditions (relative humidity = 95 % and at 60 °C) the position of the water peak was shown to shift to higher values of Q with increasing levels of sulfonation. This result was attributed to a decrease in the average distance between sulfonate groups upon increasing sulfonation. A contrast-variation study was performed to determine the origin of the water peak. The water reservoir was filled with a volumetric mixture of D_2O/H_2O of 32/68, chosen to

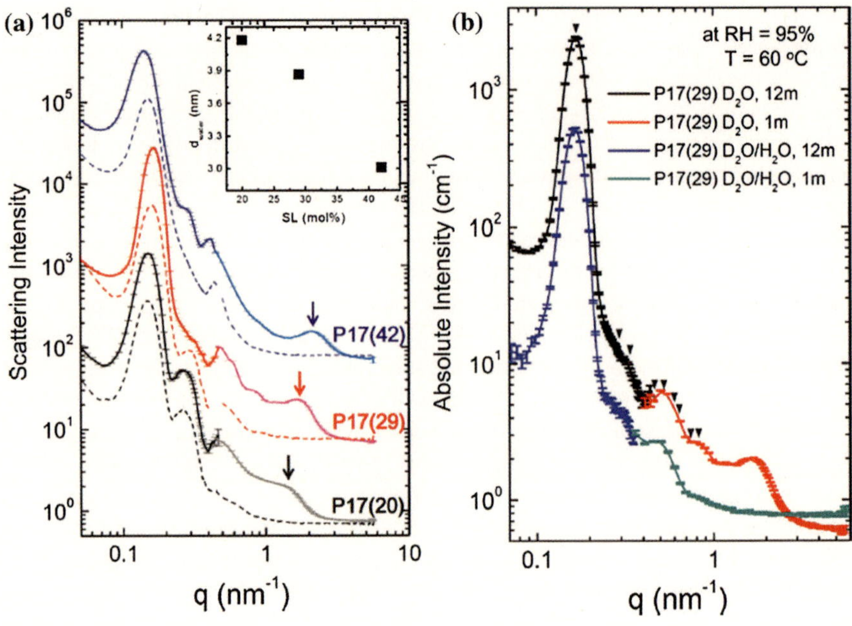

Fig. 10.2 **a** In situ SANS profiles as macroscopic cross-section versus Q (written as q) of various PSS-*b*-PMB copolymer electrolytes equilibrated at 95 % relative humidity (RH) and 60 °C (*solid lines*) and at 25 % RH and 60 °C (*dashed lines*). The P17 refers to the nominal molecular weight of the PSS block (17 kg/mol) and the number in parentheses is the level of sulfonation of the styrene units in mol.%. The *inset* shows spacing of the water domains as a function of sulfonation level. **b** In situ scattering from the P17(29) sample equilibrated in D$_2$O vapour at 95 % RH and 60 °C and at a mixture of D$_2$O/H$_2$O to match the scattering of the PSS matrix. Of note is the disappearance of the 'water peak' at higher Q values. Reprinted with permission from (S.Y. Kim, M.J. Park, N. P. Balsara, A. Jackson, Macromolecules **43**, 8128 (2010)) [33] © 2009 American Chemical Society

match the scattering-length density (SLD) of the dry PSS block. The water peak was shown to disappear when the samples were humidified with this mixture, indicating that the peak arises due to the presence of a substructure within the PSS superstructure, most likely a heterogeneous distribution of water-rich and water-poor domains as is found in most polystyrene ionomer systems.

One of the most important observations of this study was the absence of the water peak when the size of the hydrophilic domains was below a critical thickness (Fig. 10.3). For PSS-PMB copolymers this critical thickness was on the order of 6 nm to 10 nm for a sulfonation level of about 47 mol percent. It was determined that the water-rich domains were effectively homogenized due to confinement effects. This SANS work has played a critical supporting role in determining the molecular and morphological origins for the enhanced water retention and proton transport observed in the study of these copolymer systems. These results have provided a new perspective in the strategy for developing materials for use in PEM fuel cells.

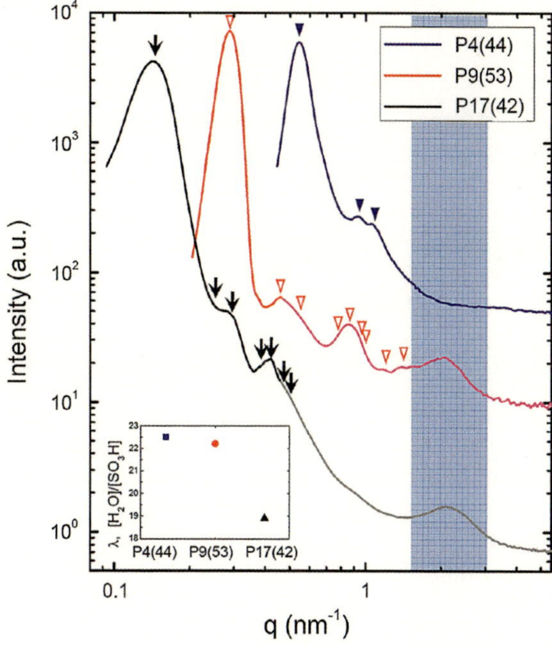

Fig. 10.3 In situ SANS profiles of various PSS-*b*-PMB copolymer electrolytes equilibrated at 95 % relative humidity at 60 °C. The number following the P refers to the nominal molecular weight of the PSS block in kg/mol and the number in parentheses is the level of sulfonation of the styrene units in mol.%. All three block-copolymer systems have comparable volume fractions of the PSS phase compared to the PMB phase and comparable levels of sulfonation. The sample with the smallest domain spacing, P4(44), shows the highest hydration level (λ) and the absence of a water peak. Reprinted with permission from (S.Y. Kim, M.J. Park, N.P. Balsara, A. Jackson, Macromolecules **43**, 8128 (2010)) [33] © 2009 American Chemical Society

10.3.2 Nanoscale Structure at Interfaces

While the bulk PEM is at the heart of a working fuel cell, it is also a critical component in the catalyst layer of the membrane electrode assembly (MEA). The polyelectrolyte is typically used as a binder in the electrode(s) where it is in contact with other components including platinum and carbon. These materials can co-exist along with pores (filled with O_2, H_2O, etc.) in the electrodes to form what is known as the triple-phase interface, or boundary. This term refers to the comingled interfaces of (i) carbon/platinum (C/Pt) particles and pores, (ii) C/Pt particles and polyelectrolyte, and (iii) polyelectrolyte and pores (Fig. 10.4). It has been shown that in these composite electrodes the polyelectrolyte is heterogeneously dispersed and can be confined to films on the order of 2–10 nm thick. It is crucial to the development of such materials for fuel-cell applications to understand how the polyelectrolyte structures at these interfaces impact water transport, proton transport, electrochemical reactions, and how certain forms of degradation occur at the

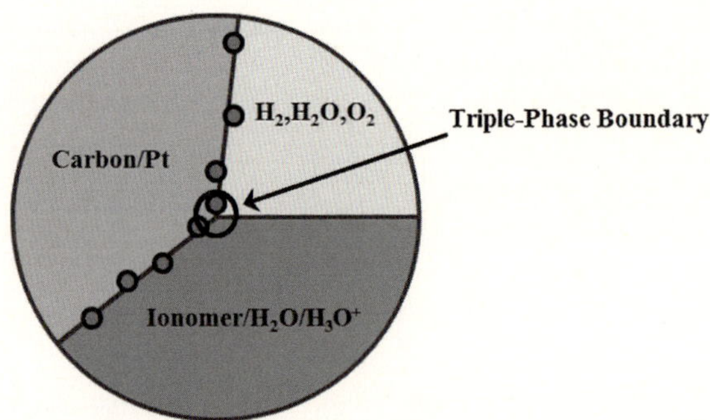

Fig. 10.4 Schematic representation of the triple-phase boundary in a PEM fuel cell where catalytically active particles (C/Pt), the proton-conducting electrolyte, and gas pores intersect

triple-phase boundaries. Furthermore, the structures at interfaces between Nafion[®] and additive nano-particles may serve to enhance, or improve, properties such as water transport and water retention. Although the structural properties of bulk PEMs have been the focus of many studies using SANS, fewer studies have focused on the thin film and interfacial structural aspects of these materials. Therefore, researchers have employed in situ NR techniques to investigate the structural characteristics of PEMs at interfaces with a variety of materials [38–42].

To date, there have been a limited number of relevant studies that have used NR to investigate the structure of PEM materials. Most of this work has been carried out on Nafion[®] thin films deposited on various substrates including smooth glassy carbon (GC) [41, 43], sputtered Pt [40, 41, 43], electrochemically oxidized Pt (PtO) [41, 43], SiO_2 [38, 44, 45], and gold (Au) [38, 44, 45].

Wood et al. report results from Nafion[®] thin films spin-coated onto glassy carbon (GC), platinum (Pt) and platinum oxide (PtO) surfaces used to experimentally model the PEMFC electrode interface and by annealing at 140 and 210 °C simulate the decal electrode-preparation method developed by Wilson and Gottesfeld [41, 43, 46, 47]. The films were exposed to 10 % relative humidity H_2O and D_2O vapour as well as saturated D_2O vapour and were found to have different multi-layer structures depending on the substrate. In composite structures of Nafion[®]/Pt/GC different behaviour was found depending on the relative humidity. At low relative humidity ($\approx$10 %) in either H_2O or D_2O the scattering results were fitted with a single-layer model consisting of hydrated Nafion[®] with thicknesses on the order of 61–62 nm. The SLD determined for films exposed to 10 % relative humidity H_2O and D_2O were relatively high ($SLD_{Nafion®H_2O} = 4.59 \times 10^{-6}\,\text{Å}^{-2}$; $SLD_{Nafion®D_2O} = 4.80 \times 10^{-6}\,\text{Å}^{-2}$) when compared to the value calculated for "dry" Nafion[®] ($SLD_{Nafion®dry} = 4.16 \times 10^{-6}\,\text{Å}^{-2}$) with a known mass density of 1.98 g.cm^{-3}. One would expect that, given the SLD of H_2O ($-0.56 \times 10^{-6}\,\text{Å}^{-2}$), the SLD of the hydrated Nafion[®] film should be lower than that of a dry film. Assuming the water content is unaffected by the isotope the two reported SLD values can be used to calculate that the water volume-fraction at 10 % relative humidity for Wood et al.'s films is approximately 3.2 % by volume and that the SLD of the dry Nafion[®] would be $4.76 \times 10^6\,\text{Å}^{-2}$.

This corresponds to a mass density of 2.27 g.cm^{-3}, which is about 15 % greater than the reported bulk density of 1.98 g.cm^{-3}. One possible explanation for this high density is that the density of the films is higher than that found in bulk Nafion® because of the thermal-processing procedure used to prepare the samples, which can increase their crystallinity. However, data for Nafion® on glassy carbon surfaces in ambient air could be fitted as a single layer with SLD $= 4.12 \times 10^{-6}$ Å^{-2}, which is much lower than for Nafion® in the same conditions on Pt. Another explanation might be the relatively narrow Q-range of the data. When Nafion® on Pt was exposed to saturated D_2O vapour the reflectivity curve was best modelled using a two-layer heterogeneous Nafion® film with an approximately 7.5 nm-thick "hydrophobic" layer at the Nafion®/ Pt interface, followed by a thicker (*ca.* 62 nm) hydrated Nafion® film. This hydrophobic layer manifests as a "dip" in the SLD profile as shown in Fig. 10.5a. This is in contrast to the work by Murthi and Dura which demonstrated that when Nafion® films that are spin-coated onto Pt or Au are exposed to H_2O vapour there is a thin water-rich layer that forms at the polymer/metal interface.

When Nafion® was in direct contact with the GC substrate a more complex scenario evolved (Fig. 10.5b). For the Nafion®/GC systems exposed to a D_2O-saturated environment, a three-layer heterogeneous model was used to describe the scattering. In this case, the researchers determined that there was a thin, rough layer (*ca.* 9 nm thick with 6 nm roughness) sandwiched between two thicker layers. The layer at the Nafion®/vapour interface was the thickest (*ca.* 57.7 nm) followed by the layer at the Nafion®/GC interface (*ca.* 26.5 nm). While the water content of each layer was not directly reported, a calculation using the SLD for each layer shows that the layer closest to the GC contained *ca.* 50 % water by volume. The thick layer at the Nafion®/vapour interface contained about 37 % water, and the middle layer was relatively water depleted at about 24 % water.

Of particular interest are the cyclic-voltammetry results obtained when the Nafion®/Pt/GC systems were electrochemically converted to Nafion®/PtO/Pt/GC (Fig. 10.5c). It was reported that although the initial potential-cycle showed no measurable Pt oxidation, subsequent cycles showed clear indications of Pt oxidation and PtO reduction. Once the PtO was formed, the structure was once again probed under saturated D_2O conditions. From analysis of the scattering curves Wood and co-workers [41, 43] found that with the PtO layer the interface became more "hydrophilic" compared to the previous Nafion®/Pt interfacial layer. Also of significance was that, after conversion of the Pt surface to PtO, there was less D_2O uptake in the Nafion®/PtO/Pt/GC system. Based on these results a vision of the development of the polymeric structure, specifically for Nafion®, near an interface was formed. Due to strong interactions of the polymer chains with the substrate it was proposed that the typically-isotropic structure reported for bulk Nafion® was modified and becomes anisotropic at the interface. According to Wood et al., the first layer acts as a template and affects the long-range structural properties of the Nafion® thin film. Murthi et al. also showed that the water uptake in Nafion® thin films on metal substrates was measurably lower than that reported for bulk films [40].

Murthi et al. [40] examined Nafion® thin films (59 nm) spin cast onto 6 nm sputtered Pt, using deposition procedures similar to Wood et al., and annealed at

Fig. 10.5 NR scattering-length density (SLD) profiles of Nafion®. **a** Nafion® on Pt/GC, **b** Nafion® on glassy carbon in saturated D_2O and ambient 10 % relative humidity environments, and **c** Nafion® on PtO in a saturated D_2O environment. Reprinted with permission from (D.L. Wood, J. Chlistunoff, J. Majewski, R.L. Borup, J. Am. Chem. Soc. **131**, 18096 (2009)) [41] © 2009 American Chemical Society

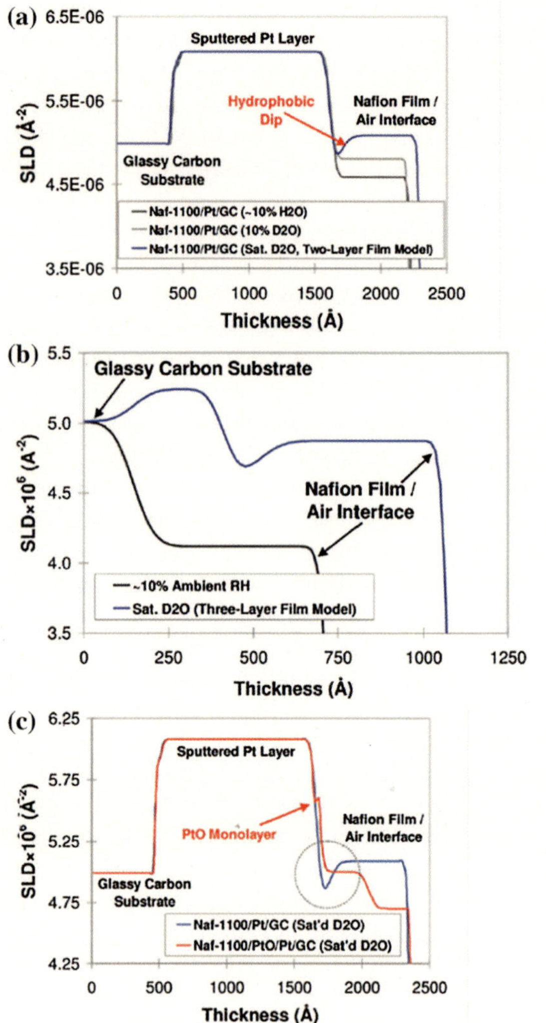

60 °C for one hour or more. From X-ray reflectivity data (to $Q_{max} = 0.7$ Å^{-1}) it was determined that prior to Nafion® deposition a 0.7 nm PtO surface-oxide layer had formed, presumably by air exposure occurring between deposition steps. Neutron reflectivity data for samples formed in controlled humidity H_2O vapour and in liquid water was obtained to measure the water uptake. H_2O was chosen over D_2O in order to provide a greater contrast between the water domains and the hydrophobic domains. The data for samples under a controlled humidity-environment at relative humidity values between 0 and 97 % were fitted using a single-layer model and the water content determined from the SLD profiles. In liquid water, a two-layer model was required to describe the data comprised of a thin 16 nm-thick

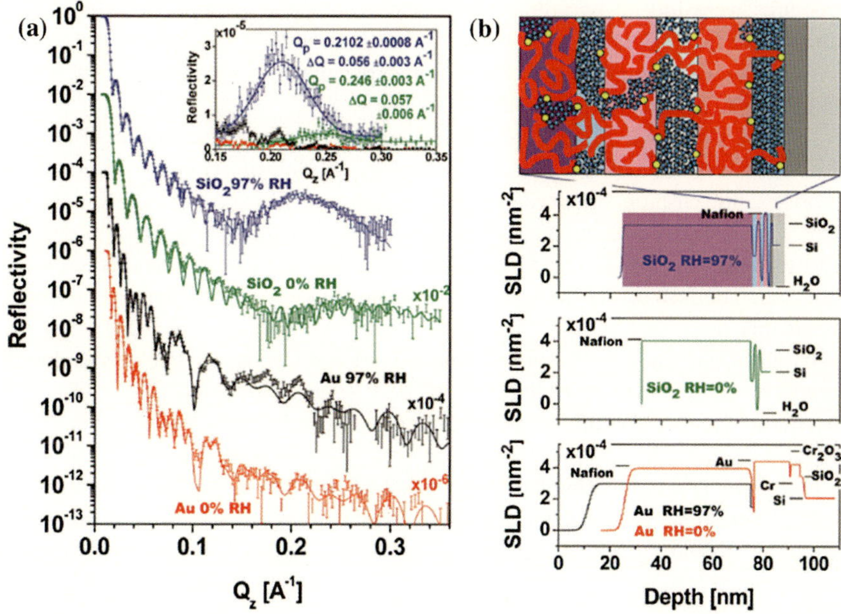

Fig. 10.6 **a** Specular NR data and model fits showing a high-Q (Q_Z) peak for SiO$_2$ at 97 % relative humidity (*blue*), a smaller high-Q peak for SiO$_2$ at 0 % relative humidity (*green*), and no high-Q peak for Au at 97 % relative humidity (*black*) or Au at 0 % relative humidity (*red*). **b** NR scattering-length density profiles and the model corresponding to SiO$_2$ at 97 % relative humidity: Nafion® fluorocarbon backbone (*red*), sulfonic acid group (*yellow*), and water (*blue*). Reprinted with permission from (J.A. Dura, V.S. Murthi, M. Hartman, S.K. Satija, C.F. Majkrzak, *Macromolecules* **42**, 4769(2009)) [38] © 2009 American Chemical Society

hydrophilic layer next to the PtO with a water content λ (moles H$_2$O per mole SO$_3^-$) of 21 and an outer layer of 76 nm with λ = 10.2. Similar results were obtained from films prepared on gold substrates.

Dura and co-workers showed a very interesting effect in Nafion® films cast on SiO$_2$ substrates. In contrast to scattering profiles fitted with simple, single-layer, or two-layer models, NR results revealed alternating lamellar layers of water-rich and Nafion®-rich domains that are induced at the interface of hydrated Nafion® films and native silicon-oxide substrates. A cartoon depiction can be seen in Fig. 10.6 which shows the silicon substrate, the native silicon-oxide, and a five-lamella structure. These structures were evidenced by the presence of a peak in the NR curve at approximately Q_z = 0.21 Å^{-1} for Nafion® on SiO$_2$ equilibrated at 97 % relative humidity. The lamellar morphology was confirmed by off-specular scattering and the location of the lamellar structures at the interface with the SiO$_2$, as opposed to the vapour-polymer interface, was confirmed by comparison of the scattering on thick thermal oxides. The position and intensity of the NR peak were shown to be highly dependent on the hydration level of the film. Detailed measurements and analysis including transverse Q-scans led to the conclusion that the structures were indeed two-dimensional sheets, or lamellae, lying parallel to the substrate surface. The first layer was found to be water-rich at nearly 100 % H_2O. The next two water rich layers were found to contain progressively less water, leading to a bulk-like swollen Nafion® layer.

He et al. [39] have used in situ NR to study the structure and kinetic absorption of water in thin films of sulfonated polyphenylene. Films of thickness ranging from 13 to 57 nm cast on oxidized-silicon substrates were exposed to D_2O vapour. Typically, NR data collection is too slow to obtain kinetic data, but by limiting the acquisition to the low-Q regime time-averaged data at 10 min intervals could be collected. The NR curves were modelled using a three-layer model. It was determined that there were D_2O-rich layers at both the vapour/polymer and polymer/ SiOx interfaces. The kinetic studies revealed that the D_2O mass uptake scaled with $time^{1/2}$ at early times and diverged at later stages. At early stages of water adsorption the effective diffusivity was found to be significantly slower compared to diffusion in the bulk polyelectrolyte.

Our group is investigating the fundamental origins of these lamellar structures in Nafion® and their potential impact on interfacial transport by using a variety of techniques including NR, grazing-incidence SAXS (GISAXS), and quartz-crystal microbalance (QCM). The aim of this work is to investigate the role that specific interactions play on lamellae formation. This is being done by using substrates with tunable chemical characteristics. From our initial studies, it appears that the interfacial structures are generally absent from hydrophobic surfaces, but that in highly hydrophilic substrates there is a strong tendency to form interfacial-lamellar structures [48]. More specifically, we have recently demonstrated that the wettability of the substrate, i.e. hydrophobic or hydrophilic, is a factor governing interfacial structure.

Although a clear and complete picture of how interfacial structure and confinement in thin polyelectrolyte films influences materials properties, such as water and proton transport, is yet to be gained, it is clear that NR techniques have a significant role to play. In summary, NR has shown that the structure at the interface between a PEM and a substrate depends largely on the surface chemistry, film processing, and even electrochemistry. These factors are certain to have an impact on the transport at these interfaces. Moreover, the confinement of the PEM to a thin film, which certainly has technological relevance, reduces the transport coefficients and can even impact the solubility of water in these materials.

10.4 Transport and Dynamics

In order to understand the performance of PEM fuel cells, it is necessary to not only understand the structure but also the water dynamics in these materials over a large range of humidities, temperatures, and processing conditions. Proper water management is critical to optimal fuel-cell performance. Many studies have focused on understanding the bulk-water transport and there has been a considerable effort to understand how this transport is related to the nanostructure of the membrane.

While bulk-water transport is an important aspect of fuel-cell operations, it is important to keep in mind that this macroscopic property is governed by the nano-structure of the membrane and, ultimately, the local-water dynamics within this structure. Researchers have used neutron spectroscopic techniques to investigate the local-water dynamics within the nanoscale ionic aggregates present in the material. In addition to understanding the dynamics and transport, the static and dynamic water concentration-gradient that is present across the MEA during fuel-cell operation is also a central piece of data for proper water management. Current neutron imaging techniques do not have the spatial resolution to map the water gradient across the thickness of the MEA during fuel-cell operation, but researchers have been able to use SANS techniques to elucidate this information in a clever way.

While the focus has mainly been on water, little effort has been spent on understanding the relationship between the water dynamics (both local and mac-roscopic) on the polymer-chain dynamics and fluctuations within, and of, Nafion®'s complex morphological features. Page and co-workers have used QENS and NSE techniques to address these issues.

The following sections summarize the work, to date, using QENS, SANS, and NSE to study the transport and dynamics in PEM materials, particularly Nafion®.

10.4.1 Water Transport and Dynamics

In the early 1980s Volino and Dianoux et al. used QENS to characterize the water dynamics in a hydrated Nafion® film [49, 50]. The Nafion® sample was 1200 EW and the water content was kept at approximately 15 % by mass. Experiments were carried out at 25 °C on the multichopper time-of-flight spectrometer (IN5) at the Institut Laue Langevin using three different neutron wavelengths of approximately 10, 11, and 13 Å, corresponding to an energy resolutions ($\Delta\omega$) of 18.5, 14, and 9 µeV, respectively. At each energy resolution, data was collected over the Q-range 0.4–1.1 Å^{-1}. Along with the hydrated films, dry films were measured as a control. In early work, the quasi-elastic broadening was characterized using a Lorentzian to describe the long-range self-diffusion of the water molecules in the hydrated membrane. The dynamic scattering-function, $S(Q,\omega)$, for diffusion of an incoherent-scattering particle inside an impermeable sphere of radius, a, was presented in the original work. Considering the broadening to be due to only long-range diffusion, i.e. a $\rightarrow \infty$, the scattering function reduces to:

$$S(Q,\omega) = \frac{1}{\pi} \frac{DQ^2}{\left(DQ^2\right)^2 + \omega^2} \tag{10.1}$$

where a Lorentzian function is used with full-width at half-maximum of $\Delta\omega = 2DQ^2$ and D is the long-range diffusion coefficient.

It was determined that this model did not sufficiently describe the QENS spectra, and therefore inadequately described the water motions in Nafion®. Given the fact that a single Lorentzian function, as used in Eq. (10.1), could not completely describe the quasi-elastic broadening, a second model was proposed that considered the rotational motions of the water molecules within the membrane. This particular model assumes that the water molecules diffuse on a sphere of radius, ρ, with a rotational-diffusion coefficient, D_r. Using the distance between the centre-of-mass and the protons (0.95 Å) as ρ, the researchers were unable to fit the data using any reasonable value for D_r. With ρ as adjustable parameter, they found reasonable fits with $\rho \approx 3$ Å, but because this value did not correspond to any reasonable length-scale in the water molecule the analysis approach was not pursued further. Therefore, it was determined that the measured QENS spectra could not be sufficiently described by the rotational diffusion of water alone.

With the failure of the two simple models described above, a third model combining translational self-diffusion with rotational diffusion was proposed. Through various initial attempts at determining D_r, ρ, and D, it was found that the general form of the model could sufficiently describe the QENS data, particularly with $\rho \approx 3$ Å. Again, given that this value did not match any reasonable length-scale for water rotation, it was determined that although the mathematical description was sufficient, the physical interpretation was lacking. Therefore, it was decided that the mathematical form of the model would be preserved and would be one in which the water molecules (*viz.* the protons) are restricted to local diffusion, D_l, inside a sphere of radius, a, combined with long-range translational diffusion, D_t, between the spheres. The best-fit curves were calculated using $D_l = 1.8 \times 10^{-5}$ cm²s⁻¹, $a = 4.25$ Å, and $D_t = 1.6 \times 10^{-6}$ cm²s⁻¹. Although this model proved to provide a reasonable explanation, and fit, to the QENS data in the Q-range 0.4–1.1 Å⁻¹, it was suggested that more sophisticated models could be proposed and were being pursued. However, this was not thought to change the overall conclusions of the work. This work showed that on a local size-scale (*ca.* 10 Å) the water molecules diffuse with a self-diffusion coefficient that is similar to bulk water ($D_{water} = 2.2 \times 10^{-5}$ cm²s⁻¹), but that the long-range self-diffusion is restricted due to the morphology of material. Essentially, water is relatively free to move within a water domain, or ionic aggregate, but motion between domains over large length-scales suffers. Subsequent measurements on oriented membranes use more sophisticated models describing diffusion within a cylinder instead of a sphere.

Pivovar and Pivovar [51] also used QENS to investigate the water dynamics in Nafion® over a range of hydration levels from $\lambda = 2$ to 16 in order to determine how the local dynamical behaviour of water is correlated with proton conductivity, particularly in the low-λ regime. The QENS data were collected at a time-of-flight instrument at NIST. The data were collected at 295 K over a Q-range from 0.4 to 2.0 Å⁻¹ with neutron wavelength of 6 Å. The energy resolution was between 55 and 70 µeV. Assuming that the QENS data were the result of both elastically and quasi-elastically scattered neutrons arising from immobile and mobile protons, respectively, the experimental results were successfully modelled using the following function:

$$S_{\exp}(Q, \omega) = I(Q)(x \cdot R(\omega) + (1 - x)R(\omega) \otimes L(\omega)) + B_o \tag{10.2}$$

where $I(Q)$ is the scaled intensity at a given value of Q, $R(\omega)$ is the instrument resolution function, $L(\omega)$ is a Lorentzian distribution, and B_o is a flat background. The quantity, x, is the elastic incoherent structure-factor (EISF), which is the ratio of the elastic scattering-intensity to the total scattering-intensity from both the quasi-elastic and elastic components. The EISF can yield information as the length-scale of the proton motions occurring within the sample. The fits to the experimental data based on Eq. 10.2 were interpreted using two different models. For $Q < 0.7\,\text{Å}^{-1}$, the Lorentzian broadening was analysed according to a model where the incoherent scattering particles, i.e. protons, were thought to undergo continuous diffusion within a sphere as in the model proposed by Volino and Dianoux [55, 56]; above this Q-value a random unconstrained jump-diffusion model was used. From these two models the characteristic length-scales over which motion occurs, along with local (D_{local}) and jump (D_{jump}) diffusion coefficients, were determined. In the context of the diffusion in a sphere model, the measured EISF was modelled with the following equation

$$x = EISF = f_{ND} + (1 - f_{ND})\left(\frac{3j_1(QR)}{QR}\right)^2 \tag{10.3}$$

where f_{ND} is the fraction of protons not diffusing in the timescale of the measurement, R is the characteristic radius of the sphere within which the protons are diffusion and j_1 is the first-order Bessel function. The results of this analysis over the entire hydration range can be seen in Fig. 10.7.

The radius of the dynamic sphere is characterized by a linear increase from about 2 to 3.5 Å in the λ range from 1 to 7. Above a λ of *ca.* 7 an asymptotic value of 3.68 Å is reached for fully saturated membranes. Correspondingly, within the time-scale of the measurement, f_{ND} decreases with increasing water content. These results indicate that at low λ values the hydrophilic domains are small and water molecules are likely to have strong interactions with the acid sites (i.e. the confining surface). These water molecules are effectively bound. As the domains swell, a smaller fraction of the protons feel the confining effects of these interactions. There is, however, inconsistency between the proposed size of these confined hydrophilic domains and the size as determined by SANS measurements. This can be accounted for by limitations in the time resolution of the instrument.

Following the work by Pivovar and Pivovar, further advancements in modelling of the QENS obtained from time-of-flight and backscattering spectrometers were made by Perrin and co-workers [52, 53]. Their efforts aimed at developing a single model to account for bound-continuous diffusion, long-range diffusion, and atomic granularity (*i.e.* jump diffusion). This model used Gaussian statistics to describe the diffusion of the scattering particles within a restricted geometry with ill-defined boundaries. Essentially, the model accounted for two populations of diffusing protons: a population of fast protons and a second population of slower-moving protons.

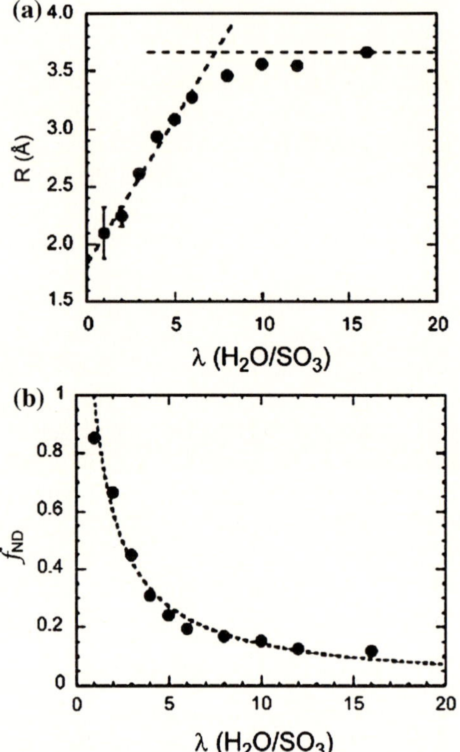

Fig. 10.7 a The dynamic sphere radius, R, from a model fit to the EISF as a function of the water content, λ, where the dashed lines demark the slope in the low-swelling regime of the asymptotic value at high hydration-level and **b** the fraction of non-diffusion hydrogen atoms, f_{ND}, as determined from the model fit as a function of water content, where the dashed line is the theoretically-calculated value of f_{ND} assuming a single, non-diffusing hydronium ion for each sulfonic acid site. Reprinted with permission from (A.A. Pivovar, B.S. Pivovar, J. Phys. Chem. B **109**, 785 (2005)) [51] © 2009 American Chemical Society

As described by Perrin, the slow protons are, for all intents and purposes, related to the hydronium ion formed when water abstracts a proton from the sulfonic acid site, but it is also likely that these are tightly bound or restricted, water molecules similar to those described by Pivovar and Pivovar [53]. The timescale of motion of this population becomes shorter with increasing hydration level and the characteristic length-scale of motions increases. No long-range diffusion was reported. With increased hydration, the "extra" water molecules comprise the fast protons whose motions are characterized over a similar length-scale as the slower protons along with long-range diffusion between adjacent confining domains. In general it was shown that for both populations of protons, the length-scale increases and the timescale of motions decrease with increasing hydration level. As with earlier studies, an asymptotic value was reached for these values at λ of about 10.

Other studies have shown similar results. Paciaroni and co-workers [54, 55] studied the dynamics of the confined water in Nafion® films at low hydration levels (*ca.* $\lambda = 6$) over a temperature range from 200 to 300 K. They characterized the motion according to random jumping inside a confining region, which they associate with the boundaries of the ionic aggregates, or clusters, of Nafion®. They also observed a transition in the water dynamics at approximately 260 K, which was reported to be related to an increase in the available degrees of freedom of the water created by a melting-like process of the bound water. Their results concerning the local-diffusion coefficient of the water are similar to those previously reported [51]. Paciaroni et al. [55] also studied a composite membrane comprised of Nafion$^{®}$ with zirconium phosphate nano-filler. Again, the QENS data were modelled according to random jump-diffusion inside a confining spherical region. Ultimately, this work demonstrated that the local water dynamics were unaffected by the presence of the nano-filler.

Researchers have used QENS to look at novel PEM materials and to investigate the fundamental relationship between structure and transport. Peterson and co-workers used QENS to investigate the water dynamics in novel plasma-polymerized PEMs [56, 57]. The membranes were synthesized using a pulsed-plasma enhanced chemical vapour-deposition system. The resulting membrane was a cross-linked polystyrene containing trifluoromethanesulfonic-like acid sites. The ion-exchange capacity of this novel membrane was shown to be somewhat less than that of Nafion$^{®}$ (0.7 mequiv/gm compared with 0.91 mequiv/gm) with a significantly lower water content. The quasi-elastic broadening was modelled using a Lorentzian. Plots of the half-width at half-maximum (HWHM) of the Lorentzian as a function of Q^2 were used to calculate the diffusion coefficients. The water in the plasma produced PEM showed similar diffusion behaviour compared to Nafion$^{®}$. However, they did observe an increase in HWHM for the plasma produced PEM at low Q values compared to that measured for Nafion$^{®}$, but no explanation was given for this observation. A later, more detailed, QENS study from this group revealed interesting behaviour in the plasma-polymerized PEM (PP-PEM) [57]. They found that the PP-PEMs studied had a proton conductivity, as measured by impedance spectroscopy that was ≈21 % higher than that of Nafion$^{®}$. A careful QENS study, and subsequent fitting, of the proton motions revealed that the quasi-elastic broadening could be described by two components, or types of protons. One component was due to motions arising from a dispersed relaxation in the frequency domain, which is equivalent to a stretched exponential in the time domain, as proposed by Bergman [58]. These protons were labelled as type 2 and considered to diffuse much like the protons in Nafion$^{®}$ membranes. The other, faster component was described by a broad Lorentzian and accounted for 41(3) % of the protons. The Q^2 dependence of the two types of protons was used to determine self-diffusion coefficients for the fast and slow diffusing protons. The fast protons, which accounted for 41(3) %, of the protons in the system, had a self-diffusion coefficient of $2.8(1) \times 10^{-4}$ cm^{2}s^{-1}. This is an order of magnitude faster than that measured for the Nafion$^{®}$-like protons ($3.0(2) \times 10^{-5}$ cm^{2}s^{-1}). These results help to support and give an explanation as to the increased proton conductivity measured for the PP-PEMs, despite having a lower water uptake. It was proposed that the superfast proton diffusion is likely due to the arrangement of the sulfonates in PP-PEM. While detailed structural information was not presented, a simple model was proposed which involve the hydronium ion being passed between adjacent trifluoromethanesulfonate groups. The local environments of these groups are thought to allow for rapid, relatively long-range proton transfer as compared to Nafion$^{®}$ [57]. Lyonnard et al. [59] have also used QENS to study the role of structure and confinement on water mobility. This study, however, was not carried out on PEMs, but on a perfluorinated surfactant (PFOS) bearing a similarity to the side-chain found in Nafion®. By varying the ratio of PFOS to water, they were able to create both lamellar and hexagonal phases. They showed that the water motions are spatially confined and, more importantly, that the geometry of the confinement affects the diffusion behaviour. In the hexagonal phase, the water dynamics were found to be almost bulk-like while for the lamellar phase there were serious restrictions in water mobility. This information could play a vital role in the rationale design of future, novel membrane materials.

While it is not in the scope of this chapter, it is worth mentioning that QENS has also been used to study the motions of hydrogen gas, the fuel in a fuel cell, on PEM fuel-cell catalyst supports. Such studies demonstrated that the interactions between H_2 and the carbon support play a significant role in reactant transport in the PEM fuel cell [60, 61].

The aforementioned studies demonstrate the efficacy of QENS for understanding the water transport in PEM materials. Several key pieces of information concerning the fundamental nature of water mobility in PEMs have emerged thanks to detailed and clever analyses carried out by the researchers mentioned. In summary, one can say that the water motions in PEMs, especially Nafion®, have the following characteristics:

1. Water motions occur in confined domains, the nature of which is largely dictated by the nanoscale morphology of the material.
2. Within this confined geometry the water motions are influenced largely by the number of water molecules present. As the number of water molecules decreases, the dynamics are increasingly restricted.
3. With a sufficient level of hydration, the local dynamics are not unlike the motions that occur in bulk water, but the long-range motions are restricted due to the material morphology.

This body of information has contributed significantly to our current understanding of these materials and will continue to illuminate the path toward the rational design of new PEM materials.

10.4.2 Water Transport

Although historically used as a tool to probe the structure of PEM materials, recently neutron techniques and sample environments have been developed to probe the transport of water in these materials by enabling structural changes to be monitored as a function of time. Kim and co-workers developed an in situ vapour sorption apparatus for SANS that is capable of controlling the vapour pressure of a given solvent and have employed it to investigate the effects of water vapour sorption in Nafion® films [62, 63]. A French group has developed an in situ, in operando SANS experiment and analysis method to observe the structure of Nafion® and determine the water profile across the thickness of the PEM [64–69]. This technique has also

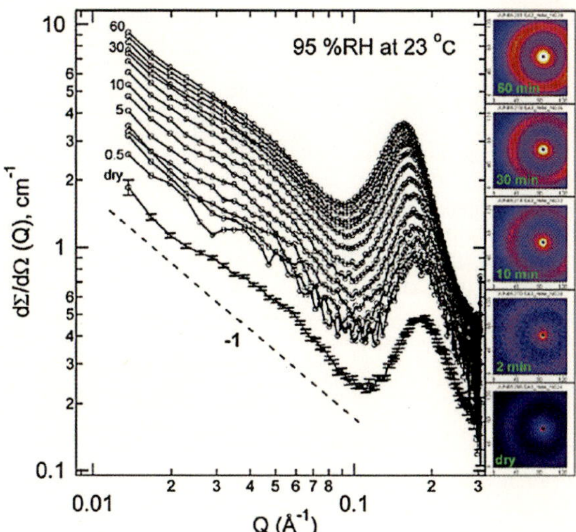

Fig. 10.8 Evolution of the macroscopic scattering cross section over 60 min from a pretreated Nafion® membrane as a function of time during hydration from 'dry' to 90 % relative humidity. The two-dimensional detector images are shown on the right. Reprinted with permission from (M.H. Kim, C.J. Glinka, S.A. Grot, W.G. Grot, Macromolecules **39**, 4775 (2006)) [63] © 2009 American Chemical Society

been used by this group and others to investigate the behaviour of water in a working fuel-cell environment and will be discussed in greater detail below [70, 71]. More recently, Gebel and co-workers [72] used a similar cell and have demonstrated the ability to measure the kinetics of water sorption in Nafion® along with the water concentration-profiles across the thickness of the membrane using neutron scattering.

Kim and co-workers were able to measure structural changes in Nafion® under various relative humidity conditions, ranging from dry to hydrated, using an in situ vapour-sorption SANS (iVSANS) cell. Scattering intensity was measured over the Q range 0.1–0.3 Å^{-1} as a function of time as shown in Fig. 10.8. The position and intensity of the ionomer peak were determined from the scattering profiles to measure the structural evolution of pretreated and as-received Nafion® films after being exposed to water vapour. The humidity changes investigated included dry to 20, 35, 50, 65, 80, and 95 % relative humidity. The results for the as-received Nafion® film can be seen in Fig. 10.9 for a humidity change from dry to 95 % relative humidity at 23 °C. Over the course of the sorption experiment, the ionomer peak increases in intensity and the position of the peak shifts to lower Q, as detailed in Fig. 10.9. The macroscopic scattering intensity (differential cross-section on an absolute scale), $d\Sigma/d\Omega$, of the ionomer peak was correlated with the water uptake (Fig. 10.9a) and increased rapidly during the early stages of sorption and levelled off upon reaching equilibrium for each of the target humidity values. The rate of water sorption and the intensity, both related to the water uptake, were found to increase with increasing relative humidity. The ionomer peak position was found to follow the same trend, with the domain spacing increasing with time during the early stages and plateauing at later times. The equilibrium spacing increases with increasing relative humidity. The time-resolved (kinetic) data of the time-evolution

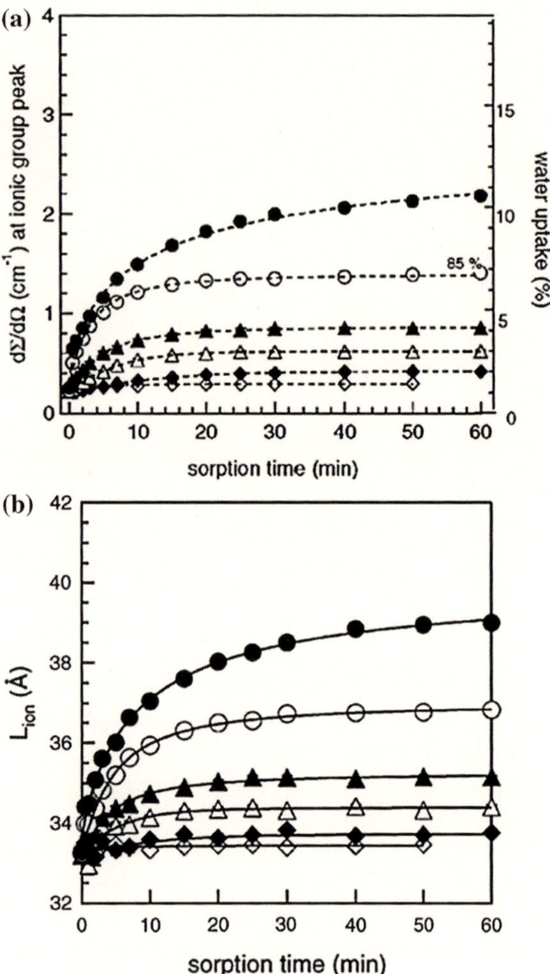

Fig. 10.9 **a** $d\Sigma/d\Omega$ at Q_{max} for the ionomer peak correlated with water uptake versus the sorption time and **b** the domain spacing of the ionic aggregates versus sorption time after changing relative humidity values. Reprinted with permission from (M.H. Kim, C.J. Glinka, S.A. Grot, W.G. Grot, Macromolecules **39**, 4775 (2006)) [63] © 2009 American Chemical Society

of the macroscopic scattering intensity was modelled with a solution to Fick's second law to determine the diffusion coefficient for both as-received and pretreated Nafion® membranes. More recently, Gebel and co-workers performed a similar set of experiments [72]. In addition to obtaining kinetic data to determine the diffusion coefficient, Gebel and co-workers used an established technique to determine the water concentration-profile across the membrane during the sorption process [64–68, 70, 71]. Scattering data from Nafion® equilibrated at various relative humidity values, were recorded and served as a reference to reconstruct the scattering obtained during the equilibration process. It is thought that the scattering data taken during the sorption process could be considered as a sum of slices with varying thickness and water contents. These slices can be considered to be represented by the recorded reference spectra and the total in situ scattering from the membrane in the operating fuel-cell can be recreated by a linear combination of said reference spectra.

One of the most innovative uses of SANS to date has been the in situ, operando technique developed by a group in France [64–68, 70, 71]. In a working PEM fuel

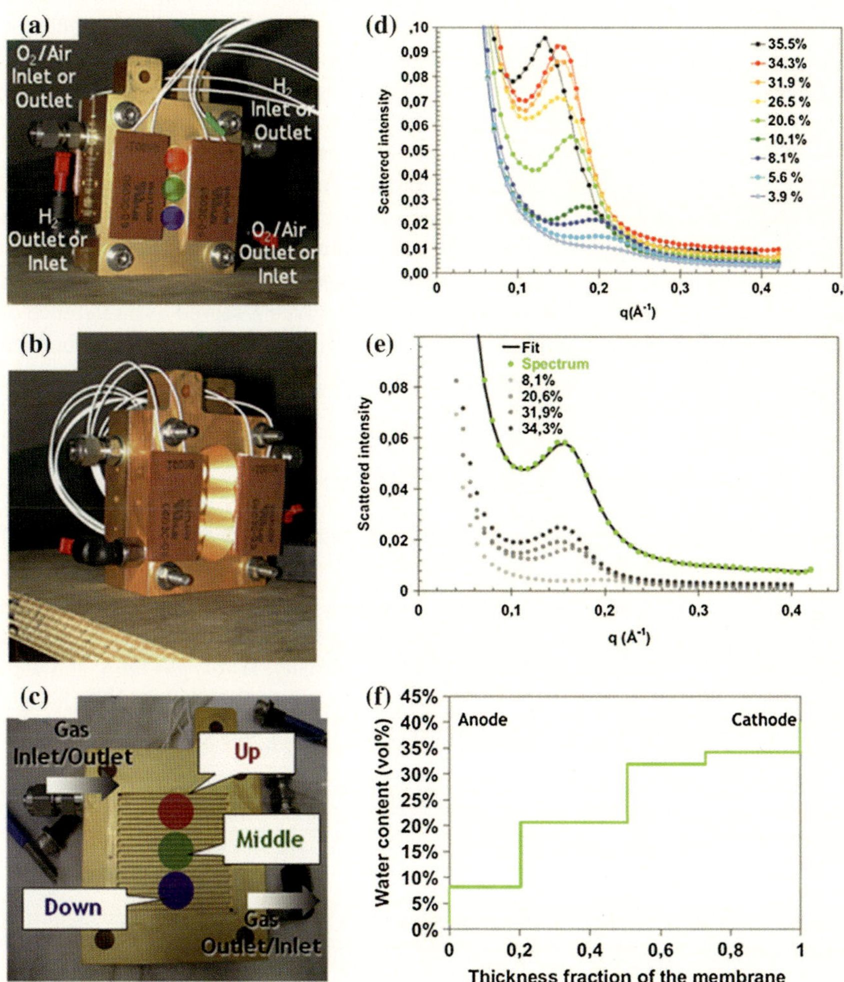

Fig. 10.10 Photographs of the cell on the cathode side (upstream side of neutron beam) (a), on the anode side (b), and of the monopolar plate with the serpentine gas channel (c). On the anode side the cell was machined such that the neutron beam could be moved along the vertical axis of the cell in order to investigate the distribution of water between the gas inlets and outlets. (d) SANS reference spectra obtained on a membrane electrode-assembly equilibrated at various relative humidity conditions. The percentages denote the water content in volume %. (e) An example of a SANS profile taken during operation and the subsequent deconvolution into the relevant reference spectra. (f) Corresponding water profile obtained from the combination of reference spectra shown in (e). Reprinted with permission from (A. Morin, F.N. Xu, G. Gebel, O. Diat, Int. J. Hydrog. Energ. 36, 3096 (2011)) [67] © 2011 International Journal of Hydrogen Energy

cell, the membrane is not uniformly hydrated across the thickness of the cell and there is usually a water gradient from the anode to the cathode, with the water concentration being higher at the cathode. Ideally, one would want to know the water concentration profile through the membrane thickness as a function of the operating conditions in order to optimize fuel-cell performance and water management. The group in France was the first to develop a fuel cell that was neutron-transparent, which enabled them to measure the scattering from the membrane during cell operation. The premise behind this technique is the water gradient across the fuel cell and the varying membrane nanostructure as a consequence of the different amounts of water. The features of the neutron-scattering data that arise from the nanostructure of Nafion® (i.e. the ionomer peak, incoherent background, etc.) are sensitive to the amount of water in the membrane. Thus, the membrane in the working fuel cell was considered to consist of a series of slices, each with different water content. Reference spectra were obtained for Nafion® membranes that were considered to be at equilibrium with respect to swelling over a range of water contents. The shape of the ionomer peak at a given λ served as a reference of the scattering for that particular water content. The scattering intensity obtained during operation, $I_{total}(Q)$, was considered to be a linear combination of the scattering intensity of the reference spectrum, $I_i^{ref}(Q)$, where the weight (or coefficient) associated with each individual reference spectrum, a_i, was directly correlated with the thickness of the corresponding hydration layer.

$$I_{total}(Q) = \sum_i a_i I_i^{ref}(Q) + k; \text{ with } \sum_i a_i = 1 \qquad (10.4)$$

The cell, the reference spectra, and a typical measured water profile can be seen in Fig. 10.10. This technique has been used to study the water gradient profiles during fuel-cell operation as a function of current density, H_2/O_2 gas ratio, and with differing gas-diffusion layers and gas-flow configurations. Details of this technique and the neutron-transparent fuel cell can be found in the literature [64–67, 73].

10.4.3 Ion and Polymer Dynamics

In addition to the studies of water dynamics in PEM materials, inelastic scattering techniques have been used to examine the dynamics of ions in neutralized membranes as well as the polymer-chain dynamics. The aim of these studies is to gain a deeper fundamental insight into the mechanisms of charge transport and to examine the correlation between ion motion and polymer-chain dynamics. In the earliest studies of this kind, Rollet and co-workers [74, 75] used QENS to probe the dynamics of $N(CH_3)_4^+$ ions in hydrated (D_2O) Nafion® membranes. Nafion® and D_2O both have very low incoherent scattering cross-sections, and hence the measured scattering in these studies is dominated by the hydrogenated counter-ion, allowing for directionality of ion motions, as a function of ion concentration, to be determined. In order to extend the time-range over which the ion dynamics could be probed, nuclear magnetic resonance and radiotracer experiments were also performed. For the given set of experimental conditions outlined, they found that at short time-scales the self-diffusion of the ions within the water domains of Nafion® was similar to that found in non-confined solutions. With increasing electrolyte concentration the self-diffusion coefficient of $N(CH_3)_4^+$ was found to decrease, a phenomenon thought to arise from viscosity effects. For long-range diffusion, the transport of ions was found to be limited by the tortuosity of the diffusion path which is in large part determined by the channels connecting the water domains.

Page and co-workers [76–79] have also used inelastic neutron methods to study the molecular dynamics of Nafion® as part of an effort to further the fundamental understanding of the role of electrostatic interactions in relaxation phenomena observed in these materials. QENS was used to measure the dynamics of the counter-ions in perfluorosulfonate ionomers (PFSIs) neutralized with various alkyl ammonium ions. While the work by Rollet and co-workers focused on measuring the counter-ion dynamics at room temperature in hydrated systems, Page and co-workers measured the counter-ion dynamics in dry systems over a range of temperatures in the vicinity of the corresponding alpha-relaxation temperature, which is highly dependent of the choice of counter-ion. Transitions in the counter-ion dynamics were correlated with bulk mechanical-relaxations in an effort to directly observe the ion-hopping process thought to be the mechanism for long-range diffusive motions of polymer chains and ions in these systems. This work explicitly showed that the counter-ion dependent alpha-relaxation, observed in thermo-mechanical analysis, is linked to the onset of mobility of the counter-ions on the length-scale of 2–3 nm. These data, taken together with other studies, demonstrate that the motions of the ions and the polymer chains are highly correlated in these systems (Fig. 10.11).

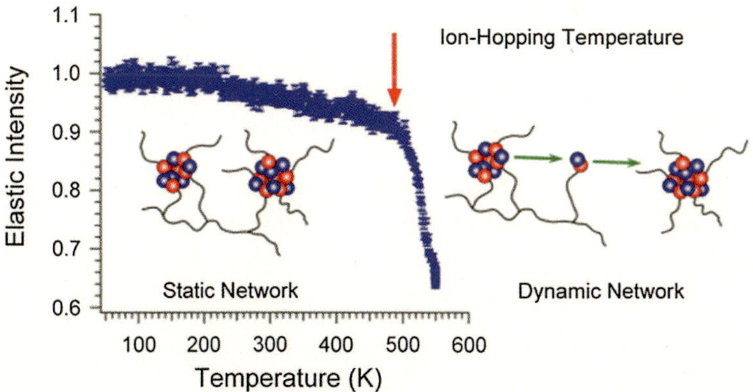

Fig. 10.11 The elastic-scattering intensity as function of temperature (at $Q = 0.25$ Å^{-1}) for a tetramethyl ammonium-neutralized Nafion® membrane. The rapid decrease in intensity at high temperature corresponds to the transition from a static, electrostatic network, to a dynamic one. Reprinted with permission from (K.A. Page, J.K. Park, R.B. Moore, V.G. Sakai, Macromolecules **42**, 2729 (2009)) [76] © 2009 American Chemical Society

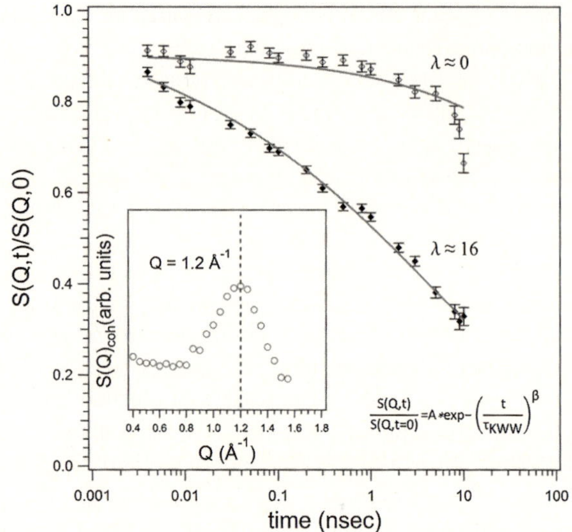

$$\frac{S(Q,t)}{S(Q,t=0)} = A * \exp - \left(\frac{t}{\tau_{KWW}}\right)^{\beta}$$

Fig. 10.12 The normalized *I(Q,t)*, *I(Q,t)/I(Q,0)* and labelled *S(Q,t)/S(Q,0)*, for a nominally dry and hydrated ($\lambda = 16$) Nafion[®] membrane at $Q = 1.2$ Å^{-1}. The Q value was chosen to coincide with the peak in the coherent structure-factor associated with inter-chain scattering of the Nafion[®] polymer backbone (inset). The *I(Q,t)* was fit with the KWW equation as shown in the graph, where τ_{KWW} is the relation time of the polymer chains and β is related to the distribution of relaxation times observed. Error bars represent one standard deviation in the measured scattering intensity

While work on dry, neutralized forms of Nafion[®] has helped to further the fundamental understanding of the complex molecular-relaxation behaviour, i.e., the relationship between electrostatic interactions, counter-ion dynamics, and polymer chain relaxations, these model studies are less useful in understanding the inter-dependence of water dynamics/transport and polymer-chain motions in the more device-relevant, acid-form hydrated Nafion[®] membranes. More recently, Page and co-workers have begun using NSE spectroscopy to probe the relationship between water content and polymer-chain dynamics in Nafion[®]. NSE enables the study of slow relaxation processes in polymeric systems [80] and for Nafion[®] samples hydrated with D_2O, the polymer-chain dynamics can be measured by monitoring the intermediate-scattering function, *I(Q,t)*, at a *Q* corresponding to the length-scale associated with the distance between the individual polymer chains. An example of the results from an ongoing study can be found in Fig. 10.12. The *I(Q,t)* decay was fit with a Kohlrausch-Williams-Watts (KWW) function (Fig. 10.12) and it was determined that with increasing water content and temperature, the relaxation time for chain motions decreased (i.e. chain motions became faster). Interestingly, the timescale associate with the polymer-chain motions plateaus at value near a λ of 6, which is also where other studies have shown a transition in the dynamics and transport of water [51]. This application of NSE indicates that there is a degree of coupling between the local water dynamics/transport and the polymer-chain dynamics. This influence of water on the polymer-chain dynamics provides a molecular-level understanding of the observed decrease in Young's modulus with increasing humidity and temperature. The increased mobility in molecular relaxations induced by the presence of water points to the molecular origins of the temperature- and humidity-dependent softening mechanisms in Nafion[®] and other poly (perfluorosulfonic acid) membrane materials.

10.5 Conclusions and Outlook

While not an extensive review of the field, the work described here shows the critical role that neutron techniques play in the research and development of PEM fuel cells. Interestingly, neutron techniques can serve researchers at several points along the process of developing a working and high-performance PEM fuel cell. At the earliest stages, neutron scattering can inform chemists of the structures that develop given the choices made in molecular architecture. In turn, it can then be determined how these structures may impact important materials properties such as water and ion transport. Finally, state-of-the-art neutron methods can be used to monitor and analyse operating PEM fuel cells and aid in determining operating conditions for optimum fuel-cell performance. Therefore, researchers have the tools necessary to correlate materials chemistry and structure to overall device performance. This can deliver critical information and serve as a powerful tool to a chemist or materials developer, by providing them with a general set of fundamental design parameters with which they can move forward in membrane development. It is reasonable to consider that neutron-based techniques will continue to serve the PEM fuel-cell community by aiding in the characterization and establishment of fundamental membrane structure-property-performance relationships.

References

1. K.A. Page, B.W. Rowe, An overview of polymer electrolyte membranes for fuel cell applications, in *Polymers for Energy Storage and Delivery: Polyelectrolytes for Batteries and Fuel Cells*. ACS Symposium Series, vol. 1096, pp. 147–164. American Chemical Society (2012)
2. Equipment and instruments or materials are identified in the paper in order to adequately specify the experimental details. Such identification does not imply recommendation by the National Institute of Standards and Technology (NIST), nor does it imply the materials are necessarily the best available for the purpose
3. T.D. Gierke, G.E. Munn, F.C. Wilson, J. Polym. Sci. Pt. B-Polym. Phys. **19**, 1687 (1981)
4. H.W. Starkweather, Macromolecules **15**, 320 (1982)
5. R.B. Moore, C.R. Martin, Macromolecules **21**, 1334 (1988)
6. M. Fujimura, T. Hashimoto, H. Kawai, Macromolecules **14**, 1309 (1981)
7. M. Fujimura, T. Hashimoto, H. Kawai, Macromolecules **15**, 136 (1982)
8. G. Gebel, J. Lambard, Macromolecules **30**, 7914 (1997)
9. R.B. Moore, C.R. Martin, Macromolecules **22**, 3594 (1989)

10. E.J. Roche, M. Pineri, R. Duplessix, A.M. Levelut, J. Polym. Sci. Pt. B Polym. Phys. **19**, 1 (1981)
11. E.J. Roche, M. Pineri, R. Duplessix, J. Polym. Sci. Pt. B Polym. Phys. **20**, 107 (1982)
12. J.A. Elliot, S. Hanna, A.M.S. Elliot, G.E. Cooley, Macromolecules **33**, 4161 (2000)
13. J.A. Elliot, P.J. James, T.J. McMaster, J.M. Newton, A.M.S. Elliot, S. Hanna, M.J. Miles, Hydrolysis of the Nafion precursor studied by X-ray scattering and in situ atomic force microscopy. http://www.e-polymers.org (2001)
14. G. Gebel, P. Aldebert, M. Pineri, Macromolecules **20**, 1425 (1987)
15. G. Gebel, R.B. Moore, Macromolecules **33**, 4850 (2000)
16. L. Rubatat, G. Gebel, O. Diat, Macromolecules **37**, 7772 (2004)
17. O. Diat, S. Lyonnard, G. Gebel, A.L. Rollet, Phys. B **350**, E959 (2004)
18. G. Gebel, Polymer **41**, 5829 (2000)
19. G. Gebel, O. Diat, Fuel Cells **5**, 261 (2005)
20. M. Pineri, E. Roche, B. Rodmaco, J. Electrochem. Soc. **127**, C404 (1980)
21. L. Rubatat, A.L. Rollet, G. Gebel, O. Diat, Macromolecules **35**, 4050 (2002)
22. T. Xie, K.A. Page, S.A. Eastman, Adv. Funct. Mater. **21**, 2057 (2011)
23. S.K. Young, S.F. Trevino, N.C.B. Tan, J. Polym. Sci. Pol. Phys. **40**, 387 (2002)
24. S. Lyonnard, G. Gebel, Eur. Phys. J. Spec. Top. **213**, 195 (2012)
25. Y.A. Elabd, M.A. Hickner, Macromolecules **44**, 1 (2011)
26. W. Essafi, G. Gebel, R. Mercier, Macromolecules **37**, 1431 (2004)
27. V. Delhorbe, C. Cailleteau, L. Chikh, A. Guillermo, G. Gebel, A. Morin, O. Fichet, J. Membr. Sci. **427**, 283 (2013)
28. G. Gebel, O. Diat, C. Stone, J. New Mat. Electr. Sys. **6**, 17 (2003)
29. H. Iwase, S. Sawada, T. Yamaki, S. Koizumi, M. Ohnuma, Y. Maekawa, Macromolecules **45**, 9121 (2012)
30. M. Yoonessi, H. Heinz, T.D. Dang, Z.W. Bai, Polymer **52**, 5615 (2011)
31. A.I.I. Sodeye, T.Z. Huang, S.R. Gido, J.W. Mays, Polymer **52**, 3201 (2011)
32. L. Rubatat, C.X. Li, H. Dietsch, A. Nykanen, J. Ruokolainen, R. Mezzenga, Macromolecules **41**, 8130 (2008)
33. S.Y. Kim, M.J. Park, N.P. Balsara, A. Jackson, Macromolecules **43**, 8128 (2010)
34. S. Balog, U. Gasser, K. Mortensen, L. Gubler, G.G. Scherer, H. Ben, Youcef. Macromol. Chem. Phys. **211**, 635 (2010)
35. A.L. Rollet, O. Diat, G. Gebel, J. Phys. Chem. B **106**, 3033 (2002)
36. S.Y. Kim, S. Kim, M.J. Park, Nat. Commun. **1**, 88 (2010)
37. M.J. Park, K.H. Downing, A. Jackson, E.D. Gomez, A.M. Minor, D. Cookson, A.Z. Weber, N.P. Balsara, Nano Lett. **7**, 3547 (2007)
38. J.A. Dura, V.S. Murthi, M. Hartman, S.K. Satija, C.F. Majkrzak, Macromolecules **42**, 4769 (2009)
39. L.L. He, H.L. Smith, J. Majewski, C.H. Fujimoto, C.J. Cornelius, D. Perahia, Macromolecules **42**, 5745 (2009)
40. V.S. Murthi, J.A. Dura, S. Satija, C.F. Majkrzak, *Water uptake and interfacial structural changes of thin film nafion (R) membranes measured by neutron reflectivity for PEM fuel cells, in Fuller*, ed. by T. Shinohara, K. Ramani, V. Shirvanian, P. Uchida, H. Cleghorn, S. Inaba, M. Mitsushima, S. Strasser, P. Nakagawa, H. Gasteiger, H.A. Zawodzinski, T. Lamy, C, Proton Exchange Membrane Fuel Cells 8, Pts 1 and 2, vol. 16. Electrochemical Society Transactions, vol. 2, pp. 1471–1485. (Electrochemical Society Inc, Pennington 2008)
41. D.L. Wood, J. Chlistunoff, J. Majewski, R.L. Borup, J. Am. Chem. Soc. **131**, 18096 (2009)
42. S. Kim, K.A. Page, C.L. Soles, Structure and properties of proton exchange membrane fuel cells at interfaces, in *Polymers for Energy Storage and Delivery: Polyelectrolytes for Batteries and Fuel Cells*, vol. 1096. ACS Symposium Series, vol. 1096, pp. 267–281. American Chemical Society (2012)
43. D.L. Wood, J. Chlistunoff, E.B. Watkins, P. Atanassov, R.L. Borup, ECS Trans. **3**, 1011 (2006)

44. S.A. Eastman, S Kim, K.A. Page, B.W Rowe, S.H. Kang, S.C. DeCaluwe, J.A. Dura, C.L. Soles, K.G. Yager, Macromolecules **46**, 571 (2013)
45. S.A. Eastman, S. Kim, K.A. Page, B.W. Rowe, S.H. Kong, C.L. Soles, Macromolecules **45**, 7920 (2012)
46. M.S. Wilson, S. Gottesfeld, J. Appl. Electrochem. **22**, 1 (1992)
47. M.S. Wilson, S. Gottesfeld, J. Electrochem. Soc. **139**, L28 (1992)
48. S. Kim, J.A. Dura, K.A. Page, B.W. Rowe, K.G. Yager, H.J. Lee, C.L. Soles, Macromolecules **46**, 5630 (2013)
49. A.J. Dianoux, M. Pineri, F. Volino, Mol. Phys. **46**, 129 (1982)
50. F. Volino, M. Pineri, A.J. Dianoux, A. Degeyer, J. Polym. Sci. Pol. Phys. **20**, 481 (1982)
51. A.A. Pivovar, B.S. Pivovar, J. Phys. Chem. B **109**, 785 (2005)
52. J.C. Perrin, S. Lyonnard, F. Volino, J. Phys. Chem. C **111**, 3393 (2007)
53. J.C. Perrin, S. Lyonnard, F. Volino, A. Guillermo, Eur. Phys. J. Spec. Top. **141**, 57 (2007)
54. A. Paciaroni, M. Casciola, E. Cornicchi, M. Marconi, G. Onori, M. Pica, R. Narducci, J. Phys. Chem. B **110**, 13769 (2006)
55. A. Paciaroni, M. Casciola, E. Cornicchi, M. Marconi, G. Onori, M. Pica, R. Narducci, A. De Francesco, A. Orecchini, J. Phys. Condens. Mat. **18**, S2029 (2006)
56. V.K. Peterson, C. Corr, G.J. Kearley, R. Boswell, Z. Izaola, Mater. Sci. Forum **654**, 2871 (2010)
57. V.K. Peterson, C.S. Corr, R.W. Boswell, Z. Izaola, G.J. Kearley, J. Phys. Chem. C **117**, 4351 (2013)
58. R. Bergman, J. Appl. Phys. **88**, 1356 (2000)
59. S. Lyonnard, Q. Berrod, B.A. Bruning, G. Gebel, A. Guillermo, H. Ftouni, J. Ollivier, B. Frick, Eur. Phys. J. Spec. Top. **189**, 205 (2010)
60. O.E. Haas, J.M. Simon, S. Kjelstrup, J. Phys. Chem. C **113**, 20281 (2009)
61. O.E. Haas, J.M. Simon, S. Kjelstrup, A.L. Ramstad, P. Fouquet, J. Phys. Chem. C **112**, 3121 (2008)
62. M.H. Kim, C.J. Glinka, R.N. Carter, Rev. Sci. Instrum. **76**, 113904 (2005)
63. M.H. Kim, C.J. Glinka, S.A. Grot, W.G. Grot, Macromolecules **39**, 4775 (2006)
64. F. Xu, O. Diat, G. Gebel, A. Morin, J. Electrochem. Soc. **154**, B1389 (2007)
65. G. Gebel, O. Diat, S. Escribano, R. Mosdale, J. Power Sources **179**, 132 (2008)
66. A. Morin, F. Xu, G. Gebel, O. Diat, Fuel Cells **12**, 156 (2012)
67. A. Morin, F.N. Xu, G. Gebel, O. Diat, Int. J. Hydrog. Energ. **36**, 3096 (2011)
68. R. Mosdale, G. Gebel, M. Pineri, J. Membr. Sci. **118**, 269 (1996)
69. S. Deabate, G. Gebel, P. Huguet, A. Morin, G. Pourcelly, Energy Environ. Sci. **5**, 8824 (2012)
70. H. Iwase, S. Koizumi, H. Iikura, M. Matsubayashi, D. Yamaguchi, Y. Maekawa, T. Hashimoto, Nucl. Instrum. Meth. A **605**, 95 (2009)
71. A. Putra, H. Iwase, D. Yamaguchi, S. Koizumi, Y. Maekawa, M. Matsubayashi, T. Hashimoto, J. Phys. Conf. Ser. **247**, 012044 (2010)
72. G. Gebel, S. Lyonnard, H. Mendil-Jakani, A. Morin, J. Phys. Condens. Mat. **23**, 234107 (2011)
73. M. Thomas, M. Escoubes, P. Esnault, M. Pineri, J. Membr. Sci. **46**, 57 (1989)
74. A.L. Rollet, M. Jardat, J.F. Dufreche, P. Turq, D. Canet, J. Mol. Liq. **92**, 53 (2001)
75. A.L. Rollet, J.P. Simonin, P. Turq, G. Gebel, R. Kahn, A. Vandais, J.P. Noel, C. Malveau, D. Canet, J. Phys. Chem. B **105**, 4503 (2001)
76. K.A. Page, J.K. Park, R.B. Moore, V.G. Sakai, Macromolecules **42**, 2729 (2009)
77. K.A. Page, K.M. Cable, R.B. Moore, Macromolecules **38**, 6472 (2005)
78. K.A. Page, W. Jarrett, R.B. Moore, J. Polym. Sci. Pt. B Polym. Phys. **45**, 2177 (2007)
79. K.A. Page, F.A. Landis, A.K. Phillips, R.B. Moore, Macromolecules **39**, 3939 (2006)
80. D. Richter, M. Monkenbusch, A. Arbe, J. Colmenero, Advances in polymer science, pp. 1–221. (Springer, Berlin, 2005)

Glossary of Abbreviations

ADP Atomic displacement parameter

AFC Alkaline fuel cell

APT Atom probe tomography

BHJ Bulk heterojunction

CCS Carbon capture and storage

CMS Carbon molecular sieve

CP Cross polarization

CT Charge transfer

CTB Conservation of tetrahedral bonds

CTI Charge-transfer integral

DA Donor acceptor

DFT Density-functional theory

DLC Discotic liquid crystal

D-LC Diamond-like carbon

DLPA Diffraction line profile analysis

DOE Department of energy

EISF Elastic incoherent structure-factor

ES Excited state

FEA Finite element analysis

FF Fill factor

FWHM Full width at half maximum

© Springer International Publishing Switzerland 2015

G.J. Kearley and V.K. Peterson (eds.), *Neutron Applications in Materials for Energy*, Neutron Scattering Applications and Techniques,

DOI 10.1007/978-3-319-06656-1

GB Grain boundary

GC Glassy carbon

GDOES Glow-discharge optical emission spectroscopy

GISAXS Grazing incidence small-angle scattering

GS Ground state

HAZ Heat-affected zone

HOMO Highest occupied molecular oribital

HT High temperature

HWHM Half width at half maximum

IAST Ideal-adsorbed solution theory

ILL Institut Laue Langevin

INS Inelastic neutron scattering

IPCC Intergovernmental panel on climate change

IPCE Incident photon-to-current efficiency

IPNS Intense-pulsed neutron source

IR Infrared

ISIS Neutron scattering at Rutherford and Appleton laboratories

ITO Indium tin oxide

LOCA Loss of coolant accident

LT Low temperature

LUMO Lowest unoccupied molecular orbital

MAS Magic-angle spinning

MCFC Molten carbonate fuel-cell

MD Molecular dynamics

MEA Membrane electrolyte assembly

MEM Maximum-entropy method

MIL Matériaux de l'Institut Lavoisier

MO Molecular orbital

MOF Metal-organic framework

MOM Metal-organic material

MPP Maximum-power point

NASA National aeronautics and space administration

ND Neutron diffraction

NIST National institute of standards and technology

NDP Neutron depth profiling

NMR Nuclear magnetic resonance

NPD Neutron powder diffraction

NR Neutron reflectometry

NSE Neutron spin echo

OCV Open-cell voltage

ODS Oxide dispersion strengthened

OPV Organic photovoltaic

PAFC Phosphoric acid fuel cell

PAH Polycyclic aromatic hydrocarbon

PCE Power-conversion efficiency

PDF Pair-distribution function

PEM Polymer-electrolyte membrane

PEMFC Polymer-electrolyte membrane fuel-cell

PES Potential energy surface

PFG NMR Pulsed field gradient NMR

PGAA Prompt-gamma activation analysis

PP-PEM Plasma polymerised PEM

PV Photovoltaic

QCM Quartz crystal microbalanace

QENS Quasielastic Neutron scattering

RDF Radial-distribution function

SANS Small-angle neutron scattering

SCWR Supercritical-water cooled reactor

SNS Spallation Neutron source

SAXS Small angle X-ray scattering

SCC Stress corrosion cracking

SEI Solid electrolyte interface

SL Scattering-length

SLD Scattering-length density

SOF Site occupancy factor

SOFC Solid oxide fuel-cell

STP Standard temperature and pressure

TOF Time of flight

USANS Ultra-small angle neutron scattering

UV Ultra-violet

WDX Wavelength dispersive X-ray

YSZ Yttria stabilized zirconia

ZIF Zeolitic imidazolate framework

ZMOF Zeolitic MOF

Printed by Printforce, the Netherlands